Ecological Studies, Vol. 195

Analysis and Synthesis

Edited by

M.M. Caldwell, Washington, USA
G. Heldmaier, Marburg, Germany
R.B. Jackson, Durham, USA
O.L. Lange, Würzburg, Germany
H.A. Mooney, Stanford, USA
E.-D. Schulze, Jena, Germany
U. Sommer, Kiel, Germany

Ecological Studies

Volumes published since 2003 are listed at the end of this book.

I.J. Gordon • H.H.T. Prins
Editors

The Ecology of Browsing and Grazing

Iain J. Gordon
CSIRO Davies Laboratory
PMB PO Aitkenvale
Qld 4814
Australia

Herbert H.T. Prins
Resource Ecology Group
Wageningen University
Droevendaalsesteeg 3a
6708 PB Wageningen
The Netherlands

Cover illustration: Zebras alerted by a predator in Mana Pools National Park in the Zambezi Valley, Zimbabwe. (Photo Iain J. Gordon)

ISSN 0070-8356
ISBN 978-3-540-72421-6   e-ISBN 978-3-540-72422-3

Library of Congress Control Number: 2007931594

This work is subject to copyright. All rights are reserved, whether the whole or part of the material is concerned, specifically the rights of translation, reprinting, reuse of illustrations, recitation, broadcasting, reproduction on microfilm or in any other way, and storage in data banks. Duplication of this publication or parts thereof is permitted only under the provisions of the German Copyright Law of September 9, 1965, in its current version, and permissions for use must always be obtained from Springer-Verlag. Violations are liable for prosecution under the German Copyright Law.

Springer is a part of Springer Science+Business Media

springer.com

© Springer-Verlag Berlin Heidelberg 2008

The use of general descriptive names, registered names, trademarks, etc. in this publication does not imply, even in the absence of a specific statement, that such names are exempt from the relevant protective laws and regulations and therefore free for general use.

Editor: Dr. Dieter Czeschlik, Heidelberg, Germany
Desk editor: Dr. Andrea Schlitzberger, Heidelberg, Germany
Cover design: WMXDesign GmbH, Heidelberg, Germany
Production and typesetting: SPi Publisher Services

Printed on acid-free paper SPIN 11382201 31/3180 5 4 3 2 1 0

Foreword

About fifty years ago, when I, as a young comparative anatomist, first looked at a wild ruminant – the European roe deer – the basic thinking concerning the ecology, behaviour, physiology and anatomy of ruminants was based on domesticated grazers, namely sheep and cattle. I could not believe the customary view that the roe deer was nothing more than a mini-cow with a choosy predilection for flowering herbs, tender leaves and shoots. Already my comparison of a red deer stomach with that of a roe deer caused me to bring to mind the different evolutionary traits of cervids as compared with bovids, of which Europe has but a few wild species left. At that time, there was no thought of integrated management of vegetation and herbivores: hunters aimed for higher game densities, foresters considered (and still do) browsers a pest, to be reduced if not eliminated from their planted forests, and advocates of animal welfare agitated against hunting. All this has negatively influenced any serious attempt to develop sustained yield concepts, certainly in Central Europe.

Thus I was overwhelmed by the living demonstration of bovid evolutionary 'explosion' and niche separation between extant species, when I came to study large herbivores in East Africa for ten years prior to decolonisation (the 'Uhuru' of 1963). When I first presented some of my morphological findings on African herbivores at a London Symposium in 1966, the audience encouraged me to extend and deepen my observations systematically. This lead in 1972, initially in collaboration with the British botanist and wildlife researcher Don Stewart, to a classification of ruminants into three feeding types – first recognising a dichotomic evolution with numerous intermediate forms, a system in common use today.

We have to remember that mammalian digestive tracts (of carnivores, omnivores and herbivores) are extremely set and conservative – the result of evolution; this is especially the case with foregut-fermenting herbivores. Ruminant evolution beyond tragulids proceeded over more than 25 million years apparently not step by step (like a ladder), but frequently in parallel fashion, like the growth pattern of a bush or baobab tree. This is why we find extant frugivorous and browsing concentrate selectors (both large and small) in dominant numbers in three of the four ruminant families, but almost no true bulk and roughage grazers (except the Père David's deer) amongst the cervids. In contrast, we see most of the grazers, stimulated by changing climate and following the spread of the grasses, amongst the Bovidae.

Browsers, irrespective of the family they belong to, have retained their archaic morphophysiological features, which evolved before grasses became the dominant plants under the then-prevailing climatic conditions. Browsers are poorly adapted to digesting the structural carbohydrates within grasses, yet browsers have successfully remained within the large herbivore spectrum for more than 10 million years. If it is the browsers that, according to the elaborate analyses and conclusions of this stimulating book, will be the prime winners of the global future, one can only hope and pray that the type of collaboration between scientists and ecosystem managers (including foresters and agronomists) which the editors appeal for will come to pass.

After a long and active life in the field of basic and applied herbivore wildlife research, I feel honoured and encouraged by the authors and especially by the editors of this future-oriented volume to contribute a foreword, with all my good wishes for a worldwide positive reception not only of this book but also of the fascinating animals which, I strongly believe, must remain the gentle modifiers of our landscapes and perhaps even of our anthropocentric view of this world's nature.

Berlin, Baruth

Reinhold R. Hofmann
Dr.med.vet.
Professor emeritus

Contents

1 Introduction: Grazers and Browsers in a Changing World 1
Herbert H.T. Prins and Iain J. Gordon
1.1 Introduction 1
1.2 Dominance of Domesticated Grazers and Browsers 3
1.3 The Last 30 Years and the Immediate Future 5
1.4 Societal Relevance 9
References 16

2 An Evolutionary History of Browsing and Grazing Ungulates 21
Christine Janis
2.1 Introduction 21
    2.1.1 Ungulate Phylogeny and Evolutionary History 22
    2.1.2 Determination of Feeding Adaptations 26
    2.1.3 Cenozoic Changes in Climate 30
2.2 Fossil Record Evidence of Dietary Evolution in Ungulates 32
    2.2.1 Early Archaic Ungulates and Ungulate-like Mammals (65 to 40 Ma) 32
    2.2.2 The Eocene Emergence of Modern Ungulates (55 to 34 Ma) 32
    2.2.3 The Lull Before the Storm: Oligocene and Early Miocene Times (34 to 20 Ma) 34
    2.2.4 The Rise of the Grasslands (20 to 10 Ma) 35
    2.2.5 The Rise of Grazing Ungulates (10 to 2 Ma) 37
    2.2.6 The Late Cenozoic Dawn of the Modern Ungulate Fauna 39
2.3 Discussion and Conclusions 41
References 42

3 The Morphophysiological Adaptations of Browsing and Grazing Mammals ... 47
Marcus Clauss, Thomas Kaiser, and Jürgen Hummel
- 3.1 Introduction ... 47
- 3.2 Grass and Browse ... 48
- 3.3 Predictions ... 52
- 3.4 Testing the Hypotheses ... 61
- 3.5 Results ... 66
- 3.6 Conclusion and Outlook ... 77
- References ... 78

4 Nutritional Ecology of Grazing and Browsing Ruminants ... 89
Alan J. Duncan and Dennis P. Poppi
- 4.1 Introduction ... 89
- 4.2 Nutritive Value of Plant Material for Ruminant Herbivores ... 90
  - 4.2.1 Digestion of Plant Material by Herbivores ... 90
  - 4.2.2 How do Browse and Grass Differ in Nutritive Value? ... 90
  - 4.2.3 Plant Secondary Metabolites ... 92
- 4.3 Do Browsers and Grazers Differ in the Way They Process Their Food? ... 93
  - 4.3.1 Reticulo-Rumen Size ... 100
  - 4.3.2 Retention Time and Passage Rate ... 100
  - 4.3.3 Mean Particle Size Escaping Rumen ... 101
  - 4.3.4 Absorptive Surface Area of Rumen ... 102
  - 4.3.5 Saliva Flow ... 103
  - 4.3.6 Fermentation Rate and Fibre Digestibility in the Rumen ... 103
  - 4.3.7 Rumen Microbial Genetic Profiles and Digestion End Products ... 105
- 4.4 Problems with the Meta-Analysis Approach ... 106
- 4.5 Detoxification of Plant Secondary Metabolites: Do Browsers and Grazers Differ? ... 107
  - 4.5.1 Salivary Tannin-Binding Proteins ... 107
  - 4.5.2 Ruminal Detoxification ... 109
  - 4.5.3 Post Absorptive Metabolism ... 110
- 4.6 Conclusions ... 111
- References ... 112

5 The Comparative Feeding Bahaviour of Large Browsing and Grazing Herbivores ... 117
Kate R. Searle and Lisa A. Shipley
- 5.1 Introduction ... 117
- 5.2 The Functional Response ... 117
  - 5.2.1 Components of the Functional Response ... 120

	5.3	Foraging in Patches	126
		5.3.1 Definition of Patches	127
	5.4	Spatial Variation in the Quality of Grass	128
	5.5	Spatial Variation in the Quality of Browse	130
	5.6	Perception of Patches	132
		5.6.1 Evidence for Patch Perceptions of Grazers and Browsers	134
		5.6.2 Patch Perceptions by Grazers	134
		5.6.3 Patch Perceptions by Browsers	136
	5.7	Summary	138
		References	139

6 The Comparative Population Dynamics of Browsing and Grazing Ungulates ... 149
Norman Owen-Smith

	6.1	Introduction	149
	6.2	Spatial and Temporal Dynamics of Grass and Browse	150
	6.3	Population Density Levels	159
	6.4	Demographic Patterns	162
	6.5	Regulation of Abundance	163
	6.6	Population Dynamics	165
	6.7	Weather Patterns and Population Fluctuations	168
	6.8	Summary and Conclusions	170
		References	171

7 Species Diversity of Browsing and Grazing Ungulates: Consequences for the Structure and Abundance of Secondary Production ... 179
Herbert H.T. Prins and Hervé Fritz

	7.1	Introduction	179
	7.2	Suggested Causes of Species Richness	181
	7.3	The Effect of Species Richness on Ecosystem Functioning: An Overview	183
	7.4	Herbivore Diversity and the Use of Primary Production	186
		7.4.1 Diet Overlap and Feeding Niches	186
		7.4.2 Postulated Advantages of Mixed-Species Feeding	189
	7.5	Mammalian Herbivore Species Richness Links to Secondary Productivity and Biomass	190
		7.5.1 Domestic Herbivore Diversity and Secondary Productivity	190
		7.5.2 Diversity–Biomass Relationship in Wild Assemblages	193
	7.6	Conclusions	195
		References	196

8 Impacts of Grazing and Browsing by Large Herbivores on Soils and Soil Biological Properties 201
Kathryn A. Harrison and Richard D. Bardgett
- 8.1 Introduction 201
- 8.2 Herbivore Effects on Nutrient Dynamics..................... 201
- 8.3 Positive Feedback Effects of Above-Ground Herbivory 202
  - 8.3.1 Urine and Dung Deposition 205
  - 8.3.2 Alterations in Plant C and N Allocation 206
  - 8.3.3 Selective Foraging on Less Nutritious Species......... 208
  - 8.3.4 Litter Deposition.................................... 208
  - 8.3.5 Increased Soil Temperature 209
- 8.4 Negative Feedback Effects of Above-Ground Herbivory......... 209
  - 8.4.1 Selective Foraging on Nutrient-Rich Tissue 210
  - 8.4.2 Production of Secondary Metabolites 211
  - 8.4.3 Impact on Soil Physical Properties 212
- 8.5 Conclusions 212
- References.. 213

9 Plant Traits, Browsing and Gazing Herbivores, and Vegetation Dynamics 217
Christina Skarpe and Alison J. Hester
- 9.1 Introduction 217
- 9.2 Plant Architecture and Herbivory........................... 218
  - 9.2.1 Introduction... 218
  - 9.2.2 Trees and Shrubs..................................... 218
  - 9.2.3 Herbaceous Plants Other Than Graminoids 220
  - 9.2.4 Graminoids 220
- 9.3 The Chemistry of Plants.................................... 222
  - 9.3.1 Photosynthesis 222
  - 9.3.2 Energy and Nutrient Reserves 223
- 9.4 Plant Resistance... 225
  - 9.4.1 Introduction... 225
  - 9.4.2 Escape Strategies 226
  - 9.4.3 Structural and Chemical Defences 227
  - 9.4.4 Plant Tolerance 229
- 9.5 Effects of Herbivory on Plants 229
  - 9.5.1 Introduction... 229
  - 9.5.2 Reserve Dynamics 230
  - 9.5.3 Sprouting and Resprouting............................ 230
  - 9.5.4 Repeated Herbivory 232
  - 9.5.5 Compensatory Growth 233
  - 9.5.6 Regeneration and Persistence 236
- 9.6 Herbivore Foraging Behaviour............................... 237
  - 9.6.1 Introduction... 237
  - 9.6.2 Hierarchical Foraging................................ 237

Contents

	9.6.3	Large Herbivores and Predators	238
	9.6.4	The Importance of Neighbours	238
9.7	Large Herbivore Effects on Vegetation		239
	9.7.1	Introduction	239
	9.7.2	Herbivory and Composition of Plant Populations	239
	9.7.3	Herbivory and the Composition of Plant Communities	240
9.8	The Theory of Vegetation Change		244
	9.8.1	Introduction	244
	9.8.2	Succession Theories	245
	9.8.3	The State-and-Transition Model	245
	9.8.4	State-and-Threshold and Catastrophe Theories	246
9.9	Conclusions		247
References			247

10 The Impact of Browsing and Grazing Herbivores on Biodiversity 263
Spike E. van Wieren and Jan P. Bakker

10.1	Biodiversity and Large Mammalian Herbivores	263
10.2	Effects on Traits	266
10.3	Effects on Plant Communities	270
10.4	Effects on Invertebrates	275
10.5	Effects on Birds	277
10.6	Effects on Mammals	280
10.7	Large Herbivores and Biodiversity	283
References		286

11 Managing Large Herbivores in Theory and Practice: Is the Game the Same for Browsing and Grazing Species? 293
Jean-Michel Gaillard, Patrick Duncan, Sip E. Van Wieren,
Anne Loison, François Klein, and Daniel Maillard

11.1	Introduction	293
11.2	The Dynamics of Grazer and Browser Populations	295
11.3	Monitoring	299
11.4	Management	302
11.5	Conclusions	303
References		304

12 Grazers and Browsers in a Changing World: Conclusions 309
Iain J. Gordon and Herbert H.T. Prins

12.1	Introduction	309
12.2	Responses in Plant Species Composition	310
12.3	Responses in Plant Chemistry	311
12.4	Responses in Terms of Population Dynamics	312
12.5	Responses in Herbivore Community Structure	313

	12.6	Ways of Managing Browsers and Grazers....................	316
		12.6.1 Managing for Variability Rather than Stability	316
		12.6.2 Gardening Versus Laisser Faire.....................	317
	12.7	Where Do We Go From Here?............................	318
		12.7.1 Landscape Scale Experiments	318
	12.8	Conclusions...	319
	References		319

Subject Index... **323**

Species Index... **327**

Contributors

Jan P. Bakker
Community and Conservation Ecology Group, University of Groningen, P.O. Box 14, 9750 AA, Haren, The Netherlands

Marcus Clauss
Division of Zoo Animals, Exotic Pets and Wildlife, Vetsuisse Faculty, University of Zurich, Winterthurerstr. 260, 8057 Zurich, Switzerland, mclauss@vetclinics.unizh.ch

Alan J. Duncan
Macaulay Institute, Craigiebuckler, Aberdeen AB15 8QH, Scotland, UK, a.duncan@macaulay.ac.uk

Patrick Duncan
Centre d'Etdudes Biologiques de Chizé, Centre National de la Recherche Scientifique, Villiers-en-Bois, 79360 Beauvoir-sur-Niort, France

Hervé Fritz
Centre d'Études Biologiques de Chizé, CNRS UPR 1934, 79360 Beauvoir-sur-Niort, France, fritzh@cebc.cnrs.fr

Jean-Michel Gaillard
Laboratoire de Biométrie et Biologie Évolutive (Unité Mixte de Recherche N° 5558), Centre National de la Recherche Scientifique, Université Lyon 1, 43 Boulevard du 11 Novembre, 69622, Villeurbanne Cedex, France, capreolus@wanadoo.fr

Iain J. Gordon
CSIRO - Davies Laboratory, PMB PO Aitkenvale, Qld 4814, Australia, iain.gordon@csiro.au

Kathryn A. Harrison
Institute of Environmental and Natural Sciences, Soil and Ecosystem Ecology Laboratory, Department of Biological Sciences, Lancaster University, Lancaster LA1 4YQ UK, k.a.harrison@lancs.ac.uk

Alison J. Hester
Macaulay Institute, Craigiebuckler, Aberdeen, AB15 8QH, UK

Jürgen Hummel
Institute of Animal Science, Department of Animal Nutrition, University of Bonn, Endenicher Allee 15, 53115 Bonn, Germany

Christine Janis
Department of Ecology and Evolutionary Biology, Brown University, Providence, RI 02912, USA, Christine_Janis@Brown.edu

Thomas Kaiser
University of Hamburg, Biozentrum Grindel and Zoological Museum, Martin-Luther-King-Platz 3, 20146 Hamburg, Germany

François Klein
Office National de la Chasse et de la Faune Sauvage, Centre National d'Étude et de Recherche Appliquée, 1 Place Exelmans, 55000 Bar-le-Duc, France

Anne Loison
Laboratoire de Biométrie et Biologie Évolutive (Unité Mixte de Recherche N° 5558), Centre National de la Recherche Scientifique, Université Lyon 1, 43 boulevard du 11 novembre, 69622, Villeurbanne Cedex, France

Daniel Maillard
Office National de la Chasse et de la Faune Sauvage, Centre National d'Étude et de Recherche Appliquée, 95 rue Pierre Flourens, BP 74267, 32098 Montpellier Cedex 05, France

Norman Owen-Smith
Centre for African Ecology, School of Animal, Plant and Environmental Sciences, University of the Witwatersrand, Wits 2050, South Africa, norman@gecko.biol.wits.ac.za

Dennis P. Poppi
Schools of Animal Studies and Veterinary Science, University of Queensland, St Lucia 4072, Brisbane, Australia

Herbert H.T. Prins
Resource Ecology Group, Wageningen University, Droevendaalsesteeg 3a, 6708 PB Wageningen, The Netherlands, herbert.prins@wur.nl

Kate R. Searle
CSIRO - Sustainable Ecosystems, Davies Laboratory, University Drive, Annandale, QLD 4814, Australia, kate.searle@csiro.au

Lisa A. Shipley
Department of Natural Resources, Washington State University, Pullman, Washington 99163, USA

Christina Skarpe
Hedmark University College, Faculty of Forestry and Wildlife Management,
2480 Koppang, Norway, Christina.Skarpe@nina.no

Sip E. Van Wieren
Resource Ecology Group, Wageningen University, Droevendaalsesteeg 3a, 6708
PB Wageningen, The Netherlands, Sip.vanWieren@wur.nl

Chapter 1
Introduction: Grazers and Browsers in a Changing World

Herbert H.T. Prins and Iain J. Gordon

1.1 Introduction

"During the second half of the 20th century, the global population explosion was the big demographic bogey ... Now that worry has evaporated, and this century is spooking itself with the opposite fear: the onset of demographic decline. The shrinkage of Russia and eastern Europe is familiar, though perhaps not the scale of it: Russia's population is expected to fall by 22% between 2005 and 2050, Ukraine by a staggering 43%. Now the phenomenon is creeping into the rich world: Japan has started to shrink and others, such as Italy and Germany, will soon follow. Even China's population will be declining by the early 2030s, according to the UN which projects that by 2050 populations will be lower than they are today in 50 countries.... People should not mind, though. What matters for economic welfare is GDP per person ..... The new demographics that are causing populations to age and to shrink are something to celebrate [not in Russia or Ukraine though HP & IG]. Humanity was once caught in the trap of high fertility and high mortality. Now it has escaped into the freedom of low fertility and low mortality. Women's control over the number of children they have is an unqualified good – as is the average person's enjoyment, in rich countries, of ten more years of life than they had in 1960. Politicians [and companies] may fear the decline of their nations' economic prowess, but people should celebrate the new demographics as heralding a golden age".

(the Economist Editorial 7 January 2006 p 12).

If humanity is at the brink of a 'Golden Age', what then is the divination for nature? In this book we will try to foretell how wild herbivores will react to the changes that take place in the world in which they live. That world has been changing since it formed, and for millions of years its plants have been consumed by herbivores. This has led to adaptations in plants in reaction to herbivory, and our present-day species assemblages and landscapes are a manifestation of the forces of natural selection that have been in operation for a very long time. Over a much shorter time, these landscapes have been heavily impacted upon by humans; firstly, by accidentally burning patches of the landscape, but later as a tool to modify that landscape either for capturing or luring game or even for changing the species composition towards a modified vegetation that yields desired produce (e.g., cultivation of hazel in the Mesolithic; Simmons et al. 1981). It is now increasingly clear that the 'primordial' Amazonian, Middle American and other rainforests have been strongly modified by local people

(Noble and Dirzo 1997); indeed when the first European colonists came to what is now called New England on the eastern seaboard of the USA the forests were so open and 'Arcadian' that settlers thought it had been created especially for them (Cronon 1983; Prins 1994; cf. Motzkin and Foster 2002; Rusell and Davis 2001). In a way that was true since the indigenous people were decimated through diseases involuntarily brought to them by the newcomers, but their hunter-gatherer imprint was still there (e.g., Douglas and Hoover 1988). The same is true for what was first named New Holland and later Australia: non-agricultural people that had settled there some 40,000 years ago had been using fire to modify that landscape for an uncounted number of generations (e.g., Hughes 1987 p 3; Lewis 2002). Wild grazers and browsers that have survived to the present day must have found a way to cope with many of these man-induced changes.

Indigenous plants that had been shaped by natural selection reacted to the new forces, and new types of vegetation emerged. Really big changes started with the emergence of agriculture; not only did plant communities get modified because the competitive interaction between the domesticated and native plant species changed, but land was cleared and often alien species were introduced for food consumption or involuntarily. Many of these species are now considered part of the native flora and only palynological and archaeological research can reveal their non-local provenance and their alien roots. This can be done where the soil archive is good enough to yield the necessary information, and especially where there are sufficient sources and academics to unravel the arcane history of local plant communities, as for example in Europe (e.g., Godwin 1975; Knörzer 1971, 1975; Opravil 1978; Pennington 1969; Van Zeist 1980). In other places, it is the scourge of introduced species that became an economic threat which has fostered a desire to find out their origin (e.g., McFadyen and Skarratt 1996).

Major changes started when early farmers domesticated sheep, goats, cattle, onagers, and donkeys in the Middle East, dromedaries in the Arabian Peninsula, horses in southern Russia, camels and yaks in Central Asia, lamas in South America, water buffalo and zebu in South Asia, gaur, gayal, and banteng in Southeast Asia, and perhaps the latest to be domesticated, reindeer in northern Scandinavia and northern Siberia (e.g., Legge 1996; Köhler-Rollefson 1996; Zeuner 1967). In many of the centres of origin, the use of domestic grazers and browsers in the already modified landscapes led to a landscape that would have been unrecognisable to earlier generations of Man. Superimposed on the changes brought about by climatic change, fire, and felling, the impact of domesticated indigenous browsers and grazers caused forests to disappear first from the Zagros Mountains (e.g., Hole 1996) or in the Lebanon, and then from across that whole range of landscapes of much of Europe (Prins 1998), the high altitude areas of South America, China, and Japan.

As people moved across the landscape in prehistoric times they took with them their domesticated browsers and grazers. The rate of this spread was about 20 km per generation; the spread of farming (or farmers) from the Levant across Europe was $1\,km.yr^{-1}$ (Cavalli-Sforza 1996) while the rate for pastoralism across Africa was $0.9\,km.yr^{-1}$ (Prins 2000). The result was that by the beginning of our Common Era across most of the Old World domesticated grazers and browsers (including so-called mixed feeders) were modifying indigenous vegetation communities, but two continents stayed free of domestic livestock, namely Australia and North America.

1.2 Dominance of Domesticated Grazers and Browsers

With the advent of European colonialism, Australia and the Americas became an open access area to the grazers and browsers from the Old World. In North America the successful newcomers were cattle, horses, and sheep; in South America cattle, zebu, water buffalo, and sheep took over much of the grasslands, while in Australia introduced sheep, cattle, water buffalo, banteng, camels, donkeys, and horses lived side-by-side, while kangaroos expanded their range as a reaction to surface water becoming available through dams and boreholes. In other places, other non-native grazers and browsers, such as deer and thar (New Zealand) or reindeer (South Georgia) were introduced for sport. The newcomers invaded niches of local mammals (e.g., Breebaart et al. 2002; Dawson et al. 1992; Edwards et al. 1996; Escobar and Gonzalez 1976; Fritz et al. 1996; Genin et al. 1994; Hubbard and Hansen 1976; Lightfoot and Posselt 1977; Prins 2000; Schwartz and Ellis 1981; Thill and Martin 1986).

By the end of the 19th century, the world as we know it took further shape: railway lines opened up the vast prairies of Canada and the United States, and steamships made it possible to start commercially transporting agricultural produce to metropolises where an industrial revolution spurred human population growth. In a series of Homestead Acts, pioneers staked out 110 million hectares of land (Anon. 2005c), which is about 30–50 times the size of countries like the Netherlands, Belgium, Denmark, or Switzerland. As part of the same industrial revolution, railway lines and steamships led to a massive exodus of people from the Ukraine, Poland, Germany, Scandinavia, Scotland, Italy, England, France, Ireland, the Low Countries, and Spain to the New World and Australia. Many of these people went to the cities, but many went also to become farmhands or farmers, cowboys or ranchers. In Argentina, Uruguay, and Paraguay native people were pushed off their land, just as they were in the United States and Canada, Australia, South Africa, and Zimbabwe. Along the Trans-Siberia railway, the Russian Far East was 'opened up'. As a consequence the human population burgeoned, demanding more and more meat, milk, and fibre. Australia and New Zealand became world leaders for the production of mutton and wool, Chicago became the beef capital of the world, and horses and cattle were exported in enormous numbers from the southern part of South America for meat. Horses were likewise exported from Poland to Western Europe for meat and draught power. The end of the 19th and the beginning of the 20th century saw an enormous demand for horses, not only for transport but also for the war machinery of clashing empires; some 8 million horses were killed at only the Western Front during the Great War. The net result of all these transformations was an enormous growth of domestic browsers and grazers, while indigenous wild ungulates declined or even went extinct, such as the Bluebuck (*Hippotragus leucophaeus*) and Quagga (*Equus quagga*) in southern Africa, the Aurochs (*Bos primigenius*) and the Wild (forest) horse (*E. caballus*) in Europe, the wild Dromedary (*Camelus dromedarius*) in Asia, and in Australia the Eastern hare wallaby (*Lagorchestes leporides*), Central hare wallaby (*L. asomatus*), Toolache (*Macropus greyi*), and Crescent nailtail wallaby (*Onychogalea lunata*). Ungulates that reached the brink of extinction during the last century were the European bison (*Bison bonasus*), American bison (*Bison bison*); Père David's deer (*Elaphurus davidianus*) in China; Swamp deer (*Cervus duvaucellii*) in India; Camel (*Camelus bactrianus*) and Przewalski horse (*Equus przewalskii*) in Central Asia; Wild Asian buffalo (*Bubalus bubalis*), Gayal (*Bos frontalis*), Kouprey (*Bos sauveli*) in South East Asia; Black wildebeest (*Connochaetes gnou*) and White rhinoceros (*Ceratotherium simum*) in South Africa. In 2005 the northern form of the white rhino was declared extinct. Many grazers and browsers became very rare and are presently listed as endangered on the IUCN's Red List. Heavy hunting is cited as the cause of these (near) extinctions, in other cases it has been clearing of indigenous vegetation, or the introduction of new predators such as Red fox (*Vulpes vulpes*) in Australia. A new cause is lawlessness associated with failing nation states or civil war, but competition with domestic stock is a seriously underrated cause. So, what about this heralded 'Golden Age'?

Natural grasslands have nearly been wiped off the face of the earth. For example, in the Ukraine only about 4% of the steppes were found to remain in the original state (Goriup 1998), the South African low veld is natural in about 11% of its extent only (Low and Rebello 1996), the North American tall-grass prairie is all but gone (Packard and Mutel 1997), while in SE Australia a stunning 99.5% of its native grasslands have been converted for agricultural use (Taylor 1998). Worldwide, forests have decreased too, from about 6.5 billion hectares to 3.5 billion since the rise of agriculture-based civilizations (Noble and Dirzo 1997). At present, about 15% of the Earth's land surface is occupied by row-crop agriculture or by urban-industrial areas, and another 6–8% has been converted to pastureland (Vitousek et al. 1997). Estimates of the fraction of land transformed or degraded by humanity, and the fraction of the land's biological production that is used by Man is about 40–50% (Vitousek et al. 1997). Managed grazings cover more than 25% of the global land surface and has a larger geographical extent than any other form of land use (Asner et al. 2004). In the beginning of the 21st century CE, nearly half the nitrogen atoms in the protein of an average human being's body came at some time or another through an ammonia factory, often using the Haber-Bosch process, invented in 1909, combining nitrogen from the air with hydrogen from coal (Anon. 2005a).

So even though landscapes were getting transformed from an often rather forested state to a landscape dominated by agriculture, the potential for increased opportunities were garnered by a small subset of browsers and especially grazers, namely by those few species that were useful to man. When artificial fertilisers were introduced on a massive scale in a number of countries, grassland productivity increased so much that now agricultural economies in the rich countries are reliant on only three domestic species, all three grazers, namely cattle, sheep, and horses.

In the United States, the European Union and also the former Eastern bloc, southern Africa, Australia and New Zealand agricultural policies then started to distort markets at a massive scale in the 1960s. World markets would have dictated a shift away from agriculture in these countries because unsubsidised agriculture became, to a large extent, not profitable if farmers in these areas had to provide their products at world market prices but had to pay labour costs as dictated by the local society's norms. Governments then started subsidising agriculture for two major reasons, namely, to maintain adequate national self-sufficiency and to guarantee farmers (and their dependent communities) a sufficiently high income so that they could continue to participate in society at large. These countries maintained a landscape geared towards maximum agricultural production and very high numbers of domestic grazers. However, there has been a substantial reduction in the numbers of horses in the world over the 20th century as horse power became transplanted by engine power. The first tractors had few advantages over the best horses, but they did not eat hay or oats. The replacement of draft animals by machines released about 25% more land for growing food for human consumption in the middle of the 20th century (Anon. 2005a). Cattle and sheep remained the dominant ungulates in temperate, semi-arid, or arid regions of the globe. In most of Africa and the Middle East goats continued to play an important role in addition to sheep, but for dromedaries there was less of a place as they too were replaced by engine power. South Asia remained dominated by zebu cattle and to a lesser extent water buffalo, while in Southeast Asia it is the other way around; but water buffalo are decreasing in numbers in Indonesia at a high rate. Worldwide, fewer and fewer ungulate species dominate the herbivore communities that modify the vegetation.

1.3 The Last 30 Years and the Immediate Future

Since about 1980 major changes in the non-urban landscapes of Europe and North America have taken place because arable- and livestock-based agriculture is no longer cost-effective; the countryside is becoming devoid of people who use the local resources while an ever increasing number of people move to urban metropolises or peri-urban neighbourhoods (Fig. 1.1). The same trend is visible in other developed countries, such as South Africa, Japan, Thailand, and Australia. Nowadays proportionally more and more people live in cities and towns: in Australia about 85% of the people are living in urban areas (Anon. 2006a), in France it is over 80% (Anon. 2005b), in the most developed part of the world it is on average 76%, and worldwide it is now 40% (United Nations 2006). Fewer people realise, however, that the absolute numbers of people living in rural landscapes is decreasing and, as a result, the number of abandoned villages in France, Italy, Spain, and Portugal is increasing. Even when British or Dutch pensioners take over property in these villages, the surrounding country side remains unused. According to preliminary results of the first national Census since 1989, more than half of Russia's 155,290 villages are abandoned or

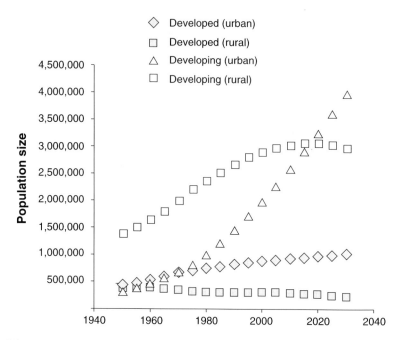

Fig. 1.1 Estimated number of people since 1940 in different parts of the world, and the predicted number up to 2040. (Population Division of the Department of Economic and Social Affairs of the United Nations Secretariat, World Population Prospects: The 2004 Revision and World Urbanization Prospects: The 2003 Revision, http://esa.un.org/unpp)

populated by 50 or fewer stragglers (Anon. 2006b). In Wyoming, Nebraska, or Colorado one can drive through ghost towns: "*Virtually all of the 20 poorest counties in America, in term of wages, are on the eastern flank of the Rockies or on the western Great Plains... The area does include several pockets of wretched Native American poverty, but in most of [these] areas the poor are as white as a prairie snowstorm. ...We are seeing a re-acceleration of the population decline. After the kids left in the 1980s and 1990s, things sort of leveled off. Now it's just plain attrition. People are dying. It is fairly common nowadays for rural counties across America to lose people: roughly one in four did in the 1990s*" (Anon. 2005c). In upstate New York and Maine people have voted with their feet: if one walks through the dense forests one stumbles over stone fences that, only a century ago, separated the potato fields of local crofters (Flinn et al. 2005; Litvaitis 1993; Motzkin and Foster 2002; Ramankutty and Foley 1999). In the drier parts of South Africa sheep farms are transformed into wildlife sanctuaries because of the high risks of stock theft or because of other factors that make it uneconomic to continue farming (e.g., Davies 2000). In Australia, one can drive through villages and towns that have been abandoned (cf. Newman 2005). In "Old Europe" (the European Union of 10 countries), the utilized agricultural area diminished by more than 3 million hectares between 1975 and 1987; between 1980 and 1990 this was 1,251,000 ha in Germany, 1,000,000 ha in France, and 307,000 ha in Italy (EC Commission data). The surface area of permanent meadows and grassland have also strongly declined in "Old Europe" (between 1980 and 1990 1 million hectares; EC Commission data). Also the number of domestic stock declines. Between 1991 and 2001, the number of cattle in Western Europe stabilised at about 90 million head, in Eastern and Central Europe it declined from 50 to 30 million head, and in the EECCA countries (the 12 countries from the Caucasus up to Russia) it showed a massive decline from about 115 to 60 million head. Similar trends are reported for sheep and goats (Anon. 2003; see also FAO 2006).

The reasons for this abandonment are not the same in every situation, but nearly always the ultimate cause is that more jobs and an easier life is available in the cities. Additional factors include lack of children, stock theft, and lack of schools, hospitals, or other services. Demography, market and neo-liberalism all lead to the same result: shops close, services disappear, jobs disappear in the villages but are offered in towns. With the ever-increasing costs of labour as compared to assets, farmers lay-off farm-hands and use machinery to do the work, which further decreases the local job markets. Farmers then get more and more dependent on banks, fodder producers, or equipment sellers for loans, which makes them more vulnerable in times of adversity. Ultimately this frequently results in bankruptcy. Politicians have tried to stop this tide by doling out hundreds of billions of dollars, pounds, liras, guilders, pesetas, yen, kronor, euros, francs, and marks in subsidies to maintain the countryside in a productive posture. For example, *"the [American Great Plains] region has plainly failed to adapt to a world in which grain and cattle are cheap. ... The crops go out by the truckload to provide jobs elsewhere. How does the region survive? For all the brouhaha about independence, it leans heavily on federal government. In 2003, the government spent an average of US$ 10,200 per person in Judith Basin county. North Dakota counties averaged $ 9,000. Most of it comes in the form of farm subsidies: federal price supports and disaster payments"* (Anon. 2005c).

This has led to an ever-increasing cost for food for the urban people, and an ever-increasing call for reducing toll barriers around 'Fortress Europe'. Agricultural subsidies have decreased already, and will decrease further. To save the last of the uneconomic farmers, governments dole out subsidies for maintaining the countryside for the sake of biodiversity, which often is a *gotzpe* because the type of biodiversity-rich agricultural landscapes of the early 1900s have been eradicated by these very same farmers when they were subsidised to meet production goals at the cost of the environment. Land abandonment leads to changes in the landscape to which wild grazers and browsers react, however, the decreasing numbers of domestic stock inhabiting the former extensive farms in developed countries offers opportunities for wild herbivores to exploit the unconsumed vegetation resource. White-tailed deer (*Odocoileus virginianus*) numbers, for example, have increased dramatically between 1980 and 2000 in many areas of North America (Riley et al. 2002).

Apart from economic drivers, sociological and demographic changes, and market distortions, there is a fourth insidious cause of change, namely climate change, especially the increasing concentrations of carbon dioxide in the atmosphere. This change may lead to changed competitive interactions between different plant groups; carbon dioxide is used in horticulture as an artificial "fertiliser" and it allows plants to grow for more hours in the day so that more carbohydrates are produced. It appears as if a number of woody species surpassed a threshold of starch deposition in their roots, making them much more fire-resistant, because with sufficiently high carbohydrate reserves these woody species can resprout and grow quickly out of the fire-trapping zone. This in turn leads to an altered balance between grasses and woody species, favouring the latter (Bond et al. 2003). This leads to changes in species composition creating decreasing opportunities for grazers and increased ones for browsers.

Under a mode of production that encourages high efficiency, land abandonment and land use intensification are two sides of one coin. "*With the [human] population growth rates falling sharply while [agricultural] yields continue to rise, even the acreage devoted to wheat may now begin to decline for the first time since the stone age.*" (Anon. 2005a). Wheat leads the field in world crops. In 2004, the worldwide area (in millions of hectares) under wheat was about 220, rice about 150, maize some 140, other cereals about 160, soybeans 90, pulses 70, roots and tubers 50 (FAO 2006). This intensification is much easier to achieve in flat areas than in montane or steeply hilly areas (Anon. 2005d). The number of farms decreases fast in OECD-countries. In Europe, most changes take place in mountainous areas (Batzing et al. 1997; MacDonald et al. 2000). For example in France the number of farms declined from about 880,000 in 1950 to 220,000 in 2005 (Anon 2005d). Poor, hilly areas in France, Italy, Spain, and Portugal lost more than half their active population since the 1970s (EC Commission data; Torrano and Valderrabano 2004). In the north of former East Germany, between 1991 and 1997 the number of people working in agriculture declined from 296,000 to 164,000 (Shaphiro and Stankevich 2000). Also in Australia the rural population is declining (Bennett et al. 2004), as it is in Patagonia where sheep stock halved between 1980 and 2000 and hundreds of farms were abandoned here (Borrelli and Cibils 2005) and in northwestern Argentina (Montero and Villalba 2005). In New Zealand's hill country woody encroachment takes place because of economic marginalisation (Popay et al. 2002), and Japan's countryside also is experiencing depopulation (Liu et al. 2003, Shimoda 2005). In a number of countries, even the absolute population numbers are falling. Russia's population is expected to fall by 22% between 2005 and 2050, the Ukraine by a staggering 43%. Japan has started to shrink in 2005 and by 2050 a 12% fall is expected, the same fall as is expected for Italy. Germany is expected to have declined in 2050 by 5%. Even China's population will be declining by the early 2030s, according to the UN, which projects that by 2050 populations will be lower than they are today in 50 countries (the Economist, Editorial, 7 January 2006). Russia's Far East is quickly depopulating (Kontorovich 2000) but so is Nova Scotia and parts of Canada's hinterland (Millward 2005).

The net results of these changes in land use, are to a large extent the focus of this book. In Portugal, for example, rural depopulation and land abandonment has led to a significant decline in agricultural land and low shrub and an increase in tall shrublands and forest, with an increase of 20–40% in fuel accumulation at a landscape level (Moreira et al. 2001; see also Debussche et al. 1999; Tinner et al. 1998; Preiss et al. 1997; Valderrabano and Torrano 2000). In Table 1.1 we have summarised different scenarios as we see them, focussing on the balance between woody species and grasses or herbs. Grazers and browsers react to these changes (Table 1.2), but they can also be used to prevent changes in vegetation composition, or to facilitate them. Herbivores, in contrast to many other organisms, can act as landscape modifiers, but they can only do this to a limited extent. Since browsers concentrate on woody species and herbs, while grazers focus on grass, the questions we are asking are:

1. What adaptations do large herbivores have to consuming browse- or grass-dominated diets?
2. What are the consequences of consuming these diets for large herbivore population ecology?
3. Are there difference between grazers and browsers in their impact on ecosystem structure and functioning?
4. How can we use this information to manage large herbivores in landscapes that are changing due to anthropogenic and climatic effects?

Land abandonment and woody encroachment are prevailing trends in the rich countries or in countries of the temperate zones. In the developing countries more typical for the tropics and sub-tropics, land abandonment is not prevalent yet although the level of urbanization is still increasing. However, after the expected peak in human numbers before the middle of the 21st century (Fig. 1.1), it is also likely that in these areas there will be an expansion of bush and forest.

1.4 Societal Relevance

Are changes in the landscape relevant? Should the public or the natural resource managers worry about a shift in the balance between woody species and grasses, or about the rapid increase of wildlife species, or about land abandonment? Is it important to know whether these shifts in the landscape mosaic can be controlled, either facilitated or suppressed? There are a number of reasons why we think this is important:

Water. A shift towards more woody cover often leads to a reduced amount of water infiltration into the soil, because trees evaporate more water than grasses, and pine trees or firs continue this evapo-transpiration even in winter (e.g., Persson 1997). This means less water is available for agricultural production and less aquifer replenishment, so less drinking water and/or process water for industry.

Table 1.1 Different combinations of change lead to different effects on herbivores and the landscape

Scenario	Population	Land use	Market/subsidies	Intensification effect	Herbivore effect	Landscape/biodiversity effect
Increased CO$_2$ levels lead to shift in balance between woody species and herbaceous layer	No depopulation of rural areas	No land abandonment	Market distortions and farm subsidies continue	Very strong intensification but no bigger farms; intensification also far from markets	No place for wild herbivores. Domestic herbivore species richness decreases and will only be kept indoors	Very intense use of fertilisers, irrigation, grass cutting; woody species only if necessary (e.g., fodder)
			Market distortions and farm subsidies stop	Very strong intensification and also bigger farms; intensification only where it pays	Place for wild herbivores on game farms. Domestic herbivore species richness can even increase if there is a market for novelty products	More interregional variation; some areas very intensively used, others much less so. Woody encroachment in less intensively used areas
		Land abandonment	Market distortions and farm subsidies continue	Intensification on bigger farms; intensification also far from markets	Increasing place for wild herbivores. 'Wild game' tolerated if incentives are paid to maintain them. Domestic herbivore species richness decreases	Governments can demand environmental schemes for biodiversity. Discouraging woody encroachment costly
			Market distortions and farm subsidies stop	Bigger farms at the cost of others; intensification only where it pays	Increasing place for wild herbivores. 'Wild game' tolerated if they provide economic return. Place for wild herbivores on game farms	Interregional variation in farm profitability will increase, and so will on-farm variation. Woody encroachment will be tolerated in many places within the (farm) landscape even if they have no function

Depopulation of rural areas	No land abandonment	Market distortions and farm subsidies continue	Increased land holdings, leading to bigger farms; further intensification	Domestic herbivore species richness may increase if there is a market for novelty products	Intermediate use of fertilisers, irrigation, grass cutting; woody species maintained on farm only if necessary (e.g., fodder)
	Land abandonment	Market distortions and farm subsidies stop	Increased and intensified holdings only where it pays (increased abandonment elsewhere)	No place for wild herbivores. Domestic herbivore species richness decreases	More interregional variation; some areas intensively used, others much less so. Woody encroachment in less intensively used areas
Increased CO_2-levels lead to shift in balance between woody species and herbaceous layer		Market distortions and farm subsidies continue	Much bigger farms are possible but hardly any intensification taking place; land holding size will also increase far from markets and hence land abandonment will be countered	No place for wild herbivores. Domestic herbivore species richness decreases	
				Little place for wild herbivores. Domestic herbivore species richness decreases, ample scope for mega dairy projects and feedlots	Importance of fertilisers decreases, but local irrigation, grass cutting; woody species maintained on farm only if necessary (e.g., as windbreak)

(continued)

Table 1.1 (continued)

Market distortions and farm subsidies stop	Much bigger farms are possible but hardly any intensification will take place, and only then where it pays (increased abandonment elsewhere)	Increasing place for wild herbivores. 'Wild game' tolerated if cost-effective. Place for wild herbivores in game parks. Domestic herbivore species richness may increase if there is a market for novelty products	Interregional variation in farm profitability will increase, and so will on-farm variation. Woody encroachment will be tolerated in many places within the (farm) landscape even if they have no function

Table 1.2 The effect of the scenarios from Table 1.1 on domestic herbivores and on wild herbivores

Scenario elements from Table 1.1 (CO_2-levels are assumed to increase in all possible scenarios)	Effect on *domestic herbivores* in the context of the ecology of grazers and browsers	Effect on *wild herbivores* in the context of the ecology of grazers and browsers
Rural areas without land abandonment or depopulation, while market distortions and farm subsidies do not allow much scope for increased land holding size due to mergers, and farmers rely heavily on further farm intensification	Very few browsers, if any, will find a place in this system. Animals will be kept indoors, and be fed with imported food and silage from intensively used (parts of) farms.	Small wild grazers, like migratory geese, can benefit but otherwise the landscape will become devoid of wild grazers and browsers.
Rural areas without land abandonment or depopulation, but cessation of market distortions and farm subsidies lead to a moderate increase in size of land holdings in places where it is economically profitable to merge. Cessation of agriculture in regions where it is no longer viable, but in regions where it is profitable most farms will further intensify	Market for novelty products can become important, such as lamas in Europe, and farmers may switch to game farming as in South Africa or New Zealand.	There will be an effect outside the region towards other regions: there abandonment will be promoted leading to woody encroachment and build-up of browser populations.
Land abandonment but no depopulation of rural areas, while market distortions and farm subsidies continue leading to much increased land holding size due to mergers even in areas where it does not make sense economically. Intensification is driven by subsidies even in places far removed from markets. There will be little land abandonment because this trend will be counteracted by subsidies	Very few browsers, if any, will find a place in this system. The industry would like to keep animals indoors, and fed on imported food and silage from intensively used farms, but the public will demand 'romantic' landscapes with domestic stock kept outdoors	Little opportunity for wild species in the landscape because abandonment is counteracted upon by government and farmers. Nature management will ask for woody species suppression by browsing, mowing, and cutting. Increased options for small wild browsers; if wild grazers are available, they will be kept fenced

(continued)

Table 1.2 (continued)

Scenario elements from Table 1.1 (CO_2-levels are assumed to increase in all possible scenarios)	Effect on *domestic herbivores* in the context of the ecology of grazers and browsers	Effect on *wild herbivores* in the context of the ecology of grazers and browsers
Land abandonment but no depopulation of rural areas, but cessation of market distortions and farm subsidies lead to an increase in size of land holdings in places where it is economically profitable but cessation of agriculture where it is not. Intensification is not necessary on farms that stay in operation	'Traditional' grazers (cattle, sheep) will be used on big farms, but market for novelty products could become important, and farmers may switch to game farming.	Reasonable opportunities for wild species in the landscape because pockets of land will be abandoned by agriculture. The opportunities are there for small and not so small browsers. Little scope for 20th century nature management. Increased options for large herds of wild grazers if they can be made economically; they will be kept fenced.
Depopulation of rural areas but without land abandonment, while market distortions and farm subsidies continue leading to increased land holding size due to mergers and further farm intensification	Very few browsers, if any, will find a place in this system. Animals will be kept both indoors and outdoors, and be fed on imported food and silage from intensively used (parts of) farms	Small wild grazers, like migratory geese, can benefit but otherwise the landscape will become devoid of wild grazers and browsers
Depopulation of rural areas but without land abandonment, but cessation of market distortions and farm subsidies lead to a moderate increase in size of land holdings in places where it is economically profitable to merge. Cessation of agriculture where it is no longer viable, but on most farms intensification necessary	Very few browsers, if any, will find a place in this system. The industry would like to see animals kept indoors but to be outdoors as much as feasible, and partly fed on silage from intensively used parts of the farms. Feedlot farming	There will be a strong effect outside the region towards other regions: there abandonment will be promoted leading to woody encroachment and build-up of browser populations
Land abandonment and depopulation of rural areas, while market distortions and farm subsidies continue to lead to increased land holding size due to mergers. Intensification is not necessary on farms that stay in operation. There will be little land abandonment because this trend will be counteracted by subsidies	Very few browsers, if any, will find a place in this system. The industry would like to see animals kept indoors but to be outdoors as much as feasible, and partly fed on silage from intensively used parts of the farms. Good options for mega-dairy projects, feedlot farming and extensive ranching	Small wild grazers, such as migratory geese, can benefit to a limited extent because of decreased fertiliser input. The landscape will become devoid of grazers and browsers, because farmers will suppress woody encroachment and will see game as a competitor for forage

(continued)

Table 1.2 (continued)

Scenario elements from Table 1.1 (CO_2-levels are assumed to increase in all possible scenarios)	Effect on *domestic herbivores* in the context of the ecology of grazers and browsers	Effect on *wild herbivores* in the context of the ecology of grazers and browsers
Land abandonment and depopulation of rural areas, but with cessation of market distortions and farm subsidies will lead to an increase in size of land holdings in places where it is economically profitable but cessation of agriculture where it is not. Intensification is not necessary on farms that stay in operation	Domestic herbivore species richness may increase if there is a market for novelty products. Low fertilizer input, coupled with large farm sizes will require the suppression of woody encroachment. Burning and stumping necessary to maintain grazing lands	Increasing place for wild herbivores. 'Wild game' tolerated if cost-effective. Place for wild herbivores on game ranches and private game parks. Burning necessary to encourage grazers. There will be a strong effect outside the region towards other regions: there abandonment will be promoted leading to woody encroachment and build-up of browser populations

Fire. More trees and less herbaceous vegetation can result in large build ups of combustible material, changing landscapes that were fire resistant in their agricultural state into fire-prone landscapes (cf. Bonazountas et al. 2005; Finney 2005; Haight et al. 2004; O'Laughlin 2005; Sturtevant et al. 2004). This leads to loss of life and property (e.g., Chen and McAneney 2004), and increased transaction costs for society because of increasing insurance premiums. Uncontrolled forest succession as consequence of the accelerated socioenvironmental change in the Mediterranean forested landscapes has shown to lead to critically enhanced risk for fires (Tabara et al. 2004).

Accidents. Land abandonment together with higher wildlife densities increase as chances for traffic accidents increase (e.g., Doerr et al. 2001). A person driving an ordinary car at 80km/hr who collides with a wild boar has a very high chance of being killed; tall-legged moose smash through wind screens in collisions. Apart from sorrow and suffering, increased insurance premiums lead to increased costs of transport.

Diseases. Increased wildlife densities, especially of deer, can lead to the closure of the life cycle of tick-borne diseases. For example, Lyme's disease needs small intermediary hosts such as mice or blackbirds but also large ones, such as roe deer or red deer. An increased number of deer leads to an increased incidence of life-endangering parasitic diseases (Jensen and Jespersen 2005; Randolph 2004; Zavaleta and Rossignol 2004).

Cultural heritage. Particularly in Japan and Europe, societies and governments have gone to great effort to stop the countryside from getting clogged-up with ungainly industry by applying zoning laws and heritage protection. In these regions, national parks have even been equated with mediaeval or pre-industrial landscapes, and not, like in Africa or North America, with pristine nature (where, admittedly, 'native peoples' had had an often unrecognised impact). Land abandonment and depopulation threaten this medium-intensity agricultural countryside that is still highly valued by the public.

By asking ourselves how grazers and browsers function in the landscape, we believe that we are able to help developing awareness of these often unwanted changes, but also that we are finding tools for the management of these landscapes. The negative aspects of landscape change that we spelled out are, however, offset by a very positive trend:

Nature. Changes in the balance between grasses and trees can have profound impact on the many other wild species that make use of the landscape. That local plant communities change is self-evident, but linked to that are the changes at lower and higher trophic levels (e.g., Litvaitis 1993; Preiss et al. 1997). Depopulation and land abandonment is leading to enormous opportunities for wildlife to re-establish itself in the countryside, and for a new natural system to develop. In many of the areas that have been intensively used for decades or even centuries, wildlife was on the brink of extinction, but now it is bouncing back. The number of roe deer in Europe or white-tailed deer in North America have not been so high for centuries, and that is true for many others species as well.

Many of the indigenous wild species have been wiped out, especially wild grazing species. Now browsers and intermediate feeders are doing well in the increasingly forested landscapes of Europe and North America. Also feral browsers do exceedingly well in some areas, for example goats and dromedaries in Australia. In the absence of wild grazers, horses and donkeys have the ability to thrive, but not sheep. Perhaps in coming years in Europe there will as much opportunity for wisent (European bison) as there is for bison in America.

Landscapes all over the world have been changing, and herbivores whether they are wild, feral or domestic, live and die at the hand of man. The different chapters in this book give clues about why some species live and others die, why some species or classes of species have selective advantages over others, and how an understanding of their ecology in the past or under 'pristine conditions' may give educated guesses about how they will fare in future. In the Third World or in the rapidly developing new economies the coming decades will be bleak, but in countries in the temperate zone we forecast a 'Golden Age' for wild browsers and to a lesser extent for wild grazers.

References

Anonymous (2003) Europe's Environment: The Third Assessment. Environmental assessment report 10. European Environment Agency, Copenhagen

Anonymous (2005a) Ears of plenty: the story of man's staple food. Economist 24 December 2005, p 26–30

Anonymous (2005b) Economist 24 December 2005, p 75

Anonymous (2005c) The poorest part of America: not here surely? Economist 10 December 2005, p 37–38

Anonymous (2005d) Europe's farm follies: why the European Union remains its strange fondness for farm subsidies. Economist 10 December 2005, p 25–27

Anonymous (2006a) http://earthtrends.wri.org/pdf_library/country_profiles/pop_cou_036.pdf

Anonymous (2006b) http://www.cdi.org/russia/johnson/7246-17.cfm

Asner GP, Elmore AJ, Olander LP, Martin RE, Harris AT (2004) Grazing systems, ecosystem responses, and global change. Annu Rev Env Resour 29:261–299

Batzing W, Perlik M, Dekleva M (1997) Urbanization and depopulation in the Alps (with 3 coloured maps). Mt Res Dev 16:335–350

Bennett J, van Bueren M Whitten S (2004) Estimating society's willingness to pay to maintain viable rural communities. Aust J Agr Resour Ec 48:487–512

Bonazountas M, Kallidromitou D, Kassomenos PA, Passas (2005) Forest fire risk analysis. Hum Ecol Risk Assess 11:617–626

Bond W J, Midgley GF, Woodward WI (2003) The importance of low atmospheric CO_2 and fire in promoting the spread of grasslands and savannas. Global Change Biol 9:973–982

Borrelli P, Cibils A (2005) Rural depopulation and grassland management in Patagonia. In: Reynolds SG, Frame J (eds) Grasslands, developments, opportunities, perspectives Science Publishers, London, pp 461–487

Breebaart L, Brikraj R, O'Connor TG (2002) Dietary overlap between Boer goats and indigenous browsers in a South African savanna. Afr J Range Forage Sci 19:13–20

Cavalli-Sforza LL (1996) The spread of agriculture and nomadic pastoralism: insights from genetics, linguistics and archaeology. In: Harris DR (ed) The origins and spread of agriculture and pastoralism in Eurasia, UCL Press, London, p 51–69

Chen KP McAneney J (2004) Quantifying bushfire penetration into urban areas in Australia. Geophys Res Lett 31:L2212, 1–4

Cronon W (1983) Changes in the land: Indians, colonists, and the ecology of New England. Hill and Wang, New York

Davies R (2000) Madikwe Game Reserve: a partnership for conservation. In: Prins HHT, Grootenhuis JG, Dolan TT (eds) Conservation of wildlife by sustainable use. Kluwer Academic, Boston, pp 439–458

Dawson TJ, Tierney PJ, Ellis BA (1992) The diet of the bridled nailtail wallaby (Onychogalea fraenata): II. Overlap in dietary niche breadth and plant preferences with the black-striped wallaby (Macropus dorsalis) and domestic cattle. Wildl Res 19:79–87

Debussche M, Lepart J, Devieux A (1999) Mediterranean landscape changes: evidence from old post cards. Global Ecol Biogeogr 8:3–15

Doerr ML, McAninch JB Wiggers EP (2001) Comparison of four methods to reduce white-tailed deer abundance in an urban community. Wildlife Soc B 29:1105–1113

Douglas JE, Hoover MD (1988) History of Coweeta. In: Swank WT, Crossley DA (eds) Forest hydrology and ecology of Coweeta, Springer, Berlin Heidelberg New York, pp 17–34

Edwards GP, Croft DB, Dawson TJ (1996) Competition between red kangaroo (Macropus rufus) and sheep (Ovis aries) in the arid rangelands of Australia. Aust J Ecol 21:165–172

Escobar A, Gonzalez JE (1976) Study on the competitive consumption of large herbivores of the flooded area of the Llanos with special reference to the capybara (Hydrochoerus hydrochaeris). Agron Trop 26:215–277

FAO (2006): http://www.fao.org/es/ess/census/default.asp/ [accessed 30/1/06]

Finney MA (2005) The challenge of quantitative risk analysis for wildland fire. Forest Ecol Manag 211:97–108

Flinn KM, Vellend M, Marks PL (2005) Environmental causes and consequences of forest clearance and agricultural abandonment in central New York, USA. J Biogeogr 32, 439–452

Fritz H, Degarinewichatitsky M Letessier G (1996) Habitat use by sympatric wild and domestic herbivores in an African savanna woodland: the influence of cattle spatial behaviour. J App Ecol 33:589–598

Genin D, Villca Z, Abasto P (1994) Diet selection and utilization by llama and sheep in high-altitude arid rangeland of Bolivia. J Range Manage 47, 245–248

Gil Montero R, Villalba R (2005) Tree rings as a surrogate for economic stress: an example from the Puna of Jujuy, Argentina in the 19th century. Dendrochronologia 22:141–147

Godwin, H (1975) History of the British flora: a factual basis for phytogeography, 2nd edn. Cambridge Univ Press, Cambridge

Goriup, P (1998) The pan-European biological and landscape diversity strategy: integration of ecological agriculture and grassland conservation. Parks 8:37–46

Haight RG, Cleland DT, Hammer RB, Radeloff VC, Rupp TS (2004) Assessing fire risk in the wildland-urban interface. J Forestry 102:41–48

Hole, F (1996) The context of caprine domestication in the Zagros region. In: DR Harris (ed) The origins and spread of agriculture and pastoralism in Eurasia. UCL Press, London, pp 263–281

Hubbard RE, Hansen RM (1976) Diets of wild horses, cattle, and mule deer in the Piceance basin, Colorado. J Range Manage 29:389–392

Hughes R (1987) The fatal shore: a history of the transportation of convicts to Australia 1787–1868. Pan Books, London

Jensen PM, Jespersen JB (2005) Five decades of tick-man interaction in Denmark: an analysis. Exp Appl Acarol 35:131–146

Knörzer KH (1971) Urgeschichtler Unkräuter im Rheinland: Ein Beitrage zur Entstehungsgeschichte der Segetalgesellschaften. Vegetatio 23:89–11

Knörzer KH (1975) Entstehung und Entwicklung der Grünlandvegetation im Rheinland. Decheniana 127:195–214

Köhler-Rollefson I (1996) The one-humped camel in Asia: origin, utilization and mechanisms of dispersal. In: Harris DR (ed) The origins and spread of agriculture and pastoralism in Eurasia. UCL Press, London, pp 282–294

Kontorovich V (2000) Can Russia resettle the Far East? Post-Communist Econ 12:365–384

Legge T (1996) The beginning of caprine domestication in Southwest Asia. In: Harris DR (ed) The origins and spread of agriculture and pastoralism in Eurasia, UCL Press, London, pp 238–262

Lewis D (2002) Slower than the eye can see: environmental change in Northern Australia's cattle lands; a case study from the Victoria River district, Northern Territory. Tropical Savannas CRC, Darwin

Lightfoot CJ, Posselt J (1977) Eland (Taurotragus oryx) as a ranching animal complementary to cattle in Rhodesia. 2. Habitat and diet selection. Rhod Agr J 74:53–61

Litvaitis JA (1993) Response of early successional vertebrates to historic changes in land-use. Conserv Biol 7:866–873

Liu YB, Nishiyama S, Kusaka T (2003) Examining landscape dynamics at a watershed scale using Landsat TM imagery for detection of wintering hooded crane decline in Yashiro, Japan. Environ Manage 31:365–376

Low B. Rebello AG (1996) Vegetation of South Africa, Lesotho and Swaziland. Department of Environmental Affairs and Tourism, Pretoria

MacDonald D, Crabtree JR, Wiesinger G, Dax T, Stamou N, Fleury P, Lazpita JG, Gibon A (2000) Agricultural abandonment in mountain areas of Europe: environmental consequences and policy response. J Environ Manage 59:47–69

McFadyen RC, Skarratt B (1996) Potential distribution of Chromolaena odorata (Siam weed) in Australia, Africa and Oceania. Agr Ecosyst Environ 59:89–96

Millward H (2005) Rural population change in Nova Scotia, 1991–2001: bivariate and multivariate analysis of key drivers. Can Geogr–Geogr Can 49:180–197

Motzkin G, Foster DR (2002) Grasslands, heathlands and shrublands in coastal New England: historical interpretations and approaches to conservation. J Biogeogr 29:1569–1590

Moreira F, Rego FC, Ferreira PG (2001) Temporal (1958–1995) pattern of change in a cultural landscape of northwestern Portugal: implications for fire occurrence. Landscape Ecol 16:557–567

Newman P (2005) The city and the bush: partnerships to reverse the population decline Australia's wheatbelt. Aust J Agr Res 56:527–535

Noble IR, R (1997) Forests as human-dominated ecosystems. Science 277:522–525

O'Laughlin J (2005) Policies for risk assessment in federal land and resource management decisions. Forest Ecol Manag 211:15–27

Opravil E (1978) Synanthrope Pflanzengesellschaften aus der Burgwallzeit (8 - 10 Jh) in der Tschechoslowakei. Berichte der deutsche Botanische Gesellschaft 91:97–106

Packard S Mutel CF (1997) The tallgrass restoration handbook for prairies, savannas and woodlands. Island Press, Washington, DC

Pennington W (1969) The history of the British vegetation. English Univ Press, London

Persson G (1997) Comparison of simulated water balance for willow, spruce, grass ley and barley. Nord Hydrol 28:85–98

Popay AI, Rahman A, James TK (2002) Future changes in New Zealand's hill country pasture weeds. New Zeal Plant Prot (Zydenbos SM ed) 55:99–105

Preiss E, Martin JL, Debussche M (1997) Rural depopulation and recent landscape changes in a Mediterranean region: consequences for the breeding avifauna. Landscape Ecol 12:51–61

Prins HEL (1994) Children of Gluskap: Wabanaki Indians on the eve of the European invasion. In: Baker EW, Churchill EA, D'Abate RS, Jones KL, Konrad VA, Prins HEL (eds) American beginnings: exploration, culture and cartography in the land of Norumbega. U Nebraska Press, Lincoln, pp 165–211

Prins HHT (1998) The origins of grassland communities in northwestern Europe. In: Wallis de Vries MF, Bakker JP, van Wieren SE (eds) Grazing and conservation management. Kluwer, Boston pp 55–105

Prins HHT (2000) Competition between wildlife and livestock. In: Prins HHT, Grootenhuis JG Dolan TT (eds) Conservation of wildlife by sustainable use. Kluwer, Boston, pp 51–80

Ramankutty N, Foley JA (1999) Estimating historical changes in land cover: North American croplands from 1850 to 1992. Global Ecol Biogeogr 8:381–396

Randolph SE (2004) Evidence that climate change has caused 'emergence' of tick-borne diseases in Europe? Int J Med Microbiol 293(Suppl):5–15

Riley SJ, Decker DJ, Enck JW, Curtis PD, Lauber TB, Brown TL (2002) Deer populations up, hunter populations down: implications for interdependence of deer and hunter population dynamics on management. Ecoscience 10:455–461

Rusell EWB, Davis RB (2001) Five centuries of changing forest vegetation in the northeastern United States. Plant Ecol 155:1–13

Schwartz CC, Ellis JE (1981) Feeding ecology and niche separation in some native and domestic ungulates on the shortgrass prairie. J Appl Ecol 18:343–353

Shaphiro S Stankevich B (2000) Structural changes in agriculture in northeastern Germany (in Russian). Vestsi Akademii Agrarnykh Navuk Respubliki Belarus 2:40–43

Shimoda M (2005) Emerged shore vegetation of irrigation ponds in western Japan. Phytocoenologia 35:305–325

Simmons IG, Dimbley GW, Grigson C (1981) The Mesolithic. In: Simmons IG, Tooley MJ (eds), The environment in British prehistory. Duckworth, London, pp 82–124

Sturtevant BR, Zollner PA, Gustafson EJ, Cleland DT (2004) Human influence on the abundance and connectivity of high-risk fuels in mixed forests of northern Wisconsin, USA. Landscape Ecol 19:235–253

Tabara D, Sauri D, Cerdan R (2004) Forest fire risk management and public participation in changing socioenvironmental conditions: a case study in a Mediterranean region. Risk Anal 23:249–260

Taylor SC (1998) South-eastern Australian temperate lowland native grasslands: protection levels and conservation. Parks 8: 21–26

Thill RE, Martin A (1986) Deer and cattle diet overlap on Louisiana pine-bluestem range. J Wildlife Manage 50:707–713

Tinner W, Conedura M, Amman B, Gaggeler HW, Gedye S, Jones R, Sagesser B (1998) Pollen and charcoal in lake sediments compared with historically documented forest fires in southern Switzerland since AD 1920. Holocene 8:31–42

Torrano L, Valderrabano J (2004) Impact of grazing on plant communities. Span J Agric Res 2:93–105

United Nations (2006) World urbanization prospects, 1999 revision. http://www.prb.org/

Valderrabano J, Torrano L (2000) The potential for using goats to control Genista scorpis shrubs in European black pine stands. Forest Ecol Manag 126:377–387

Van Zeist W (1980) Prehistorische cultuurplanten, ontstaan, verspreiding, verbouw. In: Chamaulan M, Waterbolk HT (eds), Voltooid Verleden Tijt: Een hedendaagse kijk op de prehistorie. Intermediair, Amsterdam, pp147–165

Vitousek PM, Mooney HA, Lubchenco J, Melillo JM (1997) Human domination of Earth's ecosystems. Science 277:494–499

Zavaleta JO, Rossignol PA (2004) Community-level analysis of risk of vector-borne disease. T Roy Soc Trop Med H 98:610–618

Zeuner FE (1967) Geschichte der Haustiere. Bayerische Landwirtschaftsverlag, Muenchen

Chapter 2
An Evolutionary History of Browsing and Grazing Ungulates

Christine Janis

2.1 Introduction

Browsing (i.e., eating woody and non-woody dicotyledonous plants) and grazing (i.e., eating grass) are distinctively different types of feeding behaviour among ungulates today. Ungulates with different diets have different morphologies (both craniodental ones and in aspects of the digestive system) and physiologies, although some of these differences are merely related to body size, as grazers are usually larger than browsers. There is also a difference in the foraging behaviour in terms of the relationship between resource abundance and intake rate, which is linear in browsers but asymptotic in grazers. The spatial distribution of the food resource is also different for the different types of herbage, browse being more patchily distributed than grass, and thus browsers and grazers are likely to have a very different perception of food resources in any given ecosystem (see Gordon 2003, for review).

Grass is a relatively recent type of food resource: extensive grasslands only emerged during the later Cenozoic, within the past 25 million years (Ma), while the first ungulates date back to the early Cenozoic, around 55 Ma. Browsing, probably with the incorporation of a fair amount of fruit, is thus the primitive diet of ungulates (Bodmer and Ward 2006). The general evolutionary perspective is that when the grazing species evolved later they eclipsed the browsers (Kowalevsky 1873; Matthew 1926; Pérez-Barbería et al. 2001). While this view of the later Cenozoic rise of grazing ungulates, and corresponding demise of browsing ones, is broadly correct, the actual evolutionary picture is of course much more complex. While the most familiar ungulates today (horses, cows, elephants, etc.) are mainly grazers, in fact specialised grazing is a fairly recent evolutionary adaptation: the first true grazers (i.e., animals subsisting primarily on grass year round) are no more than 10 million years old, and a predominance of tropical grazers as is familiar to us in Africa today has a history of only a couple of million years.

In order to understand the evolutionary history of diets and feeding adaptations in ungulates we need to first pose several questions. Firstly, what exactly are ungulates, and how are different ungulate groups related to each other? Secondly, how can we deduce the feeding behaviour of extinct species from their fossilised remains? And thirdly, how have differences in the earth's climate and environment in past times influenced the distribution and abundance of ungulates of different feeding behaviours?

2.1.1 Ungulate Phylogeny and Evolutionary History

While ungulates have long been considered as a clade, united by the possession of hooves (which is in fact not strictly true in any case), recent molecular phylogenies have refuted the monophyly of the 'Ungulata' (see Springer et al. 2005 for review). The paenungulates (elephants, hyraxes, and sea cows) are now united with other African endemic mammals in the basal placental clade Afrotheria. Although artiodactyls and perissodactyls are fairly closely related, they are not sister taxa: the perissodactyls are now considered to be more closely related to a clade consisting of carnivorans plus pangolins, and artiodactyls are now included with whales in the Cetartiodactyla. The name of the clade uniting this diverse assemblage is the Ferungulata.

It is difficult to state how extinct 'ungulate' groups, such as the condylarths and the endemic South American ungulates, fit within this new phylogenetic scheme. The South American orders Notoungulata and Litopterna paralleled northern ungulates in the later Cenozoic in their evolution of body morphologies (with types resembling rhinos, rodents, hyraxes, horses, and camels), and in similar evolutionary transitions to animals with skulls and teeth adapted for grazing, and limbs adapted for cursoriality (see Prothero and Schoch 2002). See the Glossary for terms of extinct ungulate groups (Box 2.1).

Box 2.1 Glossary Guide to extinct taxa. See chapters in Janis et al. (1998) and Rose (2006) for more details

ANCHITHERIINAE. A subfamily of horses (family Equidae), known from the late Eocene to late Miocene of North America and the Miocene of Eurasia, ranging from sheep-sized (earlier forms) to the size of a modern horse (later forms). More derived than the Hyracotheriinae in having more lophed teeth, indicative of a more folivorous diet (but all were brachydont, presumed browsers), and somewhat longer legs, with the loss of the fourth toe in the front foot, and no evidence of a foot pad.

BRONTOTHERIIDAE. A family of perissodactyls (also known as titanotheres), all members brachydont (indicative of browsing), known from the Eocene of North America and Eurasia. Early members were pig-sized and hornless, later forms became larger and grew forked bony nasal horns, and some of the latest forms were of a comparable size to a white rhino.

CHALICOTHEROIDEA. A superfamily of perissodactyls, all members brachydont (indicative of browsing). The earlier Eomoropidae, known from the Eocene of Asia and North America, were small (dog-sized), unspecialised forms. The later Chalicotheriidae, known from the Oligocene to Pleistocene of

Eurasia, the Miocene of North America, and the Miocene to Pleistocene of Africa, were large (camel-sized) forms that had substituted claws for hooves, and likely fed with a bipedal stance clawing down vegetation.

CONDYLARTHRA. A paraphyletic assemblage (order) of basal archaic ungulates, comprising at least eight families, known from the Palaeocene and Eocene of North America, South America and Eurasia, mostly ranging in size from the size of a squirrel to the size of a pig (some carnivorous forms reached bear size). Most forms had bunodont cheek teeth indicative of omnivory; some had more lophed teeth indicative of herbivory, while some were even carnivorous. None had cursorial (running-adapted) postcranial morphologies nor any degree of hypsodonty. The carnivorous mesonychid 'condylarths' are considered to be the sister taxon to the Cetartiodactyla, but the precise relationship of the other families to extant ungulates is not resolved.

DEINOTHERIIDAE. A family of proboscideans, known from the Miocene to Pleistocene of Africa and Eurasia. Characterised by very large size and the possession of recurved lower tusks only (no upper tusks). Cheek teeth were tapir-like, bilophodont and brachydont, indicative of a specialised browsing diet.

DINOCERATA. An order of ungulate-like (but probably not ungulate-related) mammals known from the Palaeocene and Eocene of North America and Asia, ranging from sheep-size to rhino-size, with later forms having a bizarre series of horns on their skulls and sabre-like upper canines. They had brachydont, bilophodont teeth indicative of a diet of soft browse, and later forms had postcranial adaptations indicative of a semi-aquatic mode of life.

GOMPHOTHERIIDAE. A family of Proboscidea, known from the Miocene to Pleistocene of North America, Eurasia, and Africa. Ranged from small rhino-sized to modern elephant-sized. Earlier forms had cheek teeth that were brachydont and bunodont to bunolophodont, indicative of omnivory and/or browsing. Included bizarre 'shovel-tusked' forms. Some later (Pliocene) forms developed highly lophed, hypsodont cheek teeth, indicative of mixed feeding and/or grazing.

HIPPARIONINI. Horses belonging to an extinct tribe of the subfamily Equinae (modern horses and their direct ancestors belong to the tribe Equini), known from the middle Miocene to Pliocene of North America, and late Miocene to Pliocene of Eurasia and Africa (survived into mid Pleistocene of Africa). Earlier hipparionines and equines were similar types of pony-sized animals, but hipparionines in general remained less hypsodont than later equines, and they retained a tridactyl foot. Dental microwear suggests mainly mixed-feeding diets rather than grazing (Hayek et al. 1992; Solounias and Sempebron 2002).

HYPERTRAGULIDAE. A primitive family of ruminants, known from the Eocene and Oligocene of North America and Asia. Diminutive forms, about the size of extant *Tragulus* (Tragulidae). Teeth range from brachydont to hypsodont (but not likely to have been grazing).

(continued)

Box 2.1 (continued)

HYRACOTHERIINAE. A subfamily of horses, known from the Eocene of North America. Small forms (cat to small dog-sized), with relatively bunodont cheek teeth indicative of a folivorous/frugivorous diet, and relatively short legs retaining three toes in the hind foot and four in the fore foot, with the likely presence of a tapir-like foot pad.

INDRICOTHERIINAE. A subfamily of rhinos (of the extinct family Hyracodontidae), known from the late Oligocene and early Miocene of western Asia, all with brachydont cheek teeth (indicative of browsing). Known as 'giraffe rhinos' because of the evolution of long neck and long legs, including the largest ever known land mammal, weighing up to 15 tons.

LITOPTERNA. An extinct order of South American ungulates, known from the Palaeocene to the Pleistocene, comprising forms with brachydont cheek teeth (indicative of browsing), or somewhat hypsodont teeth (indicative of mixed feeding), generally resembling horses or camels.

MAMMUTIDAE. A family of Proboscidea (mastodons), known from the Miocene to Pleistocene of North America, Eurasia, and Africa. Ranged from small rhino-sized to modern elephant-sized, and had cheek teeth that were brachydont and bunolophodont, indicative of browsing.

NOTOUNGULATA. An order of South American ungulates, known from the Palaeocene to the Pleistocene, comprising forms with high-crowned or ever-growing cheek teeth (but were not necessarily all grazers), generally resembling rodents, hyraxes, or rhinos.

OREODONTOIDEA. Extinct superfamily of (possibly) tylopod artiodactyls, pig-like or hyrax-like in appearance, known from the late Eocene to late Miocene of North America. They combined relatively derived selenodont cheek teeth (albeit in a snub-nosed skull), earlier forms all brachydont, with a primitive postcranium that retained a digitigrade stance. The earlier and more primitive Agriochoeridae (cat-sized to dog-sized) had replaced their hooves with claws and were likely secondarily arboreal. The later Merycoidodontidae (hyrax-sized to tapir-sized) were the commonest mammals in the Oligocene and early Miocene of North America. Somve later Miocene forms showed a tendency to evolve somewhat longer legs and more hypsodont teeth.

PALAEOMERYCIDAE. A family of deer-like (and deer-related) artiodactyls, known from the Miocene and Pliocene of North America (subfamily Dromomerycinae) and Eurasia. Mostly brachydont (indicative of browsing), some forms slightly hypsodont and may have been mixed feeders; many forms characterised by a third, median occipital horn in addition to a pair of frontal horns.

PALAEOTHERIIDAE. A family of horse-like, and horse-related, perissodactyls, known from the Eocene of Eurasia, ranging from dog-size to modern horse-size. All brachydont (indicative of browsing).

PANTODONTA. An order of ungulate-like (but probably not ungulate-related) mammals, known from the Palaeocene and Eocene of North America,

Eurasia, and South America, ranging from goat-sized to bison-sized, and later forms resembling ground sloths or hippos. All brachydont (indicative of browsing), and some forms were likely semi-aquatic.

SIVATHERIINAE. A subfamily of Giraffidae, known from the Miocene to Pleistocene of Africa and Eurasia. Differed from extant Giraffinae in being relatively short legged and heavily-built (often termed 'moose-like', but with shorter legs than a moose), with horns that were more lobate in form than giraffines. Cheek teeth were generally more hypsodont than giraffines, indicative of a more mixed-feeding type of diet.

TAENIODONTA. An order of ungulate-like (but probably not ungulate-related) mammals, known from the Palaeocene and Eocene of primarily North America, ranging from cat-sized to pig-sized. Later forms had hypselodont (ever-growing) cheek teeth and showed adaptations for digging, rather resembling large pigs. Their hypselodont cheek teeth were probably related to grit found on excavated roots and tubers, as their diet otherwise appears to have been omnivorous.

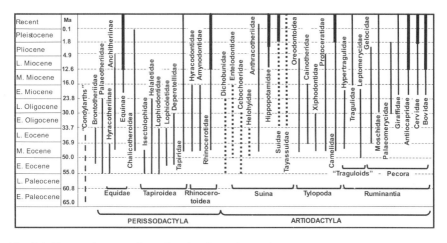

Fig. 2.1 Temporal ranges of ungulate families (adapted from range charts in Benton 1993 and Janis et al. 1998). Not all extinct families of artiodactyls shown; extinct subfamilies of Equidae included. *Dotted lines*: omnivorous groups. *Solid lines*: folivorous groups. *Thick solid lines* indicate time when there were at least some taxa in this lineage that likely took a fair amount of grass (>50%) in their diet

This chapter will mainly consider the evolution of feeding adaptations in the two major orders of extant ungulates: the Perissodactyla, or odd-toed ungulates, and the Artiodactyla, or even-toed ungulates (see Fig. 2.1). Both orders are of Northern Hemisphere origin, first appearing at the start of the Eocene (55 Ma) in North America and Eurasia. Both orders show considerable evolutionary parallelisms, including the evolution of hypsodont (high-crowned) cheek teeth, and elongated limbs with an unguligrade foot posture (i.e., standing on the tip of the phalanges, like a ballerina 'en pointe'), but they differ profoundly in their morphophysiological adaptations to folivory (the eating of leafy material, whether browse or grass). Perissodactyls are specialised hindgut fermenters while camelid and ruminant artiodactyls (in the taxonomic sense) are foregut fermenting ruminants (in the physiological sense). Such differences in digestive physiology are reflected in differences in both craniodental morphology and feeding behaviour. For example: in terms of craniodental morphology, hindgut fermenters process more food per day than ruminants (all other things such as body size, diet, etc., being equal) so they have a greater volume of masticatory musculature, which is reflected in a deeper angle of the mandible. In terms of feeding behaviour, ruminants are more limited in their daily intake, due to their longer digestive passage time, and so they have to feed in a more selective fashion than a hindgut fermenter, and use their tongue for food selection to a much greater extent than other ungulates. Figure 2.1 shows the distribution of the major families of perissodactyls and artiodactyls through time. Note that forms that have craniodental adaptations for grazing are relatively few and appear relatively late in ungulate evolutionary history.

It is worth making a brief mention of the evolutionary history of the proboscideans and hyracoids (both hindgut fermenters), even though these orders are not closely related to other ungulates (see above). Present-day hyraxes are a relict specialised clade of small-bodied forms: the extinct family Pliohyracidae comprised the dominant small-to-medium sized herbivores in the Eocene and Oligocene of Africa, with a diversity of morphologies, including a long-legged antelope-like form, but none showing any craniodental adaptations for grazing (see Prothero and Schoch 2002; Turner and Antón 2004). This radiation was largely gone by the Miocene, perhaps related to the influx of modern ungulates into Africa, but some pliohyracids persisted into the Pleistocene (including a large, hippo-like form), and were also found across southern Eurasia in the later Cenozoic. The extant family Procaviidae first appeared in the late Miocene, and the modern genus *Procavia* (the rock hyrax) is the only one that takes a large amount of grass in its diet. Proboscideans also originated in Africa, and comprised around ten separate families. Only three families (including the Miocene to Recent Elephantidae) show craniodental morphologies indicative of grazing, such as hypsodont, complexly-lophed cheek teeth. Other proboscideans have bunodont cheek teeth indicative of omnivory, or bilophodont cheek teeth indicative of browsing (see Fig. 2.2 for tooth types; hyracoids and proboscideans are not shown in Fig. 2.1, for reasons of economy of space). However, as with the other ungulates, there is no evidence of any craniodental morphology indicative of grazing until the later Miocene.

2.1.2 Determination of Feeding Adaptations

The large diversity of extant ungulates of known diet has made it possible to quantitatively determine features of cranial and dental morphology that correlate with feeding behaviour (Janis 1995; Mendoza et al. 2002). Ungulate cheek teeth (molars and premolars) are the most obvious dietary indicators. Omnivores have low-cusped molars with rounded, bumpy cusps ('bunodont'), designed to process non-brittle food such as fruit and roots (Fig. 2.2A). Many different herbivorous forms have evolved from this type of dentition a more 'lophed' or ridged type of

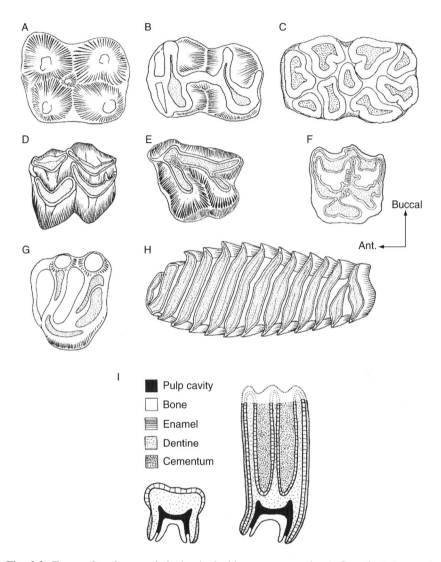

Fig. 2.2 Types of molar morphologies in herbivorous mammals. *A.* Bunodont (peccary). *B.* Bilophodont (kangaroo). *C.* Columnar (warthog). *D.* Selenodont (deer). *E.* Lophodont (rhinoceros). *F.* Plagiolophodont (horse). *G.* Bunolophodont (rodent – woodchuck). *H.* Multilophed (rodent – capybara). *I.* Longitudinal sections of molars, showing brachydont (human) form on the left and hypsodont (horse) form on the right. Modified from Janis and Fortelius 1988, with permission of Cambridge University Press

molar, where the individual cusps are thrown into higher occlusal relief, and are run together as longitudinal or horizontal ridges or lophs (Jernvall et al. 1996). These teeth have been evolved to have maximum efficiency after initial wear, so that the enamel is worn off the top of the lophs exposing a lake of dentine with enamel edges on either side, and these lophs act to shred more fibrous food, such a leaves. The different way in which the cusps have been linked into these lophs result in different patterns of occlusal anatomy: ruminant artiodactyls have 'selenodont' teeth (Fig. 2.2D), where the main pattern of the lophs is in an antero-posterior direction, while perissodactyls have 'lophodont' (Fig. 2.2E) or 'bilophodont' (Fig. 2.2B) teeth, where the main pattern of the lophs is in a labio-lingual direction. Other herbivorous mammals, such as warthogs (Fig. 2.2C) and rodents (Fig. 2.2G, H) have evolved lophed teeth independently in different fashion.

A more fibrous diet of grass rather than browse is reflected in both the level of hypsodonty (see below) and in the occlusal pattern. Highly specialised grazing ruminants and perissodactyls have more complex 'plagiolophodont' (Fig. 2.2F) occlusal enamel patterns, accomplished by cross-linking the enamel ridges. In other mammals, such as elephants, and also in many rodents and in wombats, the teeth become 'multilophed' with numerous parallel ridges of enamel that can no longer easily be homologised with the original mammalian tooth cusp pattern (Fig. 2.2H; Janis and Fortelius 1988). Any tooth in which the full crown is not visible above the gum line at eruption can be considered as 'hypsodont' to some extent, but there are varying degrees of hypsodonty, and in extreme cases the crown height may be six or seven times the width of the tooth. Hypsodont teeth usually have a layer of cementum that coats the tooth and fills in the spaces between the cusps (see Fig. 2.2I).

All dental dimensions scale isometrically: thus a simple ratio, or 'hypsodonty index' of the unworn crown height of the third molar (usually the highest-crowned tooth) to its width or length can be compared across taxa of different body sizes (Janis 1988). Fortelius et al. (2002) define a general rule of thumb for 'hypsodonty classes', based on the ratio of height to length of the second upper or lower molar. A brachydont tooth has a ratio of less than 0.8, a mesodont (partially hypsodont) tooth has a ratio of 0.8–1.2, and a hypsodont tooth has a ratio of greater than 1.2.

The traditional determinant of a browsing versus grazing diet in fossil ungulates has been the level of hypsodonty, or the degree of molar crown height, as the silica contained in grass tissue results in greater wear on the teeth. A year-round diet of grass in a (perforce) open habitat necessitates a hypsodont dentition, but other dietary or environmental factors may lead to hypsodonty. Fortelius et al. (2002) note that the factors that correlate with hypsodonty are: 'increased fibrousness, increased abrasiveness due to intracellular or extraneous dust, and decreased nutritive value'. Hypsodonty has evolved numerous times within mammals, and probably represents a fairly simple developmental change, involving delaying the closure of the tooth roots (Janis and Fortelius 1988). For an animal with an abrasive diet, and hence a high rate of tooth wear, hypsodonty is an important adaptation as a dentition that is insufficiently durable will result in a shortened life span, and hence in a reduced reproductive output (Damuth and

Janis 2005). Thus there is a strong evolutionary imperative to make teeth hypsodont if the diet is abrasive.

While almost all grazers have highly hypsodont cheek teeth, not all hypsodont ungulates are grazers, as the silica contained in grass is far from the sole abrasive element in a herbivorous diet. The exceptions to a high degree of hypsodonty in grazers include taxa such as the hippo, *Hippopotamus amphibius*, and the rock hyrax, *Procavia capensis*. These animals (both hypsodont, but at a relatively low level) both have relatively low metabolic rates for their size, resulting in less food consumed per lifetime and thus overall less dental abrasion. Grazing kangaroos are also less hypsodont than grazing ungulates, again probably due to the relatively lower metabolic rate of marsupials. Fresh grass grazers, such as the reduncine bovids, are also less hypsodont than grazers subsisting on grass in more dusty habitats. Open-habitat mixed feeders can also be highly hypsodont, approaching the level of hypsodonty of grazers in the same habitat, even if including little grass in the diet. Most gazelles fall into this category, and the pronghorn, *Antilocapra americana*, is among the most hypsodont of the ruminants despite having only about 12% of grass in its diet. The obvious interpretation of this correlation is that grit and dust accumulating on the food must also contribute to the abrasive nature of the diet (see Janis 1988; Janis et al. 2002). Hypsodont taxa in the fossil record clearly provide some form of palaeoecological signal: increasing levels hypsodonty in today's ungulate communities show a negative correlation with levels of rainfall (i.e., hypsodont ungulates are more prevalent in more arid habitats; Damuth et al. 2002; Fortelius et al. 2002).

A number of aspects of craniodental morphology can be shown to correlate with dietary behavior (see Solounias and Dawson-Saunders 1988; Janis 1995; Mendoza et al. 2002). The features distinguishing grazers from browsers relate to the different physical demands of feeding on grass versus browse. Grass is in general more fibrous and abrasive than browse, and a more fibrous and abrasive (i.e., lower quality) diet requires a greater intake and a greater degree of mastication. Grazers have bigger masseter muscles than browsers, reflected in a larger and deeper angle of the jaw, and a longer masseteric fossa on the skull. Browsers are selective feeders (as are most mixed feeders; Gordon and Illius 1988), and both these feeding types have a narrow muzzle in comparison with grazers, who have a broad muzzle for the intake of large bites. However, the majority of living ungulates are ruminant artiodactyls, and quantitative correlations of morphology and behavior derived from ruminants may not be directly applicable to other types of ungulates, although general qualitative observations may still hold true.

Dental wear (microwear, mesowear, or macrowear) is another way of determining past diets. Dental wear records the actual food preparation and mastication events that took place during the life of the animal, and thus holds the potential for recording the actual ecological history. But dental wear alone may be an insufficient guide to diet because of the continual abrasion of teeth during the life of the animal, and especially of the abrasion of the surface enamel that records microwear patterns. Solounias and Semprebon (2002), and Semprebon et al. (2004) summarise much of the current uses of dental wear in dietary determination in ungulates.

Another recently developed methodology is the use of carbon isotopes in dental enamel. Following the shift in photosynthesis in tropical grasses around 7 Ma, from a C3 carbon cycle to a C4 carbon cycle, the dental enamel of tropical grazers contains a different composition of enamel isotope from that of browsers, and mixed feeders have intermediate values between browsers and grazers. While dental isotopes are only applicable in the rather limited case of fairly recent ungulates (i.e., within the past 7 Ma or so) in tropical or subtropical settings, they can be extremely useful in such palaeoecological situations, especially in combination with morphological features (e.g., Sponheimer et al. 1999).

How does one, then, determine the diet of an extinct ungulate? To a good first approximation, molar occlusal morphology can distinguish omnivores from folivores, and hypsodonty (as well as occlusal morphology, to a certain extent) can distinguish browsers from grazers, but with a degree of caution. A low-crowned (brachydont) ungulate is extremely unlikely to be anything else but a browser, but hypsodont taxa may have a variety of diets. Within hypsodont taxa, craniodental features and/or microwear can be used to distinguish mixed feeders and grazers in most circumstances, and isotopes are useful for distinguishing diets among later Cenozoic taxa in dry, lowland tropical habitats.

2.1.3 *Cenozoic Changes in Climate*

Evolutionary transitions in ungulate diets can only be understood in the context of climatic changes on planet Earth over the past 60 million years or so. This environmental history is related to changes in the movement of the continents, which in turn have resulted in changes in ocean currents, mountain building, polar glaciation, etc. (see summaries in Janis 1993; Jacobs et al. 1999; Hooker 2000; Zachos et al. 2001). Understanding of the different environmental patterns on different continents comes from a variety of sources, including palaeotemperatures determined from deep sea foraminifera (Zachos et al. 2001) and palaeobotany (Jacobs et al. 1999). Figure 2.3 summarises the evolution of grassland habitats, and the changing global palaeotemperatures. [Note that grass phytoliths have now been isolated from the latest Cretaceous of India (Prasad et al. 2005), but there is no evidence for extensive grasslands in Asia until the later Cenozoic.]

At the start of the Cenozoic, 65 Ma, the entire globe was a fairly equable, warm place, with evidence of mainly forest cover, and tropical-like forests extending up into the high latitudes, although herbaceous (non-grassy) open habitats are recorded in places such as Central Asia. Rapid warming at the end of the Palaeocene was followed by the early Eocene climatic optimum (55–52 Ma; see Zachos et al. 2001). Following this optimum the climate cooled in the higher latitudes, with the arrival of winter frosts and more deciduous types of vegetation

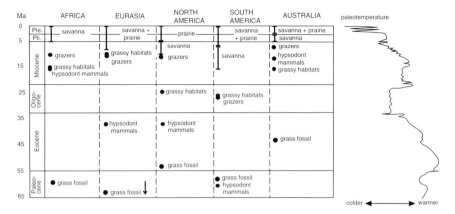

Fig. 2.3 Evolution of grasses and grazers on different continents. *Bars* indicate times of widespread grasslands: prairie (also equivalent to steppe or pampas) = treeless grassland; savanna = treed grassland. *Closed circles* record the first appearance of various events. Grassy habitats = habitats containing some grasses, probably woodland savanna or brushland. Grazers = mammals with craniodental adaptations (apart from hypsodonty) indicative of specialist grazing (i.e., >90% of grass in the diet on a year-round basis). Palaeobotanical information adapted from data in Jacobs et al. (1999) and Strömberg (2004). Palaeotemperature curve is from global deep sea oxygen isotopes (adapted from Zachos et al. 2001); as used here it represents relative temperatures only (for general comparison, mid latitude mean temperatures during the early Eocene climatic optimum were probably around 30° C). Modified from Janis et al. 2004

by the late middle Eocene (around 45 Ma). Although grass fossils are known from the Eocene, and there seem to have been areas of open habitat, such habitats have been termed 'woody savannas', not dominated by grasses, and have no analogues among modern habitat types (Leopold et al. 1992).

A million years or so after the end of the Eocene there was an episode of extreme cooling, setting the stage in the higher latitudes for the more temperate world of the Oligocene, when temperate deciduous woodlands spread in the mid latitudes along with patches of more arid habitat. Temperatures started to rise again in the late Oligocene, around 25 Ma, and after a brief fall reached a further peak at around 14 Ma, the 'mid Miocene optimum' (Zachos et al. 2001). Palaeobotanical evidence shows that grasslands spread during the Miocene in the higher latitudes, but tropical areas such as East Africa had only limited, if any, savanna regions at this time (Jacobs et al. 1999; see Fig. 2.3). The Antarctic ice cap, which had its inception in the late Eocene, was firmly in place by the Miocene.

After around 14 Ma temperatures fell in the higher latitudes, with additional evidence for increasing aridity. By the latest Miocene (around 6 Ma) the short-grass savannas of North America were replaced by tall-grass prairie (Retallack 2001), and there is evidence of woodland savannas in East Africa (Jacobs et al. 1999). Additional higher latitude cooling in the Plio-Pleistocene brought new types of cold-adapted and/or arid-adapted vegetational habitats: tundra and taiga to the high latitudes, and true deserts to tropical and subtropical regions. An extremely important event in establishing the modern grassland habitats was in the middle Pliocene, at 2.5 Ma. At this time the Isthmus of Panama appeared, linking North and South America (with subsequent animal migrations) and disrupting circum-equatorial circulation. This resulted in the aridification of East Africa, and the establishment of the extensive grasslands that support much of the diversity of grazing ungulates today. The Arctic ice cap first appeared in the Pliocene, and from the start of the Pleistocene around 2 Ma the periodic fluctuations in earth's climate, caused by the various aspects of the rotation of the earth on its own axis (Milankovitch cycles), were sufficient to regularly plunge the higher latitudes into periods of extended glaciation, which we term 'Ice Ages'.

2.2 Fossil Record Evidence of Dietary Evolution in Ungulates

2.2.1 Early Archaic Ungulates and Ungulate-like Mammals (65 to 40 Ma)

Palaeocene 'ungulates' were various taxa ascribed to the order 'Condylartha', that probably contains the ancestry of both artiodactyls and perissodactyls (see Glossary, Box 2.1). Other larger ungulate-like taxa (but probably not related to true ungulates) such as taeniodonts, pantodonts, and dinoceratans (see Glossary, Box 2.1) were also around in the Palaeocene in North America and Asia, persisting into the mid Eocene. These taxa were all apparently predominately omnivorous (taeniodonts and most condylarths) or were adapted to a diet of relatively nonfibrous browse (other condylarths, pantodonts, and dinoceratans). There was little evidence of any herbivores subsisting on fibrous vegetation (with the possible exception of a couple of smaller condylarth taxa in the latest Palaeocene).

2.2.2 The Eocene Emergence of Modern Ungulates (55 to 34 Ma)

The early Eocene saw a great diversification of artiodactyls and perissodactyls in the Northern Hemisphere, and the initial appearance of small to mid-sized mammals that appeared to be specialised folivores (primates as well as ungulates). This mammalian community shift suggests that something had happened to the structure of the vegetational habitat in the higher latitudes where these mammals were found, with leaves now available as a broad dietary resource. The early Eocene vegetation of mid to high latitudes was similar to that of modern tropical forests in terms of plant diversity (e.g., Collinson et al. 1981; Wolfe 1985). Eocene fossil localities supported a wide diversity of small-to-medium sized (5–100 kg) terrestrial herbivores, unlike the situation in present-day equatorial forest habitats, which have a paucity of small terrestrial herbivores (Hooker 2000; Janis 2000). The initial diversity of folivorous ungulates was among the perissodactyls (see Fig. 2.1). The main diversity at this time was among the 'tapiroids', mostly sheep-sized or smaller. Some of these were ancestral to modern tapirs, others to rhinos (first appearing in the middle Eocene), and others were evolutionary dead ends. Most tapiroids had bilophodont cheek teeth (see Fig. 2.2B), indicative of folivory. Hyracotheriine Equids, and the Eurasian equid-related palaeotheres were also common, especially in the early Eocene, most ranging in size from the proverbial wire-haired fox terrier to the size of a large sheep. In general, their teeth were more bunodont than those of the tapiroids and rhinos, indicative of a more omnivorous diet, although some European forms had more lophed teeth, suggestive of a greater degree of folivory. Other families of extinct browsing perissodactyls, included chalicotheres (dog-sized at this time) and brontotheres (see Glossary, Box 2.1).

Artiodactyls were also common in the early Eocene, at this time small (tragulid-sized) with bunodont cheek teeth (see Fig. 2.2A) indicative of a generalised omnivorous diet. Artiodactyls started to diversify into the modern lineages: suines (pig-related forms), tylopods (camel-related forms), and ruminants (see Fig. 2.1) in the late Eocene, coincident with the high-latitude climatic deterioration. Dental changes included more specialised bunodont cheek teeth in many of the suines, indicative of a more specialised omnivorous diet, and the evolution of more selenodont cheek teeth (see Fig 2.2D) in the tylopods and ruminants, indicative of more specialised herbivory. The common selenodont artiodactyls of the late Eocene and Oligocene were various types of now-extinct small to medium-sized traguloid ruminants and tylopods.

The declining temperatures in the higher latitudes during the late Eocene (see Fig. 2.3) basically resulted in the extinction of smaller, generalised omnivorous forms among all ungulate groups, and the rise of more specialised folivores among surviving artiodactyl and perissodactyl lineages, while the remaining condylarths and ungulate-like mammals became extinct (Janis 2000). Most higher latitude primates also disappeared during this time. The megaherbivore brontotheres also went extinct at the end of the Eocene, possibly unable to sustain their specialised browsing diet in higher latitudes at this time. Rhinos persisted through this time period, but tapiroids were badly hit, and equoids (horses and palaeotheres) became completely extinct in the Old World. The familiar story of horse evolution relates an unbroken phylogeny through the North American Eocene, but this masks the fact that the abundance of individual equid fossils decreased sharply following the early Eocene diversity of the subfamily Hyracotheriinae. Only in the late Eocene did equids return to the North American fossil record in abundance, with the first member of the subfamily Anchitheriinae, the sheep-sized *Mesohippus*. *Mesohippus* was a very different beast from the earlier hyracotheres, with more strongly-lophed teeth, indicative of committed folivory, and more cursorially-adapted limbs.

This pattern of late Eocene perissodactyl decline and artiodactyl diversification has often been held as indicative of competitive replacement, ascribed to the supposedly superior foregut system of digestion in ruminating artiodactyls (see discussion in Janis 1976), but this is not supported by the patterns in the fossil record (Janis 1989). The late Eocene fossil record shows a general shift from omnivory to folivory (in terms of adaptive dental morphologies) in rodents as well as ungulates (Collinson and Hooker 1991; Meng and McKenna 1998). Ruminating artiodactyls may have been fortuitously better-adapted than perissodactyls to cope with the changing vegetation of the later Eocene, due to their ability to subsist on smaller amounts of more selectively chosen vegetation. The climatic changes, including increased patterns of seasonality, of the later Eocene may have resulted in changes in vegetational abundance, but perhaps more importantly would probably have resulted in a greater differentiation of fibre content between plant leaf and stem, allowing for the selective feeding habits that characterise present-day ruminants (see Janis 1989). Foregut fermentation would also have been useful for the detoxification of plant secondary compounds (Bodmer and Ward 2006). Small-to-medium-sized ruminants fare better than hindgut fermenters where food is lower in quantity but higher in quality. The increased body size of many perissodactyls in the late Eocene (e.g., among the brontotheres) may also reflect an adaptation to changing patterns of food abundance and quality.

2.2.3 The Lull Before the Storm: Oligocene and Early Miocene Times (34 to 20 Ma)

The 10 million years of the Oligocene and the first few million years of the Miocene were a time of relative calm, following the high latitude temperatures of the later Eocene, the rapid plunge in temperatures just after the Eocene/Oligocene boundary (see Fig. 2.3), and the extinctions in Europe (the 'Grande Coupure') that resulted not only from climatic change but from an influx of taxa from Asia (Hooker 2000). The mid to high latitude climate would have been equable, but also temperate and seasonal, with winter frosts (see Wolfe 1985). The prominent vegetation was deciduous forest or woodland, with some areas of relative aridity in North America (Retallack 2001), but no true modern-type grasslands or deserts.

Among North American and Eurasian ungulates, tapiroids were few, and the modern family Tapiriidae was now well-established. (Note that at this time there was not yet faunal exchange between Eurasia and Africa, and India was still an isolated island.) Rhinos were represented by a variety of forms of medium to large body size, including the huge (up to 15,000 kg) indricotheres (see Glossary, Box 2.1). Anchitheriine equids underwent a moderate diversification in North America, and *Anchitherium* itself migrated to Eurasia in the early Miocene. All of these perissodactyls had skulls and dentitions suggestive of generalised browsing. The general artiodactyl diversity was primarily a continuation of the late Eocene radiation of small or medium-sized browsers and omnivores.

However, it was among the Oligocene artiodactyls that the first incidences of hypsodonty appeared among ungulates, at least in the Northern Hemisphere. (While some native South American ungulates had hypsodont teeth from Eocene times, this feature alone—as noted below—is not necessarily indicative of grazing behavior, and many hypsodont notoungulates in fact had dental microwear indicative of browsing or mixed feeding; Townsend and Croft 2005). Oligocene hypsodont ungulates include several North American taxa: the diminutive hypertragulid *Hypisodus*, rock-hyrax-like oreodonts such as *Sespia,* and gazelle-like camelids such as *Stenomylus* (see Glossary, Box 2.1). The appearance of these hypsodont ungulates has been used in support of the notion of an early spread of grasslands and grazers in North America at this time (Retallack 1983), but the cranial morphology of these animals does not support a grazing diet. These ungulates all had very narrow muzzles, typical of specialised mixed-feeders in open habitats, where grit or abrasive types of browse were the main cause of high rates of tooth wear. It is likely that they represented some sort of specialised selective feeders living in open, arid areas. Some grass may have formed some part of their diet, but they were far from being specialised grazers. All of these lineages went extinct without issue during the Miocene: they were not ancestral to any of the later groups of hypsodont mammals.

2.2.4 The Rise of the Grasslands (20 to 10 Ma)

During the late early Miocene, around 20–17 Ma, both floral and faunal change became prominent, with mid-latitude grasslands diversifying in North America and eastern Asia. Grasslands may have been established several million years earlier in southern South America, and a fauna characterised by ungulates with hypsodont teeth occurred as early as the early Oligocene (Jacobs et al. 1999; see Fig. 2.3). Sea levels fell, and a land bridge opened up between Africa and Eurasia, allowing the migration of proboscideans out of Africa and the immigration into Africa of northern types of ungulates (rhinos and various ruminants such as bovids and giraffoids; see Turner and Antón 2004). There is little evidence for grasslands in Africa at this time, although the middle Miocene (14 Ma) site of Fort Ternan in Kenya contains evidence of fossil grasses and some ungulates that were probably mixed feeders (Jacobs et al. 1999).

In North America palaeobotanical evidence exists for grasslands during the late Oligocene and earliest Miocene from both plant phytoliths (Strömberg 2004, 2006) and fossil soils (Retallack 2001). However, the soil evidence, as well as evidence from other organisms such as legumes and insects, points to initial arid bunchgrass shrubland with more savanna-like short sod grasslands developing in the late early Miocene (Retallack 2001). The advent of grasslands in North America before the evolution of hypsodont ungulates such as equids has been argued as evidence for 'adaptive lag' (Retallack 1983; Strömberg 2006)—that is, that morphological evolution had not yet caught up to behavioural evolution. This evolutionary scenario is highly unlikely: as mentioned previously, hypsodonty is essential for an animal eating an abrasive diet, and hypsodonty also seems to be an easy feature to evolve, as it has evolved so many times convergently among mammals (Janis and Fortelius 1988; Damuth and Janis 2005).

The first pecorans (horned ruminants), including early members of the modern families Bovidae, Cervidae, and Giraffidae, made their first appearance in Eurasia at around 18 Ma. The earliest definite bovid, *Eotragus*, is known from around 18 Ma in Europe and Pakistan, some cervids of a similar age are known from Europe, and some large giraffoids are known from the early Miocene of Spain (see Gentry 2000). New types of ruminant artiodactyls appeared in North America, including antilocaprids (pronghorns) and palaeomerycids (see Glossary, Box 2.1). Also in North America larger and more derived types of camelids appeared, along with a great radiation of the first member of the modern equid subfamily, Equinae, the genus *Merychippus*. However, at this point none of these ungulates had craniodental morphologies indicative of grazing, even though hypsodonty was apparent among the equids, antilocaprids, and camelids.

This initial rise in the numbers of hypsodont ungulates has long been interpreted as the radiation of a 'savanna-like' fauna (e.g., Webb 1977), but it is not at all clear that these hypsodont taxa were actually specialised grazers: the majority of the hypsodont Miocene equids have dental microwear indicating mixed feeding (Solounias and Semprebon 2002). Many lineages of camelids also became more hypsodont at this time, but again their craniodental morphology does not support the notion of a grazing habit (Dompierre and Churcher 1996), and all of these taxa display rather moderate levels of hypsodonty in comparison with modern grazers.

Until recently, the fate of the browsing ungulates during the Miocene was thought to have been one of gradual extinction, replaced by the 'better-adapted' grazers (see discussion in Janis et al. 2004). In the familiar story of horse evolution there is rarely a mention of the parallel diversification to the hypsodont forms of specialised browsing anchitheriine equids that survived into the late Miocene in both North America and Eurasia, ranging from the goat-sized *Archaeohippus* up to the horse-sized *Hypohippus* and *Megahippus* (Janis et al. 1994; Agustí and Antón 2002). This radiation of browsing horses is merely a portion of the diversity of late early and middle Miocene browsers. Others browsers included the palaeomerycids, the bizarrely-clawed chalicotheres, various types of rhinos, and a diversity of mid-sized proboscideans such as mastodons (Mammutidae) and gomphotheres (see Glossary, Box 2.1). The evolutionary pattern of these browsers has been best documented in North America (Janis et al. 2000, 2002, 2004), but similar patterns are apparent in western Europe (Fortelius et al. 1996). In North America, not only were browsing ungulates highly diverse taxonomically on a continental scale (see Fig. 2.4), but they were also exceedingly species rich at individual fossil localities. The numbers of browsing ungulates at mid-Miocene fossil sites in North America greatly exceeds those in any habitat today, and similar patterns appear to hold true for ungulates at localities in Eurasia and Africa (Janis et al. 2004), and for terrestrial herbivores in South America (Kay and Madden 1997) and Australia (Myers 2002). This abundance of browsers disappears from North America by the start of the late Miocene, around 11 Ma, and an overall decline of ungulates also occurs at this time in North America, Western Eurasia, and in Pakistan (Barry et al. 1995). These palaeocommunities clearly represent some sort of woodland savanna habitat, but with no precise modern analog.

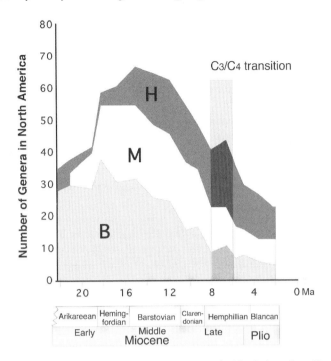

Fig. 2.4 Continent-wide generic diversity of ungulates in North America. H = hypsodont (i.e., probable grazers or mixed feeders in open habitats); M = mesodont (somewhat hypsodont; i.e., probable mixed feeders in open or closed habitats); B = brachydont (i.e., probable browsers). Modified from Janis et al. 2004

It is not clear how these mid Miocene habitats achieved a greater density of species, especially in terms of the numbers of browsers, than in their modern-day equivalents. Our (Janis et al. 2004) preferred explanation is for greater levels of atmospheric carbon dioxide than the preindustrial levels, leading to greater levels of primary productivity, but we acknowledge that this hypothesis is at odds with the current geochemical evidence. Additionally, species richness does not correlate in a simple fashion with productivity in modern ungulate faunas: Prins and Olff (1996) note that species richness of African grazers is actually highest at intermediate levels of grass productivity. Whatever the explanation, this great diversity of mid-Miocene browsers, occurring concurrently with the early diversification of more hypsodont, open-habitat forms, is an unappreciated evolutionary event.

2.2.5 The Rise of Grazing Ungulates (10 to 2 Ma)

It was not until the start of the late Miocene, around 10 Ma, that ungulates appeared with morphologies consistent with more specialised grazing adaptations. Note that, in general, the browsing lineages of the earlier Miocene went extinct without issue: although all grazers were obviously derived from browsers at some point in evolutionary history, the lineages that remained as browsers in the mid Miocene did not later transform into grazers. The remaining browsers in the late Miocene of North America were considerably bigger in body size than previously, indicating a declining quality and abundance of suitable forage (Janis et al. 1994).

A classic evolutionary pattern of distinct grazing versus browsing lineages can be seen among the North American horses. In the early Miocene the equid lineage split into the one leading to the modern Equinae (with the emergence of the genus *Parahippus*) and the lineage leading to the specialised large browsing horses of the later Miocene (derived members of the Anchitheriinae). These browsing horses (e.g., *Anchitherium, Hypohippus, Megahippus*) were a successful radiation for a good ten million years, but went extinct without issue during the climatic changes of the later Miocene. They did not change their diet at this time: instead the radiation into the grazing niche came from the Equinae line that had opted for a more mixed feeding diet back in the early Miocene. Another example is the North American dromomerycids (palaeomerycid ruminants). The final members of this predominantly browsing lineage showed some craniodental changes indicative of a more mixed-feeding strategy in the latest Miocene, but still declined in numbers and went extinct fairly rapidly, while at the same time there was an expansion of the antilocaprids that were previously more adapted for mixed feeding (Semprebon et al. 2004). The general evolutionary pattern in the late Miocene is for browsing lineages to be replaced by ones adapted for more fibrous diets (e.g., the replacement of tragulids and cervids by bovids in the Siwalik faunas of Pakistan; see, e.g., Barry 1995 and Barry et al 1995), rather than the browsers themselves undergoing evolutionary change. Likewise ungulate lineages do not transform back into browsing forms once they have evolved the more derived craniodental apparatus of grazers or mixed feeders. (A possible exception to this is the evolution of the specialised high-level browsing gerunuk, *Litocranius walleri*, that has a brachydont dentition presumably evolved from the more hypsodont dentition of other gazelles.)

In North America hypsodont horses, members of the extant tribe Equini and the extinct tribe Hipparionini (see Glossary, Box 2.1), diversified immensely in the late Miocene, and one lineage of hipparionines (or perhaps a couple) migrated into Eurasia by 10 Ma and from there into Africa. This time marks the appearance of the classic mid-latitude savanna faunas, the 'Clarendonian chronofauna' in North America and the '*Hipparion* fauna' in Eurasia. These later equines were generally larger than early members of the Equinae (such as *Merychippus*), approaching the size of modern equids (although some secondarily dwarfed lineages also existed), and had more hypsodont cheek teeth and craniodental features indicative of a more fibrous diet, such as deeper mandibles and broader muzzles. However, patterns of dental microwear suggest that these equids were still predominantly mixed feeders (Hayek et al. 1992; Solounias and Sempebron 2002).

The taxonomic diversity of early late Miocene (around 11–8 Ma) North American equids was very high: continent-wide there were ten sympatric genera, each with a large diversity of species, and individual fossil localities commonly contained half a dozen different equid species (Janis et al., 2004), making equids as taxonomically diverse as grazing bovids are today in Africa. With the exception of the short-legged rhinoceros *Teleoceras*, no other ungulates in North America appeared to challenge equids as grass specialists until the very end of the Miocene, when hypsodont proboscideans (gomphotheres) first appeared (Lambert and Shoshoni 1998). During the late Miocene there was a significant radiation of hypsodont camelids, and also of more derived antilocaprids (the antilocaprines, which include the modern pronghorn, and which were more hypsodont than the earlier merycodontine antilocaprids); but the craniodental features of these taxa (e.g., relatively narrow muzzles) are indicative of mixed feeding rather than grazing.

In the early late Miocene of Eurasia there was a reduction in numbers of browsers such as suids, tapirs, tragulids, and brachydont rhinos, suggesting a loss of forest habitat, and a radiation of bovids, equids, and more hypsodont rhinos and giraffoids, suggesting the spread of more grass-dominated habitats (Barry 1995; Fortelius et al. 1996; Agustí and Antón 2002; Costeur et al. 2004.). However, these faunas contained few undoubted grazing species, the hypsodont forms most likely being mixed feeders, and the habitat of these localities appears to have been woodland and shrubland rather than open savanna (Solounias and Dawson-Saunders 1988; Prins 1998). Few highly hypsodont bovids were present in Africa: the late Miocene fauna contained a large diversity of suids, rhinos, proboscideans, and giraffids, as well as bovids and hipparionine equids, probably representing an assemblage of browsers and mixed feeders (Turner and Antón, 2004).

2.2.6 *The Late Cenozoic Dawn of the Modern Ungulate Fauna*

Global temperatures fell dramatically in the latest Miocene (see Fig. 2.3), as did mammalian taxonomic diversity, and the mid-latitude vegetational habitats apparently became more arid (see e.g., Fortelius et al. 2002). Around 8 Ma, there was a shift in grass photosynthetic biochemistry from the C3 cycle to the C4 cycle in lower latitudes in North America and Asia (Cerling et al. 1993, 1997). It has been claimed that this event had a significant impact on the grazing mammals: a greater diversity of hypsodont artiodactyls, including bovids, appeared in Pakistan at this time while brachydont taxa such as suids and giraffids declined (Barry et al. 1995), but there was little overall impact on the hypsodonty levels of ungulates in North America (Janis et al. 2000).

Various examples of gigantism existed among ungulates at this time. There were four sympatric genera of oversized browsing or mixed-feeding camelids (i.e., ~1,500 kg) in the latest Miocene and the Pliocene of North America. The North American rhinos (one browsing and one grazing lineage) also increased in size throughout the Miocene to reach a similar size to present-day African rhinos by the end of the epoch. In the Old World, elephantids (all grazers) diversified at this time, as did large browsers such as chalicotheres, deinotheres (see Glossary, Box 2.1), and modern types of giraffids. There was also a radiation of the sivatheriine giraffids (see Glossary, Box 2.1) of similar size to the giant camelids.

In North America the rhinos were gone by the end of the Miocene, and the giant camelids did not survive past the Pliocene, but the diversity of large browsers survived in the Old World through the Pleistocene. Again, the situation of North America as an island continent probably accounts for its greater (and earlier) pattern of extinctions of large mammals. The only megaherbivores to survive into the Pleistocene in North America were the mammutid and gompthotheriid proboscideans, which were joined by the elephantid *Mammuthus* (mammoths, including the woolly mammoth and the Imperial mammoth) at this time, along with a small number of equids, antilocaprids, and camelids, and the newly immigrant bovids and cervids. The extinction of all the North American endemic ungulates and proboscideans at the end of the Pleistocene, with the exception of the pronghorn, *Antilocapra americana*, remains enigmatic, with numerous arguments for climatic change, human hunting, or some combination of the two.

African savanna habitats were first prominent in the Pliocene (Jacobs et al. 1999), with the initial radiation of specialised grazing bovids such as the hippotragines (e.g., sable) and alcelaphines (e.g., wildebeest). But the establishment of extensive open savannas of dry, secondary grasslands, versus woodland savanna or seasonally-flooded edaphic grasslands, was not apparent until the start of the Pleistocene, around 2 Ma, which is also when C4 grasses first appeared in Africa (Cerling 1992). At this time the first bovids with definitive specialised craniodental grazing adaptations appeared (Spencer 1997), evolving from earlier mixed feeding forms, and the total numbers of grazing taxa approached those of today (Reed 1997). The modern equid genus *Equus*, first evolving in North America, appeared in the Old World the late Pliocene, around 2.5 Ma; some remaining hipparionine equids (persistently three-toed forms) survived alongside *Equus* for a while, but eventually became extinct by the mid Pleistocene (Turner and Antón 2004). The grazing white rhino (genus *Ceratotherium*) also made its first appearance in the African Pliocene.

During the Pleistocene grazing megaherbivores such as the woolly mammoth (*Mammuthus*), and several types of hypsodont rhinos, including the woolly rhino (*Coelodonta*), were prevalent at mid to high latitudes, as were horses and large grazing bovids such as bison (*Bison*) and musk oxen (*Ovibos*). This array of large grazing ungulates, along with some smaller mixed-feeders [e.g., saiga antelope (*Saiga*) and reindeer (*Rangifer*)] comprised the fauna of the late Pleistocene 'steppe tundra' biome of northern Eurasia and Alaska, a habitat type with no modern analogue (Prins 1998; Agustí and Antón, 2002). In both North America and the Old World today bovids remain the prime grazers, although their diversity is much less in the Northern Hemisphere than it is in Africa. A few low latitude Asian cervids are today at least fresh-grass grazers, such as the barasinga (*Cervus duvaucelli*) and Père David's deer (*Elaphurus davidianus*; there is no equivalent radiation among cervids to bovids such as alcelaphines and hippotragines). However, it is not clear why cervids never really expanded into the grazing niche, and never colonised Africa south of the Sahara.

Ungulates such as deer, tapirs, and camelids (llamas), as well as now-extinct or extirpated forms such as gomphothere proboscideans and equids, dispersed into South America in the Pliocene, around 2.5 Ma (Webb 1991). The equids and gomphotheres diversified into the grazing niche, but no ungulate is a true grazer there today. Tapirs have remained as specialised browsers; llamas never diversified much beyond their present day high-altitude, mixed-feeding ecological role; and while cervids underwent the most profound diversification, they never evolved true grazing forms, although the marsh deer (*Blastocerus dichotomous*) is a fresh grass grazer. Bovids never reached South America until introduced by humans, and it is unclear why no cervid evolved there to be a specialised grazer.

2.3 Discussion and Conclusions

Grazing is a fairly recent evolutionary specialization among ungulates. Although extensive grasslands first appeared around 25 Ma—and along with them ungulates with hypsodont teeth designed to withstand a greater amount of abrasion—specialised grazers did not appear until around 10 Ma, and most true grazers are of Plio-Pleistocene age (Fig. 2.3). Many different lineages have evolved intermediate feeding types, but few have progressed to specialised grazing.

Although the majority of grazers today are bovids—and specialised grazing evolved a minimum of four times within this clade (within the Bovini, Caprini, Hippotragini, and Alcelaphini)—no other ruminant, past or present, has evolved a specialised grazing form (although a few modern cervids, such as *Blastocerus, Elaphurus* and species of *Cervus* and *Axis*, have a grass-dominated diet). Within other artiodactyls, grazers have appeared recently (Plio-Pleistocene) among suids (the warthog) and hippos, but were not in evidence before this time. In contrast, grazing appears to have been easier to evolve among hindgut fermenters.

Clauss et al. (2003) present an elegant argument for why ruminants, past and present, do not attain the size of rhinos and elephants. Ruminants have longer passage times of the digesta than hindgut fermenters of a comparable body size (around 40 hours for a horse versus around 60 hours for a cow), and the time for passage of the digesta increases with body size. However, if the passage time approaches 4 days, an acute problem develops with the growth of methanogenic bacteria, which convert acetic acid (the main volatile fatty acid product of cellulose fermentation) to methane and carbon dioxide, with resultant high energy losses. Elephants solve this problem by adopting relatively shorter and broader guts, thus speeding up the passage rate of the digesta; but this would not be a possible solution for a ruminant, where the rumenoreticulum portion of the stomach is specifically adapted to delay food passage. Thus ruminants are limited to body sizes where their digesta passage time is less than four days (likely under 1,500 kg).

Within the equids, several different clades in the subfamily Equinae progressed from mixed feeding to grazing (although only *Equus* survives). Grazing appeared independently within several different clades of rhinos, including the white rhino, and among extinct forms the North American *Teleoceras*, and the Old World woolly rhino (*Coelodonta*) and steppe rhino (*Elasmotherium*). Grazing also evolved three times within the Proboscidea (within the Elephantidae, Stegodontidae, and Gomphotheriidae), and once in the Hyracoidea (*Procavia*).

It is commonly thought that bovids somehow out-competed equids, but this is erroneous. Equids were predominantly a New World radiation, and their decline in taxonomic diversity in the late Cenozoic was due to being stranded in an island continent, with no tropical refuge zone when the more productive savanna grasslands turned to less productive prairie. North American equids never encountered bovids until the immigration of sheep and bison in the Pleistocene, by which time only one genus, *Equus,* remained. However, this taxon was still a very prominent part of the Pleistocene faunas, and the fossils are extremely numerous, suggesting individual abundance in life. North American equids only went extinct in the end-Pleistocene extinctions that decimated the megafauna. Equids that migrated into the Old World encountered an existing bovid-dominated faunal community, which may explain their lack of taxonomic diversification there in comparison with North America. While modern equids have diversified into a variety of different species (e.g., asses in Asia, zebras in Africa), and are individually numerous when encountered, there is still usually only one species of equid in any faunal community (this was also broadly true of Old World fossil communities), and all extant equid species belong to a single genus with essentially the same ecology (i.e., herd-living, specialised grazing).

The bias of today's world suggests a normality of the predominance of ruminant grazers, and also a diversity of grazers among other ungulates. However, the fossil record shows us that we need to think more about why grazing is so difficult to attain in non-bovid ruminants, and why in general it appears to be easy to evolve an animal that takes some grass in its diet, but not to evolve a specialised grazer.

Acknowledgements This chapter was greatly improved by comments from Herbert Prins and an anonymous reviewer.

References

Agustí J, Antón M (2002) Mammoths, sabertooths, and hominids: 65 million years of mammalian evolution in Europe. Columbia University Press, New York

Barry JC (1995) Faunal turnover and diversity in the terrestrial Neogene of Pakistan. In: Vrba ES, Denton GH, Partridge TC, Buckle LH (eds) Paleoclimate and evolution with emphasis on human origins. Yale University Press, New Haven, pp 115–134

Barry JC, Morgan ME, Flynn LJ, Pilbeam D, Jacobs LL, Lindsay EH, Raza SM, Solounias N (1995) Patterns of faunal diversity and turnover in the Neogene Siwaliks of northern Pakistan. Palaeogeogr. Palaeoclimatol 115:209–226

Benton MJ (1993) The fossil record 2. Chapman and Hall, London

Bodmer R, Ward D (2006) Frugivory in large mammalian herbivores. In: Danell K, Duncan P, Bergström R, Pastor J (eds) Large herbivore ecosystem dynamics and conservation. Cambridge University Press, Cambridge, pp 232–260

Cerling TE (1992) Development of grasslands and savannas in East Africa during the Neogene. Palaeogeogr. Palaeoclimatol 97:241–247

Cerling TE, Wang Y, Quade J (1993) Global ecological change in the late Miocene: expansion of C4 ecosystems. Nature 361:344–345

Cerling TE, Harris JM, MacFadden BJ, Leakey MG, Quade J, Eisenmann V, Ehleringer JR (1997) Global vegetation change through the Miocene-Pliocene boundary. Nature 389:153–158

Clauss M, Frey R, Kiefer B, Lechner-Doll M, Loehlein W, Polster C, Rössner GE, Streich WJ (2003) The maximum attainable body size of herbivorous mammals: morphophysiological constraints on foregut, and adaptations of hindgut fermenters. Oecologia 136:14–27

Collinson ME, Fowler MK, Boulter MC (1981) Floristic change indicates a cooling climate in the Eocene of southern England. Nature 291:315–317

Collinson ME, Hooker JJ (1991) Fossil evidence of interactions between plants and plant-eating mammals. Philos T Roy Soc B 333:197–208

Costeur L, Legendre S, Escarguel G (2004) European large mammals palaeobiogeography and biodiversity during the Neogene. Palaeogeographic and climatic impacts. Rev Paeobiol Geneve 9:99–109

Damuth JD, Fortelius M, Andrews P, Badgley C, Hadley EA, Hixon S, Janis C, Madden RH, Reed K, Smith FA, Theodor J, Van Dam JA, Van Valkenburgh B, Werdelin L (2002) Reconstructing mean annual precipitation based on mammalian dental morphology and local species richness. J Vertebr Paleontol 22(suppl):48A

Damuth J, Janis CM (2005) Paleoecological inferences using tooth wear rates, hypsodonty and life history in ungulates. J Vertebr Paleontol 25(suppl):49A

Dompierre H, Churcher CS (1996) Premaxillary shape as an indicator of the diet of seven extinct late Cenozoic New World camelids. J Vertebr Paleontol 16:141–148

Fortelius M, Werdelin L, Andrews P, Bernor RL, Gentry A, Humphrey L, Mittman H-W, Viranta S (1996) Provinciality, diversity, turnover, and paleoecology in land mammal faunas of the later Miocene of western Eurasia. In: Bernor RL, Fahlbusch V, Mittman H-W (eds) The evolution of western Eurasian Neogene mammal faunas. Columbia University Press, New York, pp 414–448

Fortelius M, Eronen J, Jernvall J, Liu L, Pushkina D, Rinne J, Tesakov A, Vislobokva I, Zhang Z, Zhou L (2002) Fossil mammals resolve regional patterns of Eurasian climatic change over 20 million years. Evol Ecol Res 4:1005–1016

Gentry AW (2000) The ruminant radiation. In: Vrba ES, Schaller GB (eds) Antelopes, deer and relatives. Yale University Press, New Haven, pp 11–25

Gordon IJ (2003) Browsing and grazing ruminants: are they different beasts? Forest Ecol Manag 181:13–21

Gordon IJ, Illius AW (1988) Incisor arcade structure and diet selection in ruminants. Funct Ecol 2:15–22

Hayek L-A, Bernor RL, Solounias N, Steigerwald P (1992) Preliminary studies of hipparionine horse diet as measured by tooth microwear. Ann Zool Fenn 28:187–200

Hooker JJ (2000) Palaeogene mammals: crises and ecological change. In: Culver SJ, Rawson PF (eds) Biotic response to global change: the last 145 million years. Cambridge University Press, Cambridge, pp 333–349

Jacobs BF, Kingston JD, Jacobs LL (1999) The origin of grass-dominated ecosystems. Ann Mo Bot Gard 86:590–643

Janis C (1976) The evolutionary strategy of the Equidae, and the origin of rumen and cecal digestion. Evolution 30:757–774

Janis CM (1988) An estimation of tooth volume and hypsodonty indices in ungulate mammals, and the correlation of these factors with dietary preferences. Mem Mus Hist Nat, Paris C 53:367–387

Janis CM (1989) A climatic explanation for patterns of evolutionary diversity in ungulate mammals. Palaeontology 32:463–481

Janis CM (1993) Tertiary mammal evolution in the context of changing climates, vegetation, and tectonic events. Annu Rev Ecol Syst 24:467–500

Janis CM (1995) Correlation between craniodental morphology and feeding behavior in ungulates: reciprocal illumination between living and fossil taxa. In: Thomason JJ (ed) Functional morphology in vertebrate paleontology. Cambridge University Press, Cambridge, pp 76–98

Janis, CM (2000) Patterns in the evolution of herbivory in large terrestrial mammals: the Paleogene of North America. In: Sues H-D, Labanderia C (eds) Origin and evolution of herbivory in terrestrial vertebrates. Cambridge University Press, Cambridge, pp 168–221

Janis CM, Fortelius M (1988) On the means whereby mammals achieve increased functional durability of their dentitions, with special reference to limiting factors. Biol Rev 63:197–230

Janis CM, Gordon I, Illius A (1994) Modelling equid/ruminant competition in the fossil record. Hist Biol 8:15–29

Janis CM, Scott KM, Jacobs LL (1998) Evolution of Tertiary mammals of North America. Cambridge University Press, Cambridge

Janis CM, Damuth J, Theodor JM (2000) Miocene ungulates and terrestrial primary productivity: Where have all the browsers gone? Proc Natl Acad Sci 97:7899–7904

Janis CM, Damuth J, Theodor JM (2002) The origins and evolution of the North American grassland biome: the story from the hooved mammals. Palaeogeogr. Palaeoclimatol 177:183–198

Janis CM, Damuth J, Theodor JM (2004) The species richness of Miocene browsers, and implications for habitat type and primary productivity in the North American grassland biome. Palaeogeogr. Palaeoclimatol 207:371–398

Jernvall J, Hunter JP, Fortelius M (1996) Molar tooth diversity, disparity, and ecology in Cenozoic ungulate radiations. Science 274:1489–1492

Kay RF, Madden RH (1997) Mammals and rainfall: paleoecology of the middle Miocene at La Venta (Colombia, South America). J Hum Evol 32:161–199

Kowalevsky W (1876) Sur l'*Anchitherium aurelianse* Cuv. et sur l'histoire paléontologie des chevaux. Mem Acad Imp Sciences de St. Petersburg, VII, 20:1–73

Lambert WD, Shoshoni J (1998) Proboscidea. In Janis CM, Scott KM, Jacobs LL (eds) Evolution of Tertiary mammals of North America. Cambridge University Press, Cambridge, pp 606–621

Leopold E, Liu G, Clay-Poole S (1992) Low biomass vegetation in the Oligocene? In: Prothero DR, Berggren WA (eds) Eocene–Oligocene climatic change and biotic evolution. Princeton University Press, Princeton, pp 399–420

Matthew WD (1926) The evolution of the horse: its record and interpretation. Quart Rev Biol 1:139–185

Mendoza M, Janis CM, Palmqvist P (2002) Characterizing complex craniodental patterns related to feeding behaviour in ungulates: a multivariate approach. J Zool Lond 58:223–246

Meng J, McKenna MC (1998) Faunal turnovers of Palaeogene mammals from the Mongolian Plateau. Nature 394:364–367

Myers TJM (2002) Paleoecology of Oligo-Miocene local faunas from Riversleigh. Unpublished PhD Thesis. University of New South Wales, Sydney

Pérez-Barbería FJ, Gordon IJ, Nores C (2001) Evolutionary transitions among feeding styles and habitats in ungulates. Evol Ecol Res 3:221–230

Prasad V, Strömberg CAE, Alimohammadian H, Sahni A (2005) Dinosaur coprolites and the early evolution of grasses and grazers. Science 310:1177–1180

Prins HHT (1998) The origins of grassland communities in northwestern Europe. In: Wallis deVries MF, Bakker JP, van Wieren SE (eds) Grazing and conservation management. Kluwer, Boston, pp 55–105

Prins HHT, Olff H (1996) Species-richness of African grazer assemblages: towards a functional explanation. In: Newbery DM, Prins HHT, Brown ND (eds) Dynamics of tropical communities. Blackwell, Oxford, pp 449–490

Prothero, DR, Schoch RM (2002) Horns, tusks and flippers: the evolution of hoofed mammals. Johns Hopkins Press, Baltimore

Reed, KE (1997) Early hominid evolution and ecological change through the Africa Plio-Pleistocene. J Hum Evol 32:289–322

Retallack GJ (1983) Late Eocene and Oligocene paleosols from Badlands National Park, South Dakota. Geol Soc Am Spec Pap 193:1–82

Retallack GJ (2001) Cenozoic expansion of grasslands and climatic cooling. J Geol 109:407–426

Rose KD (2006) The beginning of the age of mammals. Johns Hopkins Press, Baltimore

Semprebon G, Janis CM, Solounias N (2004) The diets of the Dromomerycidae (Mammalia; Artiodactyla) and their response to Miocene climatic and vegetational change. J Vertebr Paleontol 24:430–447

Solounias N, Dawson-Saunders B (1988) Dietary adaptations and paleoecology of the late Miocene ruminants from Pikermi and Samos in Greece. Palaeogeogr. Palaeoclimatol 65:149–172

Solounias N, Semprebon G (2002) Advances in the reconstruction of ungulate ecomorphology with application to early fossil equids. Am Mus Novit 3366:1–49

Spencer LM (1997) Dietary adaptations of Plio-Pleistocene Bovidae: implications for hominid habitat use. J Hum Evol 32:201–228

Sponheimer M, Reed KE, Lee-Thorpe JA (1999) Comparing isotopic and ecomorphological data to refine bovid paleodietary reconstruction. J Hum Evol 36:705–718

Springer MS, Murphy WJ, Eizirik E, O'Brien SJ (2005) Molecular evidence for major placental clades. In Rose KD, Archibald JD (eds) The rise of placental mammals: origins and relationships of major clades. John Hopkins Press, Baltimore, pp 37–49

Strömberg CAE (2004) Using phytolith assemblages to reconstruct the origin and spread of grass-dominated habitats in the great plains of North America during the late Eocene to early Miocene. Palaeogeogr. Palaeoclimatol 207:239–275

Strömberg CAE (2006) Evolution of hypsodonty in equids: testing a hypothesis of adaptation. Paleobiology 32: 236–258

Townsend KE, Croft DA (2005) Low-magnification microwear analyses of South American endemic herbivores. J Vertebr Paleontol 25 (suppl):123A

Turner A, Antón M (2004) Evolving Eden: an illustrated guide to the evolution of the African large-mammal faunas. Columbia University Press, New York

Webb SD (1977) A history of savanna vertebrates in the New World. Part I: North America. Annu Rev Ecol Syst 8:355–380

Webb SD (1991) Ecogeography and the Great American Interchange. Paleobiology 17:266–280

Wolfe JA (1985) Distribution of major vegetational types during the Tertiary. Geophys Monogr 32:357–375

Zachos J, Pagani M, Sloan L, Thomas E, Billups K (2001) Trends, rhythms, and aberrations in global climate 65 Ma to present. Science 292:686–693

Chapter 3
The Morphophysiological Adaptations of Browsing and Grazing Mammals

Marcus Clauss, Thomas Kaiser, and Jürgen Hummel

3.1 Introduction

The behaviour, physiology and morphology of animals are the outcome of adaptations to particular ecological niches they occupy or once occupied. Studying the correlation between a given set of characteristics of an ecological niche and the morphological and physiological adaptations of organisms to these characteristics is one of the most basic approaches to comparative biology, and has fuelled scientific interest for generations (Gould 2002). However, current scientific standards cannot be met by mere descriptions of both the characteristics of the niche and the organism, and a (hypothetical) intuitive explanation for the adaptive relevance of the latter; the presence or absence of a characteristic must be demonstrated in sound statistical terms (Hagen 2003)[1] ideally supported by experimental data (from in vivo, in vitro, or model assays) on its adaptive relevance.

In this chapter, we adopt an approach that first presents the relevant characteristics of the ecological niche of the 'grazer' (GR) and of the 'browser' (BR), outlines hypotheses based on these characteristics, and finally addresses examples where such hypotheses have been tested. As we consider that the discussion of morphophysiological differences between GR and BR is still unresolved, subjects of future interest, such as the particular adaptations of mixed feeders, or more elaborate classifications of feeding types (e.g., including frugivores, and differences between grasses, sedges, forbs, herbs, and woody browse), are not considered here. With respect to botanical entities, 'browse' in this chapter refers to herbs, forbs, and leaves and twigs of woody plants.

The terms 'grazer' and 'browser' have been used for a long time to characterise feeding types; however, it was Hofmann and co-workers (Hofmann and Stewart 1972; Hofmann 1973; 1988; 1989; 1991; 1999) who brought them into common use. Partly in connection with the original term 'concentrate selector', which will not be adopted in this chapter (Clauss et al. 2003b; 2003c), the term 'browser' has become synonymous with an organism feeding selectively on relatively easily digestible

---

[1] But see recent methodological work in human medicine showing that statistical significance alone cannot be used as an argument to support or falsify a hypothesis (Ioannidis 2005).

material. Hence, even mites (Siepel and de Ruiter-Dijkman 1993) and carnivorous fish (Lechanteur and Griffiths 2003) have been classified as 'grazers' and 'browsers'. In this chapter, these terms are used strictly in relation to their botanical connotation and are not used as indicators of selectivity. Demment and Longhurst (1987) proposed a classification scheme that demonstrated that there are both selective and unselective species within the GR and BR classes. Selectivity generally decreases with body size (Jarman 1974; Owen-Smith 1988), and differences between feeding type on the one hand and degree of selectivity on the other have been incorporated into a model to explain niche separation (Owen-Smith 1985).

Potential adaptations to browse or grass diets have often been compared to consequences of difference in body mass between species (Hofmann 1989; Gordon and Illius 1994; Gordon and Illius 1996). In this chapter, therefore, body mass is only included as an alternative explanation, but the influence of body mass itself on digestive processes is not reviewed.

3.2 Grass and Browse

Whereas data compilations of animal species have been published in large number (see Sects. 3.3 and 3.5 for references), there is, as far as we are aware, a surprising lack of any systematic evaluation of differences between grasses and browses in terms of their physical and chemical characteristics. In other words, the debate about differences between grazers and browsers is often based on hearsay, as far as the assumed differences between grass and browse are concerned; for example, the often quoted increased amount of grit adhering to grass forage is a conceptual cornerstone of many investigations on the hypsodont dentition of grazers (Fortelius 1985; Janis 1988; Janis and Fortelius 1988; Williams and Kay 2001), but has never been demonstrated quantitatively. Here, we only cite works that generated or at least collated comparative data (even if not statistically testing differences). When considering the literature, we think there is agreement on the forage characteristics (Table 3.1) that are of relevance for the topic of this chapter.

Growth Pattern/Location

These have the potential to influence overall body design and the food selection mechanism.

A1. It is generally assumed that grasses predominate in open landscapes, whereas browse predominates in forests or spatially more structured landscapes.

A2. It is generally assumed that grasses typically grow close to the ground (with evident exceptions such as napier grass), whereas browse grows at different heights (with forbs often at even lower growth levels than surrounding grasses, and woody browse of shrubs and trees mostly above grass level).

Table 3.1 Summary of characteristics of browse and grass used for the generation of predictions of morphophysiological differences between browsers and grazers. The functional relevance (FR) code links these predictions to the following tables. See text for more detailed explanations and references

Subject groups	FR	Characteristic	Browse	Grass
Growth pattern/location	A1	landscape	forests/spatially more structured	open
	A2	growth pattern	at different heights	mostly close to ground
	A3	nutritional homogeneity of a 'bite'	less	more
Chemical composition	B1	protein content	higher (including nitrogenous secondary compounds)	lower
	B2	fibre content	lower but more lignified	higher but less lignified
	B3	pectin content	higher	lower
	B4	secondary compounds	more	less
Physical characteristics	C1	abrasive silica	less	more
	C2	adhering grit	less	more
	C3	resistance to chewing	less	more
	C4	fracture pattern	polygonal	longish fibre-like
	C5	change in specific gravity during fermentation	less	more
Digestion/fermentation	D1	overall digestibility	lower	higher
	D2	speed of digestion	fast	slow

A3. On the scale of single bites, differences in nutritional quality are more pronounced in browse (Van Soest 1996).

Chemical Composition

These have the potential to influence overall metabolism.

B1. Grass generally contains less protein than browse (Dougall et al. 1964; Owen-Smith 1982; Codron et al. 2007a). The high protein content of browse should be regarded with some caution, since some part of the nitrogen in browse may stem from, or may be bound to, plant secondary compounds (Oftedal 1991). For the calculation of crude protein content, nitrogen content it is generally multiplied by 6.25; in contrast, Milton and Dintzis (1981) suggest that this nitrogen conversion factor should be as low as 4.4 for tropical browse.

B2. Grass contains more fibre, and a greater proportion of this fibre is cellulose, while browse has less total but more lignified fibre (Short et al. 1974; Oldemeyer et al. 1977; Owen-Smith 1982; McDowell et al. 1983; Cork and Foley 1991; Robbins 1993; Van Wieren 1996b; Iason and Van Wieren 1999; Holechek et al. 2004; Hummel et al. 2006; Codron et al. 2007a). These differences are more pronounced if C4 grasses are compared to browse (Caswell et al. 1973). The fact that no difference in fibre content between grass and browse was demonstrated in a comprehensive set of samples of East African forage plants (Dougall et al. 1964) is explained by the inclusion of twigs in the browse analysis and the use of the crude fibre method to estimate fibre content, which can considerably underestimate lignin and hemicellulose content of tropical forage, especially tropical grasses (Van Soest 1975; McCammon-Feldman et al. 1981).

B3. Although few data exist, grass and browse contain comparable low levels of easily digestible carbohydrates, such as sugar and starch (Cork and Foley 1991; Robbins 1993). This is different for pectins, an easily fermentable part of the cell wall, which is much more prominent in the browse cell wall at concentrations of 6–12 % of total forage dry matter (Robbins 1993).

B4. Browse leaves contain secondary plant compounds that can act as feeding deterrents either by poisoning or reducing plant digestibility (Freeland and Janzen 1974; Bryant et al. 1992; Iason and Van Wieren 1999). Common secondary plant compounds such as tannins occur more often in woody browse (80% of taxa) as compared to forbs (15% of taxa; Rhoades and Gates 1976) (see Duncan and Poppi in this book, Chapter 4).

Physical Characteristics

These have the potential to influence adaptations of oral food processing, and might be important drivers of the differentiation of ruminant forestomach morphology.

C1. Grasses contain abrasive silica (Dougall et al. 1964; McNaughton et al. 1985); silica is harder than tooth enamel and thus wears it down (Baker et al. 1959; but see Sanson et al. 2007).

C2. As grasses grow close to the ground, it is assumed that grass forage contains more adhering grit than browse forage, but as stated before, this has not been tested quantitatively. Herbs, typically included in the 'browse' category, should share this characteristic with grasses.

C3. Differences in the masticatory force required to comminute grass/browse have been hypothesised (e.g., Solounias and Dawson-Saunders 1988; Mendoza et al. 2002), but not described. Spalinger et al. (1986) attribute thicker cell walls to grass leaves than to forb and woody browse leaves (while cell walls of twigs were thickest). The grinding of C4 grasses needs distinctively more force than that of C3 grasses (Caswell et al. 1973), possibly due to a greater percentage of bundle sheaths in C4 grasses (Heckathorn et al. 1999).

C4. Differences in fracture patterns of grass and browse have been noted (Spalinger et al. 1986; Kay 1993; Van Wieren 1996a). Several authors have reported more polygonal particles from herbaceous forage leaves and more longish particles from grass leaves (Troelsen and Campbell 1968; Moseley and Jones 1984; Mtengeti et al. 1995). Although empirical studies are lacking, browse is thought to be a more heterogeneous material with different levels of tissue thicknesses and of resistance to breakage, whereas grass is considered more homogenous in this respect. The fibre bundles in grasses are believed to be more evenly distributed and at higher density than in most browse species (Sanson 1989).

C5. Once submitted to fermentation, different forages show different buoyancy characteristics, due to differences in fibre composition, fracture shape, hydration capacity, and bacterial attachment (Martz and Belyea 1986; Lirette et al. 1990; Wattiaux et al. 1992); particles rich in cellulose are expected to change their functional density at a slower rate. Nocek and Kohn (1987) and Bailoni et al. (1998) found that the absolute change in functional specific gravity was greater for grass than for alfalfa hays, suggesting that there may be systematic differences between forages.

Digestion/Fermentation Characteristics

These have the potential to influence overall digestive physiology.

D1. Grass should yield more energy from fermentation per unit forage (Codron et al. 2007a), which explains why both in vivo and in vitro overall digestibility (generally measured as digestibility after fermentation times >24h) are often found to be higher in grass than in woody browse (Wofford and Holechek 1982; Wilman and Riley 1993; Van Wieren 1996b; Hummel et al. 2006). However, the evidence is equivocal, as some other references give higher digestibilities for woody browse than for grass (Short et al. 1974; Blair et al. 1977; Holechek et al. 2004). All references agree that forbs have a comparatively high digestibility.

D2. Browse is fermented at a faster rate than grass during the initial stages of fermentation (Short et al. 1974; Holechek et al. 2004; Hummel et al. 2006). This characteristic distinction has been well established for alfalfa (higher fermentation rate; lower digestibility) and for grass (lower fermentation rate; higher digestibility; Waldo et al. 1972). The reasons for this might be differences in cell wall structure, but have not been explored in detail. In vivo digestion of different forages confirms the pattern found in vitro (Short 1975). In comparison to C3 grasses and legumes, C4 grasses have some characteristics, such as a more strongly attached epidermis, less loosely arranged mesophyll cells, and a parenchymal bundle sheath, which all may prolong the fermentative process (Wilson 1993), and a slower fermentation rate for a C4 than a C3 grass has been demonstrated by Wilson et al. (1989).

3.3 Predictions

Based on the knowledge of chemical and physical differences between grass and browse, we can put forward hypotheses regarding the morphological adaptations of herbivores to these feeding niches (Table 3.2). In this section, we include visual analyses of data, for example comparative graphs or tables that have not been assessed statistically; such works are considered exploratory or preliminary here. We do not infer such contributions to be of minor relevance; they simply represent an early step in hypothesis testing. Publications that actually test such predictions statistically are not included in this section but appear later in section 3.5 (Results).

It must be noted that the alternative hypothesis to most of the predictions listed here is that there is no influence of feeding type (i. e., browser or grazer) on the respective parameter, but only an influence of body mass and/or phylogenetic descent.

Chemical and Physical Ingesta Properties

B1-B4, C4-C5, D2.[2] It is to be expected that stomach contents of BR and GR differ systematically in their content of chemical, physical, and fermentation characteristics. To our knowledge, no systematic evaluation of protein or fibre content, or of fracture or buoyancy properties, has been performed of the gut contents of herbivores of different feeding types. Owen-Smith (1988) demonstrated correlations between body size and the protein or non-stem fractions in the gut contents of herbivores, thus confirming a principal correlation between body size and selectivity; however, no test for differences between feeding types was performed. It has been suggested that differences in fermentation patterns should be reflected in fermentation rates and their products between BR and GR ruminants (Hofmann 1989), or that differences in content of easily digestible carbohydrates should lead to differences in ruminal pH (Gordon and Illius 1996)—a fact confirmed for a limited number of species by Jones et al. (2001).

Overall Body Design

A1. It is expected that most GR exhibit adaptations in their limb anatomy to living in open habitats, enabling them to use the available open space by increased cursorial activity. Actually, many correlations between postcranial skeleton parameters

---

[2] These codes refer to the characteristics of grass and browse outlined in the previous section.

Table 3.2 Summary of predicted morphophysiological differences between grazers and browsers based on the plant characteristics summarised in Table 3.1 [use the functional relevance (FR) code to link plant and animal characteristics]. See text for more detailed explanations and references

Subject groups	FR	Characteristic	Browser	Grazer
Ingesta properties[a]	B1, B2	gut contents nutrient content	higher protein?, lower fibre/higher lignin	lower protein? higher fibre/lower lignin
	B3/D2	gut contents volatile fatty acid production rate	higher?	lower?
	C4	gut contents particle pattern	polygonal	longish fibre-like
	C5	gut contents buoancy characteristics	homogeneous	inhomogeneous
Overall body design	A1	limb anatomy	shorter	longer
	A2	adaptations to feeding on the ground	less pronounced: thoracic vertebrae hump, muscles supporting head and skull in grass cropping, paracondylar and glenoid attachment areas, face length	more pronounced: thoracic vertebrae hump, muscles supporting head and skull in grass cropping, paracondylar and glenoid attachment areas, face length
	A3	oral anatomy for selective feeding	more pronounced incisor differentiation, a pointed/narrow incisor arcade, a short muzzle width with long lips, a long mouth opening, pronounced lip muscles	less pronounced incisor differentiation, a square/wide incisor arcade, a long muzzle width with short lips, a short mouth opining, less flexible lip muscles
Metabolism	B1	protein requirements	higher?	lower?
	B4	adaptations against secondary plant compounds	salivary tannin-binding proteins and larger salivary glands, larger livers/increased detoxification capacity	no salivary tannin-binding proteins, smaller livers/decreased detoxification capacity

(continued)

Table 3.2 (continued)

Subject groups	FR	Characteristic	Browser	Grazer
Dental/buccal morphology	C1, C2	adaptations against abrasion	no hypsodonty; enamel adaptations	hypsodonty; enamel adaptations; consequences of hypsodonty such as deeper mandibles, posterior displacement of orbita
	C3	adaptations to different chewing forces	slender mandible, longer molar tooth row, wider palate, smaller masseter and insertion areas	robust mandible, shorter molar tooth row, narrower palate, larger masseter and insertion areas
	C4	adaptations to different fracture properties	tooth morphology; when consuming grass forage: larger ingesta particles	tooth morphology; when consuming grass forage: smaller ingesta particles
General digestive physiology	D1	daily food intake	higher	lower
	D2	adaptations to differences in digestion kinetics	shorter ingesta retention; on a comparable diet: lower digestibility; lower gut capacity; increased faecal losses (sodium); less cellulolytic activity, less diverse protozoal fauna	longer ingesta retention; on a comparable diet: higher digestibility; greater gut capacity; lower faecal losses; more cellulolytic activity; diverse protozoal fauna
Ruminant forestomach physiology	C4, C5	adaptations to the tendency of forage to stratify	weaker rumen pillars, lower reticular crests, smaller omasa, more viscous saliva, frothy contents, complete rumen papillation, more acid-producing abomasal mucosa	stronger rumen pillars, higher reticular crests, larger omasa, less viscous saliva, stratified contents with fibre mat, fluid layer and gas dome, unpapillated dorsal rumen area, less acid-producing abomasal mucosa

(continued)

Table 3.2 (continued)

Subject groups	FR	Characteristic	Browser	Grazer
Readiness to ingest forage in captivity	C1–C5	problems observed in zoo animals	less acceptance of grass hay	good acceptance of grass hay

^a Predictions about ingesta properties are made more difficult by the fact that gut contents always consist of a mixture of material recently ingested, of material that has been digested for varying lengths of time, and of symbiotic gut microbes

and habitat have been demonstrated in bovids (Scott 1985, 1987; Kappelman 1988; Köhler 1993; Plummer and Bishop 1994; Kappelmann et al. 1997; DeGusta and Vrba 2003; DeGusta and Vrba 2005a; DeGusta and Vrba 2005b; Mendoza and Palmqvist 2006b). As these represent morphological correlates of habitat rather than diet, they are not dealt with explicitly in this chapter.

A2. In accordance with comparisons using two species (Haschick and Kerley 1996; du Plessis et al. 2004), it is expected that the preferred feeding height of GR is lower than that of BR; consequently, it is expected that GR show adaptations to feeding close to the ground. A lower angle between braincase and the facial cranium should be a positional adaptation to ground feeding. The peak of the hump of the thoracic vertebrae should be correlated with the preferred feeding height, with a tall hump close to the head being advantageous for ground-level feeding (increases the moment arm of the nuchal musculature; Guthrie 1990); however, results from analyses on this hump so far are equivocal (Spencer 1995), which could be due to the fact that those browsers feeding on herbs/forbs would, by necessity, also have to feed close to the ground. Muscles supporting head and skull movements in grass cropping should be more pronounced in GR with accordingly more pronounced attachment areas (paracondlyar and glenoid). A longer face could serve to keep the eyes away from the grass, which might protect them (Janis 1995), help to maintain good visibility for predator detection during feeding (Gentry 1980), and/or enhance moment arms of the head and mandible for more efficient cropping and mastication. An interesting potential example of a skeletal feeding height adaptation is the decreasing length of the metatarsus in Sivatheres with increasing proportion of grass in their diet (Cerling et al. 2005).

A3. BR are expected to show adaptations for a more selective feeding, whereas GR are expected to show adaptations for a more unselective food intake. Browsers are expected to have a higher dental incisor index (i.e., a more pronounced size difference between the individual incisor teeth and the canine teeth; Boué 1970). BR are expected to have a lower muzzle width, a more pointed and narrower incisor arcade than GR, which have more square incisor arcades (Boué 1970; Bell 1971; Owen-Smith 1982; Bunnell and Gillingham 1985; Gordon and Illius 1988; Solounias et al. 1988; Solounias and Moelleken 1993). GR are assumed to have a small mouth opening and short lips, whereas BR should have a larger mouth opening and longer lips (Hofmann 1988). As BR are thought to need more flexible lips, a larger lip muscle attachment area has been suggested in BR (Solounias and Dawson-Saunders 1988), and a seemingly larger size of the infraorbital and stylomastoidal foramina in BR ruminants has been interpreted as indicative of more innervation of the lip muscles compared to GR (Solounias and Moelleken 1999).

Metabolism

B1. In the zoo animal literature, it has been proposed that BR have higher protein requirements for maintenance than GR. No statistical treatment of this question is known to us. However, comparisons of experimentally established protein maintenance requirements between BR and GR species (collated in Robbins 1993; Clauss et al. 2003b) do not suggest any relevant systematic difference between the feeding types.

B2. A particular adaptation of GR to the digestion of cellulose would be expected. This is subsumed under the prediction D2, as fibre content cannot be separated from fermentation characteristics.

B4. In order to counteract plant secondary compounds, it has been suggested that BR produce salivary proteins that bind to these compounds. See Clauss (2003) and Shimada (2006) for reviews of species in which such proteins have been demonstrated; however, the number of species investigated thus far makes a statistical comparison between BR and GR unfeasible. Supposedly larger salivary glands of BR ruminants (Kay et al. 1980; Kay 1987b; Hofmann 1988) have been considered to be a morphological correlate of a high production of these salivary proteins (Robbins et al. 1995). The BR that has been demonstrated to deviate from the general ruminant pattern, the greater kudu (*Tragelaphus strepsiceros*; Robbins et al. 1995), has been noted to suffer from die-offs due to tannin poisoning (Van Hoven 1991), although even kudus can include plants in their diet that are known to be poisonous for livestock (Brynard and Pienaar 1960). A larger liver, used for secondary plant compound detoxification, has been postulated in BR ruminants (Hofmann 1988; Duncan et al. 1998) and in the black rhinoceros (*Diceros bicornis*), a browser (Kock and Garnier 1993), as compared to GR ruminants and GR rhinos. It has been demonstrated, in pair-wise comparisons in rodents, macropods, and ruminants, that BR are less affected by dietary secondary compounds than GR (Iason and Palo 1991; Robbins et al. 1991; Hagerman et al. 1992; McArthur and Sanson 1993). Statistical treatments of these topics for a large range of species is still required.

Dental and Buccal Morphology

C1, C2. Silica is harder than enamel, and a grass diet should wear down teeth faster than does browse. Therefore, significant differences between GR and BR in enamel microwear or molar wear rates (Solounias et al. 1994; Solounias and

Semprebon 2002) are expected. As molar wear is a function of both attrition (tooth-to-tooth contact, which maintains sharp edges) and abrasion (tooth to food contact, which produces blunt edges), differences in the 'mesowear' pattern [the macroscopically evaluated shape and reliefs of the cusps of selected teeth first introduced by Fortelius and Solounias (2000)] between feeding types are also expected. Differences in mesowear patterns between upper and lower molars in different feeding types indicate different morphological adaptations (Franz-Oftedaal and Kaiser 2003; Kaiser and Fortelius 2003). Hypsodonty (as an adaptation to increased tooth wear due to abrasion and maybe increased attrition; C.3) has been observed in many GR groups such as marsupials, rodents, lagomorphs, and ungulates (Simpson 1953; Fortelius 1985; Janis and Fortelius 1988). Experimental work on the influence of crown height on survival is summarised in Williams and Kay (2001). Hypsodonty will have other consequences for cranial morphology, such as deeper mandibles to accommodate the hypsodont teeth (Vrba 1978). As a secondary change due to hypsodonty and the increased space requirement for the higher maxillary molars (and the masseteric insertion areas; C.3), it is assumed that in GR the orbita needs to be positioned more posteriorly than in BR (Solounias et al. 1995). This will also lead to an elongation of the whole skull. In a comparison of three macropod species, enamel hardness was greater in the GR than in the BR species (Palamara et al. 1984).

C3. The purportedly tougher consistency of grasses and higher chewing pressure needed by GR to break down grass would require a more robust mandible structure (that also accommodates hypsodont teeth; C1, C2). A possible mechanism to withstand increasing pressure/torsional forces in GR ruminants is the reduction of the premolar tooth row length (Greaves 1991), and a reduction of palatal width. In perissodactyls, premolar tooth reduction does not occur, perhaps due to the requirement for increased food intake (Janis and Constable 1993). A greater masseter muscle mass, and accordingly larger masseter insertion surfaces, are assumed for GR, which is thought to reflect the higher masticatory forces required to grind grass material (Turnbull 1970; Stöckmann 1979; Axmacher and Hofmann 1988). A larger masseter, requiring more nerve tissue for innervation, has been hypothesised to be the reason for seemingly larger foramina ovale (through which the masseter nerves pass) in GR ruminants (Solounias and Moelleken 1999).

C4. Different fracture properties of browse and grass should be reflected in differences in tooth occlusal surface morphology (Fortelius 1985; Sanson 2006); and these differences in tooth morphology should be reflected in differences in ingesta particle reduction between species on a similar food source (Lentle et al. 2003c).

General Digestive Physiology

D1. Given the higher potential digestibility of grass, one would either expect BR to have lower basal metabolic rates at similar intake levels, or similar BMRs at higher intake levels, or GR to have lower BMRs at relatively low intake levels.

The BMR data available for ruminants (Williams et al. 2001) does not suggest a systematic difference in BMR between the feeding types, and hence it would be expected that BR have higher intakes.

D2. Optimal digestion theory (Sibly 1981) predicts that animals adapted to a forage that yields energy and nutrients quickly would have short retention times, and those that ingest a forage that yields energy and nutrients more slowly would have longer retention times. This has been postulated for ruminants (Hanley 1982; Kay 1987a; Hofmann 1989; Clauss and Lechner-Doll 2001; Behrend et al. 2004; Hummel et al. 2005a), however, a comprehensive dataset based on comparable measurements is still lacking.

Differences in retention time would have far-reaching consequences: on comparable diets, BR should achieve lower digestion coefficients than GR (for ruminants: Owen-Smith 1982; Prins et al. 1983; Demment and Longhurst 1987). A higher food intake in BR, due to the supposedly shorter passage times, has been suggested (Owen-Smith 1982; Baker and Hobbs 1987; Prins and Kreulen 1991). Forage that ferments faster and is retained shorter should also be ingested in shorter time intervals, and a higher feeding bout frequency has been suggested in browsing ruminants (Hofmann 1989; Hummel et al. 2006). A combination of higher food intake and lower digestibility should theoretically result in a comparatively higher faecal output in BR, which could be expected to have further consequences. For example, Robbins (1993) states that sodium losses are a function of faecal bulk, and higher faecal sodium losses have been observed in the browsing black rhinoceros (*Diceros bicornis*) than in the domestic horse (Clauss et al. 2006a).

Cellulolytic activity in the rumen of BR is expected to be lower than that of GR (Prins et al. 1984; Deutsch et al. 1998). GR have a more diverse protozoal fauna, whereas BR protozoa are mostly *Entodinium sp.*; as these are particularly fast-growing ciliates, it has been suggested that other protozoa cannot establish viable populations in the reticulorumen (RR) of BR (Prins et al. 1984; Dehority 1995; Dehority et al. 1999; Clauss and Lechner-Doll 2001; Dehority and Odenyo 2003; Behrend et al. 2004). Other parameters indicative of shorter ingesta retention times (Clauss and Lechner-Doll 2001; Behrend et al. 2004) are a lower degree of unsaturated fatty acid hydrogenation in the RR, a greater number of glucose transporters, and a higher amylase activity in the small intestine of BR,[3] as well as larger faecal particles in BR. In order to maintain intake levels whilst still having prolonged ingesta retention in the rumen, GR should have more capacious rumens (Prins and Geelen 1971; Giesecke and Van

---

[3] The hypothesis that BR maintain a functional reticular groove throughout their adult lives (Hofmann 1989) has not been tested directly (Ditchkoff 2000); however, a comparison of fluid retention data from two different trials on roe deer (*Capreolus capreolus*) indicates that bypass of soluble substances from the rumen via the reticular groove is probably not a quantitative factor, at least in this species (Behrend et al. 2004).

Gylswyk 1975; Drescher-Kaden 1976; Hoppe 1977; Kay et al. 1980; Owen-Smith 1982; Bunnell and Gillingham 1985; Van Soest 1994). It has been suggested that BR have more capacious hindguts (Hofmann 1988), but this view has recently been modified (Clauss et al. 2003a; Clauss et al. 2004).

Macropods

C1C3. Macropods have been classified in feeding types based on their dental morphology (Sanson 1989). Macropod teeth have a crushing action over a relatively large occlusal contact area in BR, with dentine basins making up a large percentage of this surface area. As macropods have evolved to feed more on grass, the area of occlusal contact has been decreased by increasing the complexity of the enamel ridges and increasing the curvature of the tooth row (Janis and Fortelius 1988; Sanson 1989; Lentle et al. 2003b; Lentle et al. 2003a). The macropod feeding type classification of Sanson (1989) was tested for two macropod species by Sprent and McArthur (2002); the results were in accord with Sanson's prediction. Differences in ingesta particle size distribution between four macropod species, tested in sets of both free-ranging and captive animals, support the notion that teeth of GR are more suited to the fine-grinding of grass material (Lentle et al. 2003c).

C4-C5. It has been postulated that grazing macropods have a lower proportion of the sacciform relative to the tubiform forestomach (Freudenberger et al. 1989). This could be due to the fact that macropods do not ruminate, and as a result, ingesta stratification in a larger sacciform forestomach would not be beneficial to GR. At the same time, comparisons indicate that, in GR macropods, the length of the large intestine is greater than in BR, indicating additional fermentation of the slower-fermenting grasses in this site (Freudenberger et al. 1989).

Ruminant Forestomach Morphophysiology[4]

C4-C5. Recently, it has been hypothesised that a key to the understanding of ruminant forestomach physiology is the presence or absence of a stratification of RR contents (Clauss and Lechner-Doll 2001; Clauss et al. 2001, 2002, 2003c).

---

[4] We feel that a major basis of the discussion on potential differences in digestive morphophysiology between GR and BR ruminants has been to confirm or refute Hofmann's original observations and hypotheses, rather than necessarily to understand the functional relevance of his findings. Here, we present a new, complex interpretation of ruminant forestomach physiology; this is not a "refutation" of Hofmann's hypotheses but a refinement and readjustment, based on his anatomical observations, the validity of which is not drawn into question (but should be submitted to statistical evaluation).

Physical and chemical characteristics of grass are thought to enhance the development of this stratification, with particle separation occurring according to flotation/sedimentation, and a 'fibre raft' or 'fibre mat' on top of a liquid layer; to these, the animal has to adapt, for example with stronger rumen pillars (to work against the tough consistency of the 'mat'), with deeper reticular honeycomb cells (traps for the sedimenting particles in the liquid phase), or larger omasa (for water re-absorption from the liquid outflow out of the RR). As the stratification enhances particle retention and, hence, fibre digestion (Beaumont and Deswysen 1991; Lechner-Doll et al. 1991), adaptations encouraging stratification would be expected to have evolved in GR. The most controversial part of the traditional concept of wild ruminant RR physiology is that the larger salivary glands of BR (Hofmann 1988) are assumed to translate into a higher saliva production rate and, hence, a higher fluid throughput through the RR. If this was the case, then the comparatively smaller omasa of BR (Hofmann 1988; Clauss et al. 206c) would not make sense; the primary function of the omasum is fluid re-absorption. If fluid throughput was particularly high in BR, then they, not GR, should have the larger omasa. The concept of higher salivary flow rate in BR was challenged by Robbins et al. (1995) on the basis of a three-species comparison. It has been proposed that the difference between particle and fluid retention in the RR is a distinguishing characteristic of the different feeding types, with the difference between the phases being large in GR—indicative of ingesta stratification—and small in BR (Clauss and Lechner-Doll 2001; Hummel et al. 2005a). However, the question whether this difference stemmed from longer particle or shorter fluid retention times in GR, or both, has not been addressed comprehensively to date. Comparisons between the few species on which data exist indicate that GR have both a shorter fluid, and a longer particle retention time in the RR than do BR (Clauss et al. 2006c). We propose that the production of a large amount of non-viscous saliva is a particular adaptation of GR and supports the development of RR contents stratification, whereas the more viscous saliva of BR delays any separation of ingesta particles by flotation/sedimentation. Thus, the size of the salivary glands is probably not correlated with saliva production but with saliva protein content (c.f. B4) and hence viscosity (Robbins et al. 1995). Results of Jones et al. (2001) who found that several browsers had a higher dry matter content in rumen fluid as compared to grazers supports the concept of a higher fluid viscosity in browsers. In a low-viscosity medium, the fermentation gases, carbon dioxide and methane, can easily rise and gather in the dorsal RR. Rumen papillae development is stimulated by the presence of volatile fatty acids (in particular, of butyrate) (Warner et al. 1956). The continuous presence of a gas dome of CO_2 and methane in the dorsal rumen will prevent any significant concentration of volatile fatty acids in this region, leading to unpapillated dorsal rumen surfaces in GR (Hofmann 1973). In a more viscous medium, fermentation gases cannot dissociate from food particles to rise and gather in the dorsal rumen as easily, which results in the typical 'frothy' appearance of BR RR contents (Clauss et al. 2001) and an even RR papillation (Hofmann 1973). The escape of a more a viscous ingesta that traps CO_2 into the abomasum would also explain why browsers have a thicker layer of acid-producing abomasal mucosa (Hofmann 1988), inasmuch as the presence of CO_2 would increase the buffering capacity of the ingesta.

Captive Animals: Readiness to Ingest Forage

Based both on dental and, in the case of ruminants, forestomach characteristics, GR should have less problems in eating browse than BR in eating grass (Clauss et al. 2003c). This hypothesis finds support in reports for captive wild ruminants (Clauss et al. 2003b), for macropods (Lentle et al. 2003c), suids (Leus and MacDonald 1997), and from the well-known reluctance of captive tapirs to ingest grass hay (Foose 1982). Captive BR ruminants ingest less hay and have a higher incidence of rumen acidosis (due to a lack of fibre) than GR ruminants (Clauss et al. 2003b). Commercially available pelleted feeds for captive wild herbivores in general or GR in particular have a distinctively lower fibre content than diets designed especially for BR, again indicating that one cannot rely on a sufficient fibre intake via hay consumption by BR (Clauss and Dierenfeld 2007). Due to the high abrasiveness that characterizes the diets of both free-ranging grazers and that fed to captive animals in general, the tooth wear pattern of captive browsers resembles that of free-ranging grazers (Clauss et al. 2007a).

3.4 Testing the Hypotheses

Testing these hypotheses involves problems that have, historically, been addressed in different ways in different publications. The following discussion raises issues about the approaches used to test for morphophysiological adaptations to diet in herbivores.

Body Mass

Body mass (BM) is the single most influential factor on the absolute size of any anatomical, and of most (but not necessarily all) physiological parameters (Schmidt-Nielsen 1984; Peters 1986; Calder 1996). Therefore, the inclusion of BM in statistical evaluations is self-evident. Ideally, datasets should cover similar BM ranges for all feeding types investigated. On average, grazing ruminants are larger than browsing ruminants (Bell 1971; Case 1979; Bodmer 1990; Van Wieren 1996b; Pérez-Barbería and Gordon 2001). However, there is either no correlation (Van Wieren 1996b; Clauss et al. 2003c; Sponheimer et al. 2003), or it is very weak (Gagnon and Chew 2000), between BM and the proportion of grass in the natural diet of ruminants. This is because browsers are found across the body size range (Sponheimer et al. 2003). The largest extant ruminant, the giraffe (*Giraffa camelopardalis*), is a browser, the largest marsupial herbivores were browsers (Johnson and Prideaux 2004), and the largest known terrestrial mammalian herbivore ever, the Indricotherium (Fortelius and Kappelman 1993), was also a browser; extensive grasslands did not exist when this species inhabited the earth (Janis 1993). Large BM, therefore, does not preclude a browsing lifestyle (Hofmann 1989).

The question of whether there are different upper and lower body size thresholds for the feasibility of grazing or browsing in ruminants has been addressed by Demment and Van Soest (1985) and by Clauss et al. (2003a). Assuming feeding type independent general relationships between body mass and forestomach capacity, and between body mass and ingesta retention, Demment and Van Soest (1985) demonstrated that, theoretically, grazing by ruminants is feasible at greater body masses than browsing; using, in contrast, feeding type specific relationships between body mass and forestomach capacity and ingesta retention, Clauss et al. (2003a) demonstrated that browsing is theoretically feasible at larger body masses—a result seemingly in better accord with the extant and fossil ruminant record.

Two examples illustrate the importance of choosing a specific BM value when evaluating morphological and physiological data: Hofmann (1973) gives data on the length of the curvature of the omasum (which he claims is larger in GR than in BR) for the giraffe (52–71 cm) and the African buffalo (*Syncerus caffer*) (72 cm). The BM data given in Hofmann (1973) are derived from the literature (giraffe, 750 kg; buffalo, a range of 447–751 kg). Should one chose to compare both measurements on the basis of the maximal BM (750 vs. 752 kg), hardly any difference between the species would be evident; should one chose to use the averages/medians of the given BM data (750 kg vs. 599 kg), then the GR buffalo would be assumed to display a relatively larger omasum. Another example is the greater kudu in Hofmann's (1973) dataset. The BM range, again taken from the literature, is 170–257 kg. However, the handwritten notes in Hofmann's archive record estimated BM of the animals investigated to range from 220 to 350 kg (M. Clauss, pers. obs.). These actual BM data would link the anatomical measurements to a higher average BM, thus reinforcing potential differences between the feeding types. The importance of measurements of morphophysiological traits and BM from the same individuals, therefore, cannot be overemphasised. In studies that use measurements on museum skeleton specimens (for which live BM data is usually missing), the use of a parameter that can be measured on the museum specimen and is known to correlate closely with BM or BM-independent ratios are alternatives (for example Janis 1988; Spencer 1995; Archer and Sanson 2002).

Definition of Feeding Type

In order to compare adaptations of different species, the species have to be classified according to niches that are relevant to the adaptations under investigation. This has mostly been done by allocating feeding type labels, such as 'BR' or 'GR', as discrete variables. The available information on diet composition on which such a classification is based differs between species (Gagnon and Chew 2000). A common practice has been to collect published data on the botanical composition of a species' diet, calculate an average value for the different reports, and then use pre-defined thresholds to allocate a feeding type. These thresholds have not been used consistently in the literature; in particular, some publications allocate species with >75 % of the respective forage to the BR or GR category (Pérez-Barbería and Gordon 1999; Pérez-Barbería et al. 2001a; Mendoza et al. 2002), whereas other publications reserve these categories only for species consuming >90 % of the respective forage (Janis 1990; Pérez-Barbería et al. 2001b). The impact of the choice of allocation of species to feeding types is demonstrated for example by Gordon and Illius (1994), who showed that results differed depending on the classification used. A more consistent approach (Janis 1995; Clauss et al. 2003c; Sponheimer et al. 2003; Pérez-Barbería et al. 2004) does not use a discrete variable, but uses the percentage of grass and/or browse in the natural diet as a continuous variable. But while such an approach overcomes the need to make arbitrary 'threshold decisions', it should be remembered that the information contained in such a continuous variable is not perfect since there can be enormous geographical and seasonal variation in diet composition in some species (Owen-Smith 1997). An important limitation of the description of 'natural' diets is explained by Sprent and McArthur (2002): in any natural setting, the 'typical' forage preference pattern is evidently modified by the available forage. Ideally, a selected diet should always be expressed in terms of the available diet.

It should be borne in mind that allocating feeding types on the basis of actual observations does not provide full information on the nutritional adaptation of species. Although it is generally viewed that species diversification followed the sequence of BR/closed habitat, mixed feeder, GR/open habitat (Pérez-Barbería et al. 2001b), the reverse has been suggested or noted occasionally for extant and extinct species (Solounias and Dawson-Saunders 1988; Thenius 1992; Cerling et al. 1999; MacFadden et al. 1999). The morphology of species that are in a transition/regression state in this respect may not be completely correlated with dietary behaviour yet. The different evolutionary directions that led species to their present state can potentially make convergent evolutionary traits more difficult to discern (Gould 2002).

Phylogenetic Descendence

If values for individual species are used in statistical tests, these values cannot be viewed as independent because the species are phylogenetically related (Harvey and Pagel 1991; Martins and Hansen 1996). In recent years, phylogenetic control in statistical tests has become standard procedure for evaluating differences between or correlations with feeding types (c.f. the work of Pérez-Barberìa et al.). Published results can be classified into those that do not remove phylogenetic effects in the analysis (generally earlier studies) and those that do. This leads to the dilemma that results from earlier studies cannot be quoted with confidence, but direct replication of tests are rarely performed on the same datasets.

The method of phylogenetic control has been criticised (Westoby et al. 1995), but this discussion shall not be reviewed here. The most informative approach is to conduct two analyses, without and with phylogenetic control. If, for example, a certain measure shows a difference between feeding types, after controlling for body mass alone, this indicates that it represents either (1) a case of convergent evolution between lineages or (2) evolution within a certain lineage that dominates the dataset. If, in a second step, no difference between feeding types is found, when phylogeny is controlled for, then the hypothesis of convergent evolution between lineages can be rejected, but not necessarily the hypothesis of evolution within a certain lineage, nor the adaptive value of the trait as such. The rejection of the hypothesis of convergent evolution should not be confused with a rejection of the hypothesis of adaptive value, which can only be tested experimentally.

Two important choices have to be made when phylogenetic control is applied. The phylogenetic tree should, ideally, be based on characters unrelated to the character that is submitted to the test. In this respect, one should, for example, note that many of the dental characters understood to be adaptations to feeding niches (Janis 1990) have also been used to establish phylogenetic relations in ungulates (Janis and Scott 1987). The case of the more recently discovered mammalian taxonomic clades of the Afrotheria, Laurasiatheria, and Euarchontoglires implies a widespread accumulation of homoplasious morphological features in various placental clades and thus exemplifies the difficulty of basing phylogenetic trees on morphological characters (Robinson and Seiffert 2004). The other choice refers to the spectrum of species included in the analysis, that is, the level at which convergent evolution is to be assessed. On the one hand, if there is some trait showing convergent evolution within the ruminants, then this trait will be more difficult to detect in a dataset comprising only bovids than in a dataset that comprises both bovids and cervids. On the other hand, an expansion of the species dataset beyond certain phylogenetic borders may appear unreasonable. When rumen capacity or rumination activity is compared between feeding types, only ruminants (and maybe camelids) will be included but not other ungulates. Similarly, ingesta retention should be analysed separately for ruminants and hindgut fermenters, due to the differences in digestive physiology (Illius and Gordon 1992). Other examples include systematic differences in cranial morphology. For example, in perissodactyls, the lower premolar row increases in length in grazing species whereas it decreases in grazing ruminants and macropods (Janis 1990; Mendoza et al. 2002), and grazing horses have relatively smaller muzzles than corresponding grazing ruminants (Janis and Ehrhardt 1988). Therefore, inclusion of phylogenetically distant groups in one analysis might yield different results from an analysis within each of these groups; in this respect, the finding that a parameter shows no convergent evolution in GR or BR *ungulates* does not falsify the finding that such convergent evolution occurs within GR or BR *ruminants*. This latter question would have to be addressed by an analysis using only ruminant data. The power of a variable to predict the feeding type correctly, hence, usually increases as the taxonomic level is narrowed from ungulates to either artiodactyls or perissodactyls (Janis 1995).

Statistical Procedure

Generally one can distinguish uni- or bivariate statistics (testing one trait, for example a ratio, or two if control for BM is included, between feeding types), or a multivariate approach. The advantage of a multivariate approach is that a number of characteristics are included that will, each and together, contribute to the adaptation of a feeding type. A potential disadvantage of a multivariate approach might be the temptation to include as many data as possible without giving attention to the functional relevance of the different parameters. In this respect, multivariate analysis can be considered to be exploratory, unless it is followed or preceded by further detailed investigations (Spencer 1995; Archer and Sanson 2002). In the selection of data for a multivariate approach, partly repetitive information—for example, both volumetric and linear measurements of the same organ, or both the length of the molar tooth row and the length of the individual molars—has either been included (Pérez-Barbería et al. 2001a) or excluded (Mendoza et al. 2002).

Parameter Choice and Measurement Resolution

All previous considerations notwithstanding, the most important prerequisite for a meaningful analysis is that the parameters investigated have a functional relevance, and that they are measured with appropriate resolution so that meaningful differences can be detected. The necessary degree of resolution is intuitively more evident in anatomical than in physiological studies. No one would conceive of comparing incisor breadths measured to the closest 0.1 meter. Yet, for example, the method employed by Foose (1982) for the estimation of ingesta retention (using one or two pooled faecal samples per day) can be expected to be of sufficient resolution to differentiate between hindgut and foregut fermenters (the aim of Foose's study), but hardly so between browsers and grazers of either category. If data compilations from different publications are used for an analysis, the consistency of the methods applied, therefore, is of prime importance; an additional solution is to include data source as a random variation in the analysis (Pérez-Barbería et al. 2004). In many studies, the functional significance of parameters tested has not only *not* been proven by experimentation, but is sometimes not even defined in logical terms. For example, the concept stated by Hofmann (1988) and repeated many times by others that 'openings' in the ruminant forestomach such as the ostium intraruminale or the ostium ruminoreticulare, the diameters of which are by magnitudes greater than the particles that leave the rumen, should have any influence on ingesta retention is not self-evident and would have to be backed by engineering models (or rapidly dismissed).

3.5 Results

In this section, publications are summarised that generated or collated comparative data and submitted these data to statistical tests for differences between the feeding types. The individual morphological and physiological traits are listed in tables 3.3–5, and comprehensive multivariate analyses are explained in the text. In general, the lack of phylogenetically controlled studies on different taxonomic groups does not allow definite conclusions.

Macropods

Results of morphological craniodental comparisons within the macropods are given in Table 3.3. No phylogenetically controlled approach has been used thus far. Some, but not all of the predictions regarding craniodental design seem to be met. As with ruminants, GR macropods also show a shorter premolar row length, but have smaller muzzles than do BR macropods.

Table 3.3 Statistical tests for craniodental differences between grazing (GR) and browsing (BR) macropods. From (Janis 1990); $n = 52$, original data (not given), discrete feeding type allocation. FR = functional relevance, FT = feeding types (significance without/with phylogenetic control). BM = body mass

Parameter	FR	FT	Direction
Basicranial angle	A2	*/-	GR < BR
Anterior jaw length	A2	ns/-	
Width of central incisor	A3	ns/-	
Width of lateral incisor	A3	*/-	GR > BR
Muzzle width	A3	*/-	GR < BR
Hypsodonty index	C1-C2	*/-	GR > BR
Distance orbita tooth row	C1-C2	*/-	GR > BR
Lower premolar row length	C3	*/-	GR < BR
Lower molar row length	C3	ns/-	
Depth of mandibular angle	C3	ns/-	
Maximum width of mandibular angle	C3	ns/-	
Length of coronoid process	C3	ns/-	
Length of masseteric fossa	C3	ns/-	
Palatal width	C3	ns/-	

Rodents

The only analysis pertaining to rodents indicates that GR have a higher hypsodonty index (after phylogenetic control) than do BR (Williams and Kay 2001).

Ungulates

Results of uni- and bivariate analyses are given in Table 3.4. Additionally, a multivariate analysis (without phylogenetic control) of a set of 22 craniodental variables in 115 species (Mendoza et al. 2002) indicated that differences exist between the feeding types. As in macropods and rodents, hypsodonty is identified as a primary distinction between BR and GR, being greater in GR than in BR. For several parameters (skull length, muzzle width, occlusal surface, tooth row lengths), opposing trends between artiodactyls and perissodactyls might have led to nonsignificant results in the ungulate comparison. Similarly, masseter parameters could be significant simply because equids—all GR—have larger masseter muscles than ruminants (Turnbull 1970). Thus, phylogenetic control leaves only very few parameters of convergent evolution within the ungulates, either indicating that no differences between feeding types exist, or that they should be looked for at lower taxonomic levels. Recently, in an explorative analysis including phylogenetic control, without a specific hypothesis, Pérez-Barbería and Gordon (2005) did not find any relevant correlation between feeding type and brain size in ungulates.

Perissodactyls

The dental anatomy of browsing rhinoceroses (*D. bicornis, Dicerorhinus sumatrensis*) differs distinctively from that of the grazing *Ceratotherium simum* but not *Rhinoceros unicornis* (Palmqvist et al. 2003); similarly, production of salivary tannin-binding proteins differ between *C.simum* on one hand, and *D. bicornis* and *R. unicornis* on the other (Clauss et al. 2005b). This could indicate that *R. unicornis* ingests more browse or fruit than previously reported, or that this species has switched to a grass-dominated diet recently. There is a significant difference in tooth sharpness, with particularly blunt tips in the grazing *C. simum* as compared to *D. bicornis* (Popowics and Fortelius 1997). It has been demonstrated that ingesta retention in the browsing *D. bicornis* was shorter than expected for its body mass when compared to grazing rhinoceroses and equids (Clauss et al. 2005a). Digestion coefficients achieved by the two browsing rhinoceros species *D. bicornis* and *D. sumatrensis* are lower than those of grazing rhinoceroses at comparable levels of dietary fibre content (Clauss et al. 2006b).

Table 3.4 Statistical tests for morphological differences between grazing (GR) and browsing (BR) ungulates. FR = functional relevance, n = number of species, dg = data given (yes/no), od = original data (yes/no), FTC = feeding-type classification (d = discrete; c = continuous), FT = difference between grazers/browsers significant without/with phylogenetic control, BM = correlation with body mass significant without/with phylogenetic control. * = significant, ns = not significant, ds = difference in slope, - = not done

Parameter	FR	n	dg	od	FTC	FT	BM	Direction	Source
Basicranial angle	A2	136	n	y	d	*/-		GR < BR	(Janis 1990)
Total skull length	A2, C1, C2	136	n	y	d	ns[a]/-			(Janis 1990)
Posterior skull length	A2, C1, C2	136	n	y	d	*/-		GR > BR	(Janis 1990)
Rel. muzzle width (palate/muzzle)	A3	95	y	y	d	*/-		GR > BR	(Janis and Ehrhardt 1988)
Muzzle width	A3	136	n	y	d	ns[a]/-			(Janis 1990)
Muzzle width	A3	104	y	n	d	*/ns	*/*		(Pérez-Barbería and Gordon 2001)
Width central incisor	A3	136	n	y	d	*/-		GR > BR	(Janis 1990)
Width lateral incisor	A3	136	n	y	d	*/-		GR > BR	(Janis 1990)
Rel. incisor width (I1/I3)	A3	70	y	y	d	*/-		GR < BR	(Janis and Ehrhardt 1988)
Rel. incisor width (I1/I3)	A3	66	y	n	d	ns/ns	*/*		(Pérez-Barbería and Gordon 2001)
Incisor protrusion	A3	25	y	n	d	*/ns	*/*		(Pérez-Barbería and Gordon 2001)
Hypsodonty index	C1, C2	128	y	y	d	*/-		GR > BR	(Janis 1988)
Hypsodonty index	C1, C2	136	n	y	d	*/-		GR > BR	(Janis 1990)
Hypsodonty index	C1, C2	79	n	y	c	*/-		GR > BR	(Janis 1995)
Hypsodonty index	C1, C2	57	n	n	d	-/*		GR > BR	(Williams and Kay 2001)
Hypsodonty index	C1, C2	19	n	n	c	*/-		GR > BR	(Codron et al. 2007b)
M3 height	C1, C2	121[b]	y	y	d	*/-	*/-	GR > BR	(Janis 1988)
M3 height	C1, C2	113	y	n	d	*/*	*/*	GR > BR	(Pérez-Barbería and Gordon 2001)

3 The Morphophysiological Adaptations of Browsing and Grazing Mammals 69

M3 volume	C1, C2	121	y	y	d	*/-	*/-	GR > BR	(Janis 1988)
M3 volume	C1, C2	113	y	n	d	(ds)	*/*	GR > BR	(Pérez-Barbería and Gordon 2001)
Distance orbita tooth row	C1-C2	136	n	y	d	*/-		GR > BR	(Janis 1990)
Postcanine tooth row volume	C1, C3	121	y	y	d	*/-	*/-	GR > BR	(Janis 1988)
Molar row volume	C1, C3	113	y	n	d	*/*	*/*	GR > BR	(Pérez-Barbería and Gordon 2001)
Occlusal surface	C3	92	y	n	d	ns^a/ns	*/*		(Pérez-Barbería and Gordon 2001)
Lower M2 area	C3	136	n	y	d	*/-		GR < BR	(Janis 1990)
Depth mandibular angle	C3	136	n	y	d	*/-		GR > BR	(Janis 1990)
Max. width mandibular angle	C3	136	n	y	d	*/-		GR > BR	(Janis 1990)
Length coronoid process	C3	136	n	y	d	*/-		GR > BR	(Janis 1990)
Length masseteric fossa	C3	136	n	y	d	*/-		GR > BR	(Janis 1990)
Palatal width	C3	136	n	y	d	ns/-			(Janis 1990)
L. premol. tooth row length	C3	136	n	y	d	ns^a/-			(Janis 1990)
L. molar tooth row length	C3	136	n	y	d	ns^a/-			(Janis 1990)
17 jaw and 4 skull traits, including most of the parameters listed above [except those used in Pérez-Barbería and Gordon (2001)]		94	n	y	d	*/ns			(Pérez-Barbería and Gordon 1999)

^{a}opposing trends in artiodactyls and equids
^{b}excluding equids from the dataset

While extant equids are uniformly classified as GR, there is a large number of BR equids in the fossil record (MacFadden 1992). The feeding type of fossil horses is generally determined by a combination of isotope, hypsodonty, and microwear data (MacFadden et al. 1999). In theory, it would be feasible to compare feeding types classified in this manner for other osteological measurements. MacFadden (1992, pp 241–242) gives the example of an equid with a low hypsodonty index and a pointed muzzle shape as would be expected of a BR, and another one of an equid with a high hypsodonty index and a broad muzzle as would be expected of a GR. However, a quantitative approach to such correlations in the fossil record is lacking.

Proboscids

Extant elephants are intermediate feeders with a preference for browse, but isotopic investigations show that both lineages were once grazers and are in a transition back to browsing (Cerling et al. 1999). Isotopic evidence suggests that the Asian elephant (*Elephas maximus*) might ingest a higher proportion of grass than the African elephant (*Loxodonta africana*). This difference is confirmed by microwear results (Solounias and Semprebon 2002). The differences in molar structure between the two species (with *L. africana* having less enamel ridges than *E. maximus*; Maglio 1973) could be interpreted as a higher degree of adaptation for grass forage in *E. maximus*. Elephants differ in their digestive physiology from other ungulates due to their very short retention times and low digestion coefficients (Clauss et al. 2003d; Loehlein et al. 2003). Hackenberger (1987) found significantly longer ingesta retention times in *E. maximus* compared to *L. africana* and correspondingly higher digestion coefficients for *E. maximus* than *L. africana* when both species were fed hay diets. Data from Foose (1982) confirms this pattern. Anatomical data compilations suggest that *E. maximus* has a longer gastrointestinal tract, a larger masseteric insertion area, and smaller parotid glands than *L. africana* (Clauss et al. 2007b) —seemingly in parallel to similar differences between GR and BR ruminants.

Hyraxes

The GR *Procavia capensis* and the more browsing *Heterohyrax brucei* and *Dendrohyrax dorsalis* have similar differences in microwear pattern as found in other herbivore taxons (Walker et al. 1978), but they have the same hypsodonty index (Janis 1990). No differences in tooth sharpness were observed between GR and BR hyraxes (Popowics and Fortelius 1997).

Suids

The suids comprise the GR warthog (*Phacochoerus aethipicus*) or the browsing/omnivorous babyrousa (*Babyrousa babyrussa*) and bushpig (*Potamochoerus porcus*), and the BR forest hog (*Hylochoerus meinertzhageni*; Harris and Cerling 2002; Mendoza et al. 2002; Cerling and Viehl 2004). Differences in hypsodonty (Harris and Cerling 2002; Mendoza et al. 2002) accord with observations in other taxonomic groups. Preliminary results indicate that *B. babyrussa* digests grass fibre less efficiently than domestic pigs (Leus and MacDonald 1997).

Ruminants

Results of uni- and bivariate analyses are given in Table 3.5. A multivariate analysis by Solounias and Dawson-Saunders (1988) using 13 traits mainly related to masseter muscle insertion in 27 species, and another multivariate analysis by Sponheimer et al. (1999) using four craniodental traits in 23 species, both found significant differences between the feeding types. In a multivariate, stepwise discriminant analysis of data from 72 bovid species, Mendoza and Palmqvist (2006a) demonstrated systematic differences in craniodental morphology between feeding types. With four exceptions, none of the studies listed in Table 3.5 were performed with phylogenetic control; therefore, although many characters do differ according to the predictions, it cannot be determined whether this represents a case of true convergent evolution. For physiological measurements such as digestibility (Robbins et al. 1995; Iason and Van Wieren 1999) or particle retention time (Gordon and Illius 1994; Hummel et al. 2006), larger datasets yielded different results than previous studies on more limited datasets. The fact that GR digest fibre more efficiently than BR (Pérez-Barbería et al. 2004) supports the general concept of feeding type differentiation. With respect to anatomical measurements of the forestomach, very few have been submitted to tests, and the basic dataset (Hofmann 1973) has hardly been expanded. In the multivariate analysis of Pérez-Barbería et al. (2001a), no specific functionality of the traits analysed was addressed; instead, the study tested whether Hofmann's conclusions could be supported or derived from the majority of the data given in Hofmann (1973). It was found that, after controlling for both body mass and phylogeny, the forestomach structures of BR and GR are similar, whereas those of mixed feeders differ; this finding is difficult to reconcile with the result that mixed feeders represent an intermediate evolutionary state between BR and GR (Pérez-Barbería et al. 2001b). Recently, it has been shown that GR have larger omasal laminal surface areas than BR (Clauss et al. 2006c), supporting Hofmann's (1968) observation that GR have larger omasa, and that BR have larger salivary glands (Hofmann et al., in press), confirming the indications of earlier studies. Even though the respective forages differ in fermentation rate (D2), it is difficult to predict in what way RR ingesta samples, which do not represent fresh forage but forage in varying states of fermentation (that is, proportions of which

Table 3.5 Statistical tests for morphological differences between grazing (GR) and browsing (BR) ruminants. FR = functional relevance, n = number of species, dg = data given (yes/no), od = original data (yes/no), FTC = feeding-type classification (d = discrete; c = continuous), FT = difference between grazers/browsers significant without/with phylogenetic control, BM = correlation with body mass significant without/with phylogenetic control. * = significant, ns = not significant, ds = difference in slope, - = not done

Parameter	FR	n	dg	od	FTC	FT	BM	Direction	Source
Braincase angle	A2	33	y	y	d	*/-		GR < BR	(Spencer 1995)
Glenoid height	A2	33	y	y	d	*/-		GR > BR	(Spencer 1995)
Paracondylar process	A2	33	y	y	d	*/-		GR > BR	(Spencer 1995)
Diastema length	A2, C3	33	y	y	d	*/-		GR > BR	(Spencer 1995)
Length of skull	A2, C3	33	y	y	d	*/-		GR > BR	(Spencer 1995)
Predental length	A2, C3	33	y	y	d	*/-		GR > BR	(Spencer 1995)
Incisor arcade breadth	A3	88	y	y	d	*/-	*/-	GR > BR	(Gordon and Illius 1988)
Incisor arcade breadth	A3	33	y	y	d	*/-		GR > BR	(Spencer 1995)
Incisor arcade breadth	A3	79	y	n	d	*/ns	*/*		(Pérez-Barbería and Gordon 2001)
Incisor arcade shape	A3	72	y	n	d	(ds)	ns/ns		(Pérez-Barbería and Gordon 2001)
Muzzle height (lip muscle attachment)	A3	27	y[a]	y	d	*/-		GR < BR	(Solounias and Dawson-Saunders 1988)
Parotid gland size	B4	22[b]	n	yn	d	*/-	*/-	GR < BR[c]	(Robbins et al. 1995)
Parotid gland size	B4	20	y	n	d	*/(-)		GR < BR	(Jiang and Takatsuki 1999)
Parotid gland size	B4	62	y	y	c	*/*	*/-	GR < BR	(Hofmann et al., in press)
Mandibular gland size	B4	61	y	y	c	*/*	*/-	GR < BR	(Hofmann et al., in press)
Size of the ventral buccal glands	B4	44	y	y	c	*/*	*/-	GR < BR	(Hofmann et al., in press)
Sublingual gland size	B4	30	y	y	c	*/*	*/-	GR < BR	(Hofmann et al., in press)
Hypsodonty index	C1-C2	27	y	n	c	*/-		GR > BR	(Sponheimer et al. 2003)
Hypsodonty index	C1-C2	37	y	n	c	*/-		GR > BR	(Cerling et al. 2003)
Hypsodonty index	C1-C2	13	n	n	c	*/-		GR > BR	(Codron et al. 2007b)
Molar wear rates	C1-C2	9	y	n	d	*/-	*/-	GR > BR	(Solounias et al. 1994)
Distance orbita tooth row	C1-C2	22	y	y	d	*/-		GR > BR	(Solounias et al. 1995)
Mandible depth	C1-C3	33	y	y	d	*/-		GR > BR	(Spencer 1995)
Mandible depth	C1-C3	27	n	n	d	*/-		GR > BR	(Sponheimer et al. 2003)
Mandible width	C1-C3	33	y	y	d	*/-		GR > BR	(Spencer 1995)

3 The Morphophysiological Adaptations of Browsing and Grazing Mammals

Feature								Reference
Tooth blade sharpness	C1-C2, C4	14	n	y	d	ns/-		(Popowics and Fortelius 1997)
Length of premolar tooth row	C3	33	y	y	d	*/-	GR < BR	(Spencer 1995)
Length of premolar tooth row	C3	27	n	n	c	*/-	GR < BR	(Sponheimer et al. 2003)
Masseter weight	C3	22	y	y	d	ns/-	*/-	(Axmacher and Hofmann 1988)
Masseter insertion area	C3	22	y	y	d	*/-	GR > BR	(Solounias et al. 1995)
Palatal width	C3	33	y	y	d	*/-	GR < BR	(Spencer 1995)
Molar cavity complexity	C4	27	y	y	d	*/-	GR > BR	(Solounias and Dawson-Saunders 1988)
Molar enamel ridge pattern/occlusal surface complexity	C4	26	y	y	d	*/-	GR > BR	(Archer and Sanson 2002)
Faecal particle size	C4, D2	81	y	y	d	*^{d}/-	GR < BR	(Clauss et al. 2002)
RR capacity (water fill volume)	C4, C5, D2	25	n	n	d	*/-	GR > BR	(Demment and Longhurst 1987)
RR capacity (water fill volume)	C4, C5, D2	25	n	n	d	ns/-	*/-	(Van Wieren 1996b)
RR capacity (water fill volume)	C4, C5, D2	36	y	n	d	*/(−)	GR > BR	(Jiang and Takatsuki 1999)
RR capacity (water fill volume)	C4, C5, D2	21	y	n	d	ns^e/-	*/-	(Gordon and Illius 1994)
RR contents (wet weight)	C4, C5, D2	21	y	n	d	*/-	GR > BR	(Gordon and Illius 1994)
RR contents (wet weight)	C4, C5, D2	47	y	n	d	*(−)	GR > BR	(Jiang and Takatsuki 1999)
RR contents (wet weight)	C4, C5, D2	29	y	n	d	*/-	GR > BR	(Clauss et al. 2003c)
RR contents (wet weight)	C4, C5, D2	21	y	n	d/c	ns^e/-	*/-	(Gordon and Illius 1994)
RR contents (dry)	?	27	y	n	d/c	*/-	GR > BR	(Clauss et al. 2003c)
Rumen pillar thickness	C4, C5, D2	45	y	y	d	*(ds)^g/-	GR < BR	(Demment and Longhurst 1987)
Rumen total surface area	C4, C5	25	n	n	d	*^{h}/-	*/-	(Van Wieren 1996b)
Density papillation ventr. rumen	?	25	n	n	d	ns/-	*/-	(Van Wieren 1996b)
Maximum SEF (of any RR location)	?	25	n	n	d	-/-	*/-	(Demment and Longhurst 1987)
Ostium rumino-reticulare^i	?	25	y	y	c	*/*	*/*	(Clauss et al. 2006c)
Omasal laminar surface area	C4, C5	34	n^j	n	d	ns/-	*/-	(Robbins et al. 1995)
RR liquid flow rate	D2	8	n	yn	d	?^{k}		(Robbins et al. 1995)
RR liquid retention time	D2	8	n	yn	d	ns/-	*/-	(Clauss et al. 2006c)
RR liquid retention time	D2, C5	14	n	n	d	ns^m/-	ns^l/-	(Clauss et al. 2006c)
RR particle retention time	D2, C5	10	n	n	d	ns^m/-	ns^m/-	(Clauss et al. 2006c)

(continued)

Table 3.5 (continued)

Parameter	FR	n	dg	od	FTC	FT	BM	Direction	Source
RR VFA concentrations	B2, B3	16	y	y	d	*n/-	*/-	GR < BR	(Clemens and Maloiy 1983)
RR contents crude fibre content	B2	16	y	y	d	ns/-			(Woodall 1992)
RR fermentation rates (mols per DM and d)	B2, B3, D2	21	y	n	d	ns/-	*/-		(Gordon and Illius 1994)
Energy by VFA in RR per day	D2	21	y	n	d	*o/-	*/-	GR > BR	(Gordon and Illius 1994)
Energy by VFA in RR per day	D2	21	y	n	d	nsp/-	*/-		(Gordon and Illius 1994)
RR acetate:propionate ratio	B2-B3	16	y	y	d	ns/-	*/-		(Clemens et al. 1983)
Energy content of RR VFA (J per mol)	B2, B3q	21	y	n	d	ns/-	ns/-		(Gordon and Illius 1994)
RR acetate proportions	B2-B3	16	y	y	d	ns/-	ns/-		(Clemens et al. 1983)
RR propionate propotion	B2-B3	16	y	y	d	*/-	*/-	GR < BR	(Clemens et al. 1983)
RR butyrate proportion	B2-B3	16	n	y	d	ns/-	ns/-		(Clemens et al. 1983)
Dry matter RR (% of wet weight)	?	16	y	y	d	ns/-	*/-		(Clemens and Maloiy 1983)
Volume omasum-abomasum	C4, C5	25	n	n	d	ns/-	ns/-		(Van Wieren 1996b)
Ratio Vol. RR:Omas-Abomas	?	25	n	n	d	ns/-	ns/-		(Van Wieren 1996b)
DM abomasum (%)	C4, C5, D2	16	y	y	d	*/-	ns/-	GR > BR	(Clemens and Maloiy 1983)
DM small intestine (%)	C4, C5, D2	16	y	y	d	*/-	*/-	GR > BR	(Clemens and Maloiy 1983)
Hindgut contents (wet weight)	D2	15	y	n	d	*r/-	*/-	GR < BR	(Gordon and Illius 1994)
Caecum VFA concentration	B2, B3, D2	18	n	y	d	ns/-	-/-		(Maloiy and Clemens 1991)
Colonic VFA concentration	B2, B3, D2	16	y	y	d	*/-	ns/-	GR < BR	(Clemens and Maloiy 1983)
Hindgut VFA reabsorption	B2, B3, D2	16	y	y	d	*/-	ns/-	GR < BR	(Clemens and Maloiy 1984)
Hindgut fermentation rates	B2, B3, D2	15	y	n	d	ns/-	*/-		(Gordon and Illius 1994)
Energy by VFA in hindgut per day	D2	15	y	n	d	ns/-	*/-		(Gordon and Illius 1994)
Hindgut acetate:propionate ratio	B2, B3, D2	16	y	y	d	ns (*)/-	ns/-	GR > BR	(Clemens et al. 1983)
Hindgut acetate proportions	B2, B3, D2	16	y	y	d	*/-	*/-	GR > BR	(Clemens et al. 1983)
Hindgut propionate propotion	B2, B3, D2	16	y	y	d	ns/-	ns/-		(Clemens et al. 1983)
Hindgut butyrate proportion	B2, B3, D2	16	n	y	d	*/-			(Clemens et al. 1983)
Lactic acid in caecum and colon	B2, B3, D2	16	y	y	d	*/-	*/-	GR < BR	(Clemens and Maloiy 1983)
Fiber digestion	D2	15	n	n	d	ns/-		GR < BR	(Robbins et al. 1995)

Variable		n						Reference
Fiber digestion	D2	20	n	d		*/-	ns/-	GR > BR	(Van Wieren 1996b)
Fiber digestion	D2	20	n	d		*/-	*/-	GR > BR	(Iason and Van Wieren 1999)
Fibre digestion	D2	24	n	c		-/*	-/ns	GR > BR	(Pérez-Barbería et al. 2004)
Total GIT particle mean retention time	D2	26	y	d		ns^u/-	*/-		(Gordon and Illius 1994)
Total GIT particle mean retention time	D2	33	n	d		*^{u}/-	*/-	GR > BR	(Hummel et al. 2006)
Colonic fluid absorption	?	16	y	d		ns/-	ns/-		(Clemens and Maloiy 1984)
Dry matter distal colon (% of wet weight)	?	16	y	d		*/-		GR < BR	(Clemens and Maloiy 1983)
Dry matter faeces (% of wet weight)	?	81	y	d		ns^v/-	*^{w}/-		(Clauss et al. 2004)

^{a}data not measured but given as discrete categorical variables

^{b}not all species stated in the methods can be found on the graph in the results

^{c}the tragelaphinae, especially the greater kudu, were outliers to this pattern

^{d}difference in slopes, GR < BR only in species >80 kg; only captive animals on zoo winter diets used

^{e}see Clauss et al. (2003c); data compared to gut-content-free BM, using the authors' own feeding type classification

^{f}using Hofmann's classification; not mentioned in results, only in discussion

^{g}as observed (Pérez-Barbería et al. 2001a), the claim of statistical significance is not justified due to the difference in slopes

^{h}significance disappeared when three small species were excluded; note that according tests (exclusion of certain species) were not performed in other analyses

^{i}misquoted as the ostium reticulo-omasale (Illius and Gordon 1999; Gordon 2003); in contrast to the claim by these authors, no test for difference between feeding types is found in the original work

^{j}not all data from the publications cited in methods section visible in the graph

^{k}no evident pattern, unclear whether statistical test was performed, not all data from the publications cited in methods section visible in the graph

^{l}number of species too small to allow confirmation or refutation of hypotheses; note that in pair-wise comparison, cattle had shorter RR fluid retention than large browsers

^{m}number of species too small to allow confirmation or refutation of hypotheses

Table 3.5 (continued)

[n] statistical significance could not be confirmed (Streich, pers. comm.)
[o] using their own feeding type classification; note that the individual factors used to calculate this factor (RR DM load, VFA production rate, VFA energy content) did not differ significantly
[p] using Hofmann's classification
[q] due to chemical composition of forages, no difference is to be expected
[r] using Hofmann's classification
[s] $p < 0.045$ is stated in the paper
[t] correlation with BM weak but significant
[u] see comments in text and in Clauss and Lechner-Doll (2001)
[v] wider ranges in faecal DM observed in GR than in BR
[w] decreasing faecal DM with BM

do not yield further volatile fatty acids), will reflect this. In particular, a large set of truly comparable ingesta passage measurements is missing.

3.6 Conclusion and Outlook

Although many of the morphophysiological adaptations expected in BR and GR (Table 3.2) seem to be met in tests using conventional statistical methods, the number of such characteristics tested with a phylogenetic control remains low to date (Tables 3.3-3.5). Most of the studies including phylogenetic controls have addressed ungulates as a phylogenetic group, whereas the according parameters have hardly been investigated within lower taxonomic units, such as the ruminants, or, also for lack of extant species, the hindgut-fermenting ungulates. In this respect, the findings of Pérez-Barberia et al. (2004), Clauss et al. (2006b), and Hackenberger (1987) on differences in digestive efficiency between feeding types in ruminants, rhinoceroses, and elephants should incite more comparative investigations in mechanisms involved in digestive physiology. Differences in the correlation between food intake and ingesta retention (Clauss et al. 2007c) could be particularly revealing for the differentiation of BR and GR. Controlled studies that measure retention time and digestibility, on both ad libitum intake of natural forages (which is hypothesised to result in different intake levels) and on similar intake levels of the same forages, could be performed in a multitude of species. Effects of forages should be tested by feeding the same amounts of different forages to a species. The variety of macropod, rodent (Kay and Madden 1997), suid and ruminant species offers an ideal research area for such future studies.

The craniodental patterns found in ruminants with conventional statistical methods should be evaluated within the ruminant guild by phylogenetically controlled studies. More experimental (engineering or biomechanical) approaches are warranted to corroborate the proposed adaptive relevance of craniodental parameters. Amongst others, the pioneering studies of Witzel and Preuschoft (1999, 2002, 2005) could serve as an example here. In order to test for potential differences in soft tissue morphology, new data on both macropods and ruminants is desperately needed, and until more cervid species are included in a phylogenetically controlled study, tests for convergent evolution within the ruminants must remain tentative. In terms of functional relevance, we think that soft tissue anatomy poses the most challenging questions that have to be addressed by engineering models, such as the one created by Langer and Takács (2004) addressing the functional relevance of taeniae and haustra in mammalian intestines, or by invasive experiments such as the one fashioned by Kaske and Midasch (1997) who impeded the motility of the reticulum in sheep in order to elucidate the role of this forestomach compartment for the digestive process in ruminants.

While a comparison of measurements of physiological processes, such as digestive efficiency, ingesta retention time, particle size reduction, amount of food ingested, etc., represent valuable bases for the discussion of different feeding types and their ecological impact on their feeding niches, we believe that only progress in the quantitative understanding of both the function of hard and soft tissue anatomical features and functionally relevant forage properties will allow meaningful interpretations of potential morphological adaptations. It is only by such knowledge that questions like whether different sets of adaptations, with an overexpression of one morphological feature compensating for the underexpression of another feature or vice versa, will facilitate the exploitation of the same niche; or whether the evolution of a particular anatomical feature exclusively allows the use of a new niche or broadens the range of niches available to the species, leaving open the path to a back-switching to formerly used niches. We want to conclude this chapter with the puzzling example of the ruminant reticulum, well aware that differences in this organ have not been statistically demonstrated between the feeding types. The reticular honeycomb cells of many grazing ruminants are particularly pronounced and deep (Hofmann 1988). In domestic ruminants, their function as sedimentation traps for small particles which are subsequently transported into the next forestomach compartment has been determined experimentally (Kaske and Midasch 1997). In contrast, many browsing ruminants, deemed representatives of evolutionary older ruminants, have extremely shallow reticular crests (Neuville and Derscheid 1929), the mechanical function of which is beyond imagination so far. Their shape could be explained if they could be considered 'atavisms', an interpretation ruled out by the common understanding of the evolutionary sequence of feeding types to date. Therefore, the shallow crests of roe deer, moose, and giraffe remain a challenging example for the fact that the evolution of morphological characters can only be understood by their functional relevance.

Acknowledgements MC thanks R.R. Hofmann for years of support and hospitality. We thank B. Schneider for tireless help with literature acquisition. This contribution is dedicated to all those who relish the beauty of molar enamel ridges and reticular honeycomb cells, and whose hearts beat faster at the smell of acetate.

References

Archer D, Sanson G (2002) Form and function of the selenodont molar in southern African ruminants in relation to their feeding habits. J Zool 257:13–26
Axmacher H, Hofmann RR (1988) Morphological characteristics of the masseter muscle of 22 ruminant species. J Zool 215:463–473
Bailoni L, Ramanzin M, Simonetto A, Obalakov N, Schiavon S, Bittan G (1998) The effect of in vitro fermentation on specific gravity and sedimentation measurements of forage particles. J Anim Sci 76:3095–3103
Baker DL, Hobbs NT (1987) Strategies of digestion: digestive efficiency and retention times of forage diets in montane ungulates. Can J Zool 65:1978–1984
Baker G, Jones LHP, Wardrop ID (1959) Cause of wear in sheep's teeth. Nature 184:1583–1584
Beaumont R, Deswysen AG (1991) Mélange et propulsion du contenu du réticulo-rumen. Reprod Nutr Dev 31:335–359
Behrend A, Lechner-Doll M, Streich WJ, Clauss M (2004) Seasonal faecal excretion, gut fill, liquid and particle marker retention in mouflon (*Ovis ammon musimon*), and a comparison with roe deer (*Capreolus capreolus*). Acta Theriol 49:503–515

Bell RHV (1971) A grazing ecosystem in the Serengeti. Sci Am 225:86–93
Blair RM, Short HL, Epps EA (1977) Seasonal nutrient yield and digestibility of deer forage from a young pine plantation. J Wildl Manage 41:667–676
Bodmer RE (1990) Ungulate frugivores and the browser–grazer continuum. Oikos 57:319–325
Boué C (1970) Morphologie fonctionelle des dents labiales chez les ruminants. Mammalia 34:696–711
Bryant JP, Reichardt PB, Clausen TP, Provenza FD, Kuropat PJ (1992) Woody plant – mammal interactions. In: Rosenthal GA, Berenbaum MR (eds) Herbivores: their interaction with secondary plant compounds, vol 2. Academic Press, San Diego, pp 343–370
Brynard A, Pienaar VdV (1960) Annual report of the biologist, 1958/1959, Koedoe. 3:1–205
Bunnell FL, Gillingham MP (1985) Foraging behavior: the dynamics of dining out. In: Hudson RJ, White RG (eds) Bioenergetics of wild herbivores. CRC Press, Boca Raton, FL, pp 53–59
Calder WA (1996) Size, function and life history. Harvard University Press, Cambridge, MA
Case TJ (1979) Optimal body size and an animal's diet. Acta Biotheor 28:54–69
Caswell H, Reed F, Stephenson SN, Werner PA (1973) Photosynthetic pathways and selective herbivory: a hypothesis. Am Nat 107:465–480
Cerling TE, Viehl K (2004) Seasonal diet changes of the forest hog (*Hylochoerus meinertzhageni*) based on the carbon isotopic composition of hair. Afr J Ecol 42:88–92
Cerling TE, Harris JM, Leakey MG (1999) Browsing and grazing in elephants: the isotope record of modern and fossil proboscideans. Oecologia 120:364–374
Cerling TE, Harris JM, Passey BH (2003) Diets of East African bovidae based on stable isotope analysis. J Mammal 84:456–470
Cerling TE, Harris JM, Leakey MG (2005) Environmentally driven dietary adaptations in African mammals. In: Ehleringer JR, Cerling TE, Dearing MD (eds) A history of atmospheric CO_2 and its effects on plants, animals and ecosystems. Springer, New York, pp 258–272
Clauss M (2003) Tannins in the nutrition of captive wild animals. In: Fidgett A, Clauss M, Ganslosser U, Hatt JM, Nijboer J (eds) Zoo animal nutrition, vol 2. Filander, Fuerth, Germany, pp 53–89
Clauss M, Lechner-Doll M (2001) Differences in selective reticulo-ruminal particle retention as a key factor in ruminant diversification. Oecologia 129:321–327
Clauss M, Dierenfeld ES (2007) The nutrition of browsers. In: Fowler ME, Miller RE (eds) Zoo and wild animal medicine, vol VI. Saunders, Philadelphia (in press)
Clauss M, Lechner-Doll M, Behrend A, Lason K, Lang D, Streich WJ (2001) Particle retention in the forestomach of a browsing ruminant, the roe deer (*Capreolus capreolus*). Acta Theriol 46:103–107
Clauss M, Lechner-Doll M, Streich WJ (2002) Faecal particle size distribution in captive wild ruminants: an approach to the browser/grazer-dichotomy from the other end. Oecologia 131:343–349
Clauss M, Frey R, Kiefer B, Lechner-Doll M, Loehlein W, Polster C, Rössner GE, Streich WJ (2003a) The maximum attainable body size of herbivorous mammals: morphophysiological constraints on foregut, and adaptations of hindgut fermenters. Oecologia 136:14–27
Clauss M, Kienzle E, Hatt JM (2003b) Feeding practice in captive wild ruminants: peculiarities in the nutrition of browsers/concentrate selectors and intermediate feeders. A review. In: Fidgett A, Clauss M, Ganslosser U, Hatt JM, Nijboer J (eds) Zoo animal nutrition, vol 2. Filander, Fuerth, Germany, pp 27–52
Clauss M, Lechner-Doll M, Streich WJ (2003c) Ruminant diversification as an adaptation to the physicomechanical characteristics of forage. A reevaluation of an old debate and a new hypothesis. Oikos 102:253–262
Clauss M, Löhlein W, Kienzle E, Wiesner H (2003d) Studies on feed digestibilities in captive Asian elephants (*Elephas maximus*). J Anim Physiol Anim Nutr 87:160–173
Clauss M, Lechner-Doll M, Streich WJ (2004) Differences in the range of faecal dry matter content between feeding types of captive wild ruminants. Acta Theriol 49:259–267
Clauss M, Froeschle T, Castell J, Hummel J, Hatt JM, Ortmann S, Streich WJ (2005a) Fluid and particle retention times in the black rhinoceros (*Diceros bicornis*), a large hindgut-fermenting browser. Acta Theriol 50:367–376

Clauss M, Gehrke J, Hatt JM, Dierenfeld ES, Flach EJ, Hermes R, Castell J, Streich WJ, Fickel J (2005b) Tannin-binding salivary proteins in three captive rhinoceros species. Comp Biochem Physiol A 140:67–72

Clauss M, Castell J, Kienzle E, Dierenfeld ES, Flach EJ, Behlert O, Ortmann S, Hatt JM, Streich WJ, Hummel J (2006a) Macromineral absorption in the black rhinoceros (*Diceros bicornis*) as compared to the domestic horse. J Nutr 136:2017S–2020S

Clauss M, Castell JC, Kienzle E, Dierenfeld ES, Flach EJ, Behlert O, Ortmann S, Streich WJ, Hummel J, Hatt JM (2006b) Digestion coefficients achieved by the black rhinoceros (*Diceros bicornis*), a large browsing hindgut fermenter. J Anim Physiol Anim Nutr 90:325–334

Clauss M, Hofmann RR, Hummel J, Adamczewski J, Nygren K, Pitra C, Reese S (2006c) The macroscopic anatomy of the omasum of free-ranging moose (*Alces alces*) and muskoxen (*Ovibos moschatus*) and a comparison of the omasal laminal surface area in 34 ruminant species. J Zool 270:346–358

Clauss M, Hummel J, Streich WJ (2006d) The dissociation of the fluid and particle phase in the forestomach as a physiological characteristic of large grazing ruminants: an evaluation of available, comparable ruminant passage data. Eur J Wildl Res 52:88–98

Clauss M, Franz-Odendaal TA, Brasch J, Castell JC, Kaiser TM (2007a) Tooth wear in captive giraffes (*Giraffa camelopardalis*): mesowear analysis classifies free-ranging specimens as browsers but captive ones as grazers. J Zoo Wildl Med (in press)

Clauss M, Steinmetz H, Eulenberger U, Ossent P, Zingg R, Hummel J, Hatt JM (2007b) Observations on the length of the intestinal tract of African (*Loxodonta africana*) and Asian elephants (*Elephas maximus*). Eur J Wildl Res 53:68–72

Clauss M, Streich WJ, Schwarm A, Ortmann S, Hummel J (2007c) The relationship of food intake and ingesta passage predicts feeding ecology in two different megaherbivore groups. Oikos 116:209–216

Clemens ET, Maloiy GMO (1983) Digestive physiology of East African wild ruminants. Comp Biochem Physiol A 76:319–333

Clemens ET, Maloiy GMO (1984) Colonic absorption and secretion of fluids, electrolytes and organic acids in East African wild ruminants. Comp Biochem Physiol A 77:51–56

Clemens ET, Maloiy GMO, Sutton JD (1983) Molar proportions of volatile fatty acids in the gastrointestinal tract of East African wild ruminants. Comp Biochem Physiol A 76:217–224

Codron J, Lee-Thorp JA, Sponheimer M, Codron D, Grant RC, De Ruiter DJ (2006) Elephant (*Loxodonta africana*) diets in Kruger National Park, South Africa: spatial and landscape differences. J Mammal 87:27–34

Codron D, Lee-Thorp JA, Sponheimer M, Codron J (2007a) Nutritional content of savanna plant foods: implications for browse/grazer models of ungulate diversification. Eur J Wildl Res 53:100–111

Codron D, Lee-Thorp JA, Sponheimer M, Codron J, de Ruiter D, Brink JS (2007b) Significance of diet type and diet quality for ecological diversity of African ungulates. J Anim Ecol 76:526–537

Cork SJ, Foley WJ (1991) Digestive and metabolic strategies of arboreal mammalian folivores in relation to chemical defenses in temperate and tropical forests. In: Palo RT, Robbins CT (eds) Plant defenses against mammalian herbivores. CRC Press, Boca Raton, pp 133–166

DeGusta D, Vrba E (2003) A method for inferring paleohabitats from the functional morphology of bovid astagali. J Archaeol Sci 30:1009–1022

DeGusta D, Vrba E (2005a) Methods for inferring paleohabitats from discrete traits of the bovid postcranial skeleton. J Archaeol Sci 32:1115–1123

DeGusta D, Vrba E (2005b) Methods for inferring paleohabitats from the functional morphology of bovid phalanges. J Archaeol Sci 32:1099–1113

Dehority BA (1995) Rumen ciliates of the pronghorn antelope (*Antilocapra americana*), mule deer (*Odocoileus hemionus*), white-tailed deer (*Odocoileus virginianus*) and elk (*Cervus canadensis*) in the Northwestern United States. Arch Protistenkd 146:29–36

Dehority BA, Odenyo AA (2003) Influence of diet on the rumen protozoal fauna of indigenous African wild ruminants. J Eukaryot Microbiol 50:220–223

Dehority BA, Demarais S, Osborn DA (1999) Rumen ciliates of white-tailed deer (*Odocoileus virginianus*), axis deer (*Axis axis*), sika deer (*Cervus nippon*) and fallow deer (*Dama dama*) from Texas. J Eukaryot Microbiol 46:125–131

Demment MW, Van Soest PJ (1985) A nutritional explanation for body-size patterns of ruminant and nonruminant herbivores. Am Nat 125:641–672

Demment MW, Longhurst WH (1987) Browsers and grazers: constraints on feeding ecology imposed by gut morphology and body size. Proceedings of the IVth International Conference on Goats, Brazilia, Brazil, 989–1004

Deutsch A, Lechner-Doll M, Wolf AG (1998) Activity of cellulolytic enzymes in the contents of reticulorumen and caecocolon of roe deer (*Capreolus capreolus*). Comp Biochem Physiol A 119:925–930

Ditchkoff SS (2000) A decade since "diversification of ruminants": has our knowledge improved? Oecologia 125:82–84

Dougall HW, Drysdale VM, Glover PE (1964) The chemical composition of Kenya browse and pasture herbage. E Afr Wildl J 2:86–121

Drescher-Kaden U (1976) Tests on the digestive system of roe deer, fallow deer and mouflon. Report 1: Weight statistics and capacity measurements on the digestive system, particularly of the rumenreticulum. Z Jagdwiss 22:184–190

du Plessis I, van der Waal C, Webb EC (2004) A comparison of plant form and browsing height selection of four small stock breeds - preliminary results. S Afr J Anim Sci 34(Suppl. 1):31–34

Duncan P, Tixier H, Hofmann RR, Lechner-Doll M (1998) Feeding strategies and the physiology of digestion in roe deer. In: Andersen R, Duncan P, Linell JDC (eds) The European roe deer: the biology of success. Scandinavian University Press, Oslo, pp 91–116

Foose TJ (1982) Trophic strategies of ruminant versus nonruminant ungulates. Dissertation, University of Chicago, Chicago

Fortelius M (1985) Ungulate cheek teeth: developmental, functional, and evolutionary interrelations. Acta Zool Fenn 180:1–76

Fortelius M, Kappelman J (1993) The largest land mammal ever imagined. Zool J Linn Soc–Lond 107:85–101

Fortelius M, Solounias N (2000) Functional characterization of ungulate molars using the abrasion–attrition wear gradient: a new method for reconstructing paleodiets. Am Mus Novit 3301:1–36

Franz-Oftedaal TA, Kaiser TM (2003) Differential mesowear in the maxillary and mandibular cheek dentition of some ruminants (Artiodactyla). Ann Zool Fenn 40:395–410

Freeland WJ, Janzen DH (1974) Strategies in herbivory by mammals: The role of secondary compounds. Am Nat 108:269–289

Freudenberger DO, Wallis IR, Hume ID (1989) Digestive adaptations of kangaroos, wallabies and rat-kangaroos. In: Grigg G, Jarman P, Hume I (eds) Kangaroos, wallabies and rat-kangaroos. Surrey-Beatty, Sydney, pp 151–168

Gagnon M, Chew AE (2000) Dietary preferences in extant African bovidae. J Mammal 81:490–511

Gentry AW (1980) Fossil bovidae (Mammalia) from Langebaanweg. Ann S Afr Mus 79:213–337

Giesecke D, Van Gylswyk NO (1975) A study of feeding types and certain rumen functions in six species of South African wild ruminants. J Agr Sci 85:75–83

Gordon IJ (2003) Browsing and grazing ruminants: Are they different beasts? Forest Ecol Manag 181:13–21

Gordon IJ, Illius AW (1988) Incisor arcade structure and diet selection in ruminants. Funct Ecol 2:15–22

Gordon IJ, Illius AW (1994) The functional significance of the browser–grazer dichotomy in African ruminants. Oecologia 98:167–175

Gordon IJ, Illius AW (1996) The nutritional ecology of African ruminants: a reinterpretation. J Anim Ecol 65:18–28

Gould SJ (2002) The structure of evolutionary theory. Harvard University Press, Cambridge, MA
Greaves W (1991) A relationship between premolar loss and jaw elongation in selenodont artiodactyls. Zool J Linn Soc–Lond 101:121–129
Guthrie RD (1990) Frozen fauna of mammoth steppe: the story of blue babe. University of Chicago Press, Chicago
Hackenberger MK (1987) Diet digestibilities and ingesta transit times of captive Asian and African elephants. University of Guelph, Guelph
Hagen J (2003) The statistical frame of mind in systematic biology from *Quantitative zoology* to *Biometry*. J Hist Biol 36:353–384
Hagerman AE, Robbins CT, Weerasuriya Y, Wilson TC, McArthur C (1992) Tannin chemistry in relation to digestion. J Range Manage 45:57–62
Hanley TA (1982) The nutritional basis for food selection by ungulates. J Range Manage 35:146–151
Harris JM, Cerling TE (2002) Dietary adaptations of extant and Neogene African suids. J Zool 256:45–54
Harvey PH, Pagel MD (1991) The comparative method in evolutionary biology. Oxford University Press, Oxford
Haschick SL, Kerley GIH (1996) Experimentally determined foraging heights of buchbuck (*Tragelaphus scriptus*) and Boer goats (*Capra hircus*). S Afr J Wildl Res 26:64–65
Heckathorn SA, McNaughton SJ, Coleman JS (1999) C4 plants and herbivory. In: Sage RF, Monson RK (eds) C4 plant biology. Academic Press, San Diego, pp 285–312
Hofmann RR (1968) Comparison of the rumen and omasum structure in East African game ruminants in relation to their feeding habits. Sym Zool Soc Lond 21:179–194
Hofmann RR (1973) The ruminant stomach. East African Literature Bureau, Nairobi
Hofmann RR (1988) Morphophysiological evolutionary adaptations of the ruminant digestive system. In: Dobson A, Dobson MJ (eds) Aspects of digestive physiology in ruminants. Cornell University Press, Ithaca, NY, pp 1–20
Hofmann RR (1989) Evolutionary steps of ecophysiological adaptation and diversification of ruminants: a comparative view of their digestive system. Oecologia 78:443–457
Hofmann RR (1991) Endangered tropical herbivores - their nutritional requirements and habitat demands. In: Ho YW, Wong HK, Abdullah N, Tajuddin ZA (eds) Recent advances on the nutrition of herbivores. Malaysia Society of Animal Production, UPM Serdang, pp 27–34
Hofmann RR (1999) Functional and comparative digestive system anatomy of Arctic ungulates. Rangifer 20:71–81
Hofmann RR, Stewart DRM (1972) Grazer or browser: a classification based on the stomach-structure and feeding habit of East African ruminants. Mammalia 36:226–240
Hofmann RR, Streich WJ, Fickel J, Hummel J, Clauss M. Convergent evolution in feeding types: salivary gland mass differences in wild ruminant species. J Morphal (in press)
Holechek JL, Pieper RD, Herbel CH (2004) Range management. Principles and practices, 5th edn. Pearson/Prentice Hall, Upper Saddle River, NJ
Hoppe PP (1977) Rumen fermentation and body weight in African ruminants. In: Peterle TJ (ed) 13th Congress of Game Biology, vol 13. The Wildlife Society, Washington, DC, pp 141–150
Hummel J, Clauss M, Zimmermann W, Johanson K, Norgaard C, Pfeffer E (2005) Fluid and article retention in captive okapi (*Okapia johnstoni*). Comp Biochem Physiol A 140:436–444
Hummel J, Südekum KH, Streich WJ, Clauss M (2006) Forage fermentation patterns and their implications for herbivore ingesta retention times. Funct Ecol 20:989–1002
Iason G, Palo RT (1991) The effects of birch phenolics on a grazing and a browsing mammal. A comparison of hares. J Chem Ecol 17:1733–1743
Iason GR, Van Wieren SE (1999) Digestive and ingestive adaptations of mammalian herbivores to low-quality forage. In: Olff H, Brown VK, Drent RH (eds) Herbivores: between plants and predators. 38th Symp Brit Ecol Soc Blackwell, Oxford, pp 337–369
Illius AW, Gordon IJ (1992) Modelling the nutritional ecology of ungulate herbivores: evolution of body size and competitive interactions. Oecologia 89:428–434

Illius AW, Gordon IJ (1999) The physiological ecology of mammalian herbivory. In: Jung HJG, Fahey GC (eds) Nutritional ecology of herbivores. The American Society of Animal Science, Savoy, IL, pp 71–96

Ioannidis JPA (2005) Why most published research findings are false. PLos Med 2:e124 (696–701)

Janis CH (1988) An estimation of tooth volume and hypsodonty indices in ungulate mammals and the correlation of these factors with dietary preference. Teeth revisited. Proceedings of the VIIth International Symposium on Dental Morphology. Mem Mus Hist Naturelle Paris C 53:367–387

Janis CM (1990) Correlation of cranial and dental variables with dietary preferences in mammals: a comparison of macropodoids and ungulates. Mem Queensland Mus 28:349–366

Janis CM (1993) Tertiary mammal evolution in the context of changing climates, vegetation, and tectonic events. Annu Rev Ecol Syst 24:467–500

Janis CM (1995) Correlations between craniodental morphology and feeding behavior in ungulates: reciprocal illumination between living and fossil taxa. In: Thomason JJ (ed) Functional morphology in vertebrate paleontology. Cambridge Univ. Press, New York, pp 76–98

Janis CM, Scott KM (1987) The interrelationships of higher ruminant families with special emphasis on the members of the cervoidea. Am Mus Novit 2893:1–85

Janis CM, Ehrhardt D (1988) Correlation of the relative muzzle width and relative incisor width with dietary preferences in ungulates. Zool J Linn Soc–Lond 92:267–284

Janis CM, Fortelius M (1988) The means whereby mammals achieve increased functional durability of their dentitions, with special reference to limiting factors. Biol Rev 63:197–230

Janis CM, Constable E (1993) Can ungulate craniodental features determine digestive physiology? J Vertebr Paleontol 13:abstract

Jarman PJ (1974) The social organization of antelope in relation to their ecology. Behaviour 48:215–266

Jiang Z, Takatsuki S (1999) Constraints on feeding type in ruminants: a case for morphology over phylogeny. Mammal Study 24:79–89

Johnson CN, Prideaux GJ (2004) Extinctions of herbivorous mammals in the late Pleistocene of Australia in relation to their feeding ecology: no evidence for environmental change as cause of extinction. Austr Ecol 29:553–557

Jones RJ, Meyer JHF, Bechaz FM, Stolzt MA, Palmer B, van der Merwe G (2001) Comparison of rumen fluid from South African game species and from sheep to digest tanniniferous browse. Aust J Agr Res 52:453–460

Kaiser TM, Fortelius M (2003) Differential mesowear in occluding upper and lower molars: opening mesowear analysis for lower molars and premolars in hypsodont horses. J Morphol 258:67–83

Kappelman J (1988) Morphology and locomotor adaptations of the bovid femur in relation to habitat. J Morphol 198:119–130

Kappelmann J, Plummer T, Bishop L, Duncan A, Appleton S (1997) Bovids as indicators of plio-pleistocene paleoenvironments in East Africa. J Hum Evol 32:229–256

Kaske M, Midasch A (1997) Effects of experimentally impaired reticular contractions on digesta passage in sheep. Brit J Nutr 78:97–110

Kay RF, Madden RH (1997) Mammals and rainfall: paleoecology of the middle Miocene at La Venta (Columbia, South America). J Hum Evol 32:161–199

Kay RNB (1987a) Comparative studies of food propulsion in ruminants. In: Ooms LAA, Degtyse AD, Van Miert ASJ (eds) Physiological and pharmacological aspects of the reticulo-rumen. Marinus Nijhoff, Boston, pp 155–170

Kay RNB (1987b) Weights of salivary glands in some ruminant animals. J Zool 211:431–436

Kay RNB (1993) Digestion in ruminants at pasture. World Conference on Animal Production Edmonton, Canada, pp 461–474

Kay RNB, Engelhardt Wv, White RG (1980) The digestive physiology of wild ruminants. In: Ruckebush Y, Thivend P (eds) Digestive physiology and metabolism in ruminants. MTP Press, Lancaster, pp 743–761

Kock RA, Garnier J (1993) Veterinary management of three species of rhinoceros in zoological collections. In: Ryder OA (ed) Rhinoceros biology and conservation. Zoological Society of San Diego, San Diego, pp 325–338

Köhler M (1993) Skeleton and habitat of recent and fossil ruminants. Münchner Geowiss Abh A: Geol Paläontol 25:1–88

Langer P, Takács A (2004) Why are taeniae, haustra, and semilunar folds differentiated in the gastrointestinal tract of mammals, including man? J Morphol 259:308–315

Lechanteur YARG, Griffiths CL (2003) Diets of common suprabenthic reef fish in False Bay, South Africa. Afr Zool 38:213–227

Lechner-Doll M, Kaske M, Engelhardt Wv (1991) Factors affecting the mean retention time of particles in the forestomach of ruminants and camelids. Proc Internat Sym Ruminant Physiol 7:455–482

Lentle RG, Hume ID, Stafford KJ, Kennedy M, Haslett S, Springett BP (2003a) Comparisons of indices of molar progression and dental function of brush-tailed rock-wallabies (*Petrogale penicillata*) with tammar (*Macropus eugenii*) and parma (*Macropus parma*) wallabies. Aust J Zool 51:259–269

Lentle RG, Hume ID, Stafford KJ, Kennedy M, Haslett S, Springett BP (2003b) Molar progression and tooth wear in tammar (*Macropus eugenii*) and parma (*Macropus parma*) wallabies. Aust J Zool 51:137–151

Lentle RG, Hume ID, Stafford KJ, Kennedy M, Springett BP, Haslett S (2003c) Observations on fresh forage intake, ingesta particle size and nutrient digestibility in four species of macropod. Aust J Zool 51:627–636

Leus K, MacDonald A (1997) From barbirusa (*Babyrousa babyrussa*) to domestic pig: the nutrition of swine. Proc Nutr Soc 56:1001–1012

Lirette A, Milligan LP, Cyr N, Elofson RM (1990) Buoyancy separation of particles of forages, feces, and ruminal contents and nuclear magnetic resonance examination. Can J Anim Sci 70:1099–1108

Loehlein W, Kienzle E, Wiesner H, Clauss M (2003) Investigations on the use of chromium oxide as an inert, external marker in captive Asian elephants (*Elephas maximus*): passage and recovery rates. In: Fidgett A, Clauss M, Ganslosser U, Hatt JM, Nijboer J (eds) Zoo animal nutrition, vol 2. Filander, Fuerth, Germany, pp 223–232

MacFadden BJ (1992) Fossil horses. Cambridge University Press, Cambridge

MacFadden BJ, Solounias N, Cerling TE (1999) Ancient diets, ecology, and extinction of 5-million-year-old horses from Florida. Science 283:824–827

Maglio V (1973) Origin and evolution of the elephantidae. Trans Am Philos Soc 63:1–149

Maloiy GMO, Clemens ET (1991) Aspects of digestion and in vitro fermentation in the caecum of some East African herbivores. J Zool 224:293–300

Martins EP, Hansen TF (1996) The statistical analysis of interspecific data: a review and evaluation of phylogenetic comparative methods. In: Martins EP (ed) Phylogenies and the comparative method in animal behavior. Oxford University Press, Oxford, pp 22–75

Martz FA, Belyea RL (1986) Role of particle size and forage quality in digestion and passage by cattle and sheep. J Dairy Sci 69:1996–2008

McArthur C, Sanson GD (1993) Nutritional effects and costs of a tannin in a grazing and a browsing macropodid marsupial herbivore. Funct Ecol 7:690–969

McCammon-Feldman B, van Soest PJ, Horvath P, McDowell RE (1981) Feeding strategy of the goat. Cornell University, Ithaca, NY

McDowell RE, Sisler DG, Schermerhorn EC, Reed JD, Bauer RP (1983) Game or cattle for meat production on Kenya rangelands? Cornell International Agriculture Monograph, Ithaca, NY

McNaughton SJ, Tarrants JL, MacNaughton MM, Davis RH (1985) Silica as a defense against herbivory and a growth promotor in African grasses. Ecology 66:528–535

Mendoza M, Palmqvist P (2006a) Characterizing adaptive morphological patterns related to diet in bovidae. Acta Zool Sin 52:988–1008

Mendoza M, Palmqvist P (2006b) Characterizing adaptive morphological patterns related to habitat use and body mass in bovidae. Acta Zool Sin 52:971–987

Mendoza M, Janis CM, Palmqvist P (2002) Characterizing complex craniodental patterns related to feeding behaviour in ungulates: a multivariate approach. J Zool 258:223–246

Milton K, Dintzis FR (1981) Nitrogen-to-protein conversion factors for tropical plant samples. Biotropica 13:177–181

Moseley G, Jones JR (1984) The physical digestion of perennial ryegrass (*Lolium perenne*) and white clover (*Trifolium repens*) in the foregut of sheep. Brit J Nutr 52:381–390

Mtengeti EJ, Wilman D, Moseley G (1995) Physical structure of white clover, rape, spurrey and perennial ryegrass in relation to rate of intake by sheep, chewing activity and particle breakdown. J Agr Sci 125:43–50

Neuville H, Derscheid JM (1929) Recherches anatomiques sur l'okapi. IV. L'estomac. Rev Zool Afr 16:373–419

Nocek JE, Kohn RA (1987) Initial particle form and size on change in functional specific gravity of alfalfa and timothy hay. J Dairy Sci 70:1850–1863

Oftedal O (1991) The nutritional consequences of foraging in primates: the relationship of nutrient intakes to nutrient requirements. Philos Trans R Soc B 334:161–170

Oldemeyer JL, Franzmann AW, Brundage AL, Arneson PD, Flynn A (1977) Browse quality and the Kenai moose population. J Wildlife Manage 41:533–542

Owen-Smith N (1982) Factors influencing the consumption of plant products by large herbivores. In: Huntley BJ, Walker BH (eds) Ecology of tropical savannas. Springer, Berlin Heidelberg New York, pp 359–404

Owen-Smith N (1985) Niche separation among African ungulates. In: Vrba ES (ed) Species and Speciation, vol 4. Transvaal Museum, Pretoria, pp 167–171

Owen-Smith N (1988) Megaherbivores - the influence of very large body size on ecology. Cambridge University Press, Cambridge

Owen-Smith N (1997) Distinctive features of the nutritional ecology of browsing versus grazing ruminants. Z Säugetierkd 62 (Suppl. 2):176–191

Palamara J, Phakey PP, Rachinger WA, Sanson GD, Orams HJ (1984) On the nature of the opaque and translucent enamel regions of some macropodinae (*Macropus giganteus*, *Wallabia bicolor* and *Peradorcas concinna*). Cell Tissue Res 238:329–337

Palmqvist P, Groecke DR, Arribas A, Farina RA (2003) Paleoecological reconstruction of a lower Pleistocene large mammal community using biogeochemical and ecomorphological approaches. Paleobiology 29:205–229

Pérez-Barbería FJ, Gordon IJ (1999) The functional relationship between feeding type and jaw and cranial morphology in ungulates. Oecologia 118:157–165

Pérez-Barbería FJ, Gordon IJ (2001) Relationships between oral morphology and feeding style in the Ungulata: a phylogenetically controlled evaluation. Proc R Soc Lond B Bio 268:1023–1032

Pérez-Barbería FJ, Gordon IJ (2005) Gregariousness increases brain size in ungulates. Oecologia 145:41–52

Pérez-Barbería FJ, Gordon IJ, Illius A (2001a) Phylogenetic analysis of stomach adaptation in digestive strategies in African ruminants. Oecologia 129:498–508

Pérez-Barbería FJ, Gordon IJ, Nores C (2001b) Evolutionary transitions among feeding styles and habitats in ungulates. Evol Ecol Res 3:221–230

Pérez-Barbería FJ, Elston DA, Gordon IJ, Illius AW (2004) The evolution of phylogenetic differences in the efficiency of digestion in ruminants. Proc Roy Soc Lond B Bio 271:1081–1090

Peters RH (1986) The ecological implications of body size. Cambridge University Press, Cambridge

Plummer TW, Bishop LC (1994) Hominid paleoecology at Olduvai Gorge, Tanzania as indicated by antelope remains. J Hum Evol 27:47–75

Popowics TE, Fortelius M (1997) On the cutting edge: tooth blade sharpness in herbivorous and faunivorous mammals. Ann Zool Fenn 34:73–88

Prins RA, Geelen MJH (1971) Rumen characteristics of red deer, fallow deer and roe deer. J Wildlife Manage 35:673–680

Prins RA, Kreulen DA (1991) Comparative aspects of plant cell wall digestion in mammals. In: Hoshino S, Onodera R, Minoto H, Itabashi H (eds) The rumen ecosystem. Japan Scientific Society Press, Tokyo, pp 109–120

Prins RA, Rooymans TP, Veldhuizen M, Domhof MA, Cliné-Theil W (1983) Extent of plant cell wall digestion in several species of wild ruminants kept in the zoo. Zool Garten NF 53:393–403

Prins RA, Lankhorst A, Van Hoven W (1984) Gastro-intestinal fermentation in herbivores and the extent of plant cell wall digestion. In: Gilchrist FMC, Mackie RI (eds) Herbivore nutrition in the subtropics and tropics. Science Press, Craighall, South Africa, pp 408–434

Rhoades DF, Gates RG (1976) Towards a general theory of plant antiherbivore chemistry. Recent Adv Phytochem 10:168–213

Robbins C, Hagerman A, Austin P, McArthur C, Hanley T (1991) Variation in mammalian physiological responses to a condensed tannin and its ecological implications. J Mammal 72:480–486

Robbins CT (1993) Wildlife feeding and nutrition. Academic Press, San Diego

Robbins CT, Spalinger DE, Van Hoven W (1995) Adaptations of ruminants to browse and grass diets: are anatomical-based browser–grazer interpretations valid? Oecologia 103:208–213

Sanson GD (1989) Morphological adaptations of teeth to diets and feeding in the macropodoidea. In: Grigg G, Jarman P, Hume I (eds) Kangaroos, wallabies and rat-kangaroos. Surrey-Beatty, Sydney, pp 151–168

Sanson GD (2006) The biomechanics of browsing and grazing. Am J Bot 93:1531–1545

Sanson GD, Kerr SA, Gross KA (2007) Do silica phytoliths really wear mammalian teeth? J Archaeol Sci 34:526–531

Schmidt-Nielsen K (1984) Scaling: Why is animal size so important? Cambridge University Press, Cambridge

Scott KM (1985) Allometric trends and locomotor adaptations in the Bovidae. Bull Am Mus Nat Hist 179:197–288

Scott KM (1987) Allometry and habitat-related adaptations in the postcranial skeleton of cervidae. In: Wemmer CM (ed) Biology and management of the cervidae. Smithsonian Press, Washington, DC, pp 65–79

Short HL (1975) Nutrition of southern deer in different seasons. J Wildlife Manage 39:321–329

Short HL, Blair RM, Segelquist CA (1974) Fiber composition and forage digestibility by small ruminants. J Wildl Manage 38:197–209

Sibly RM (1981) Strategies of digestion and defecation. In: Townsend C, Calow P (eds) Physiological ecology: an evolutionary approach to resource utilization. Blackwell, Oxford, pp 109–139

Siepel H, de Ruiter-Dijkman EM (1993) Feeding guilds of oribatid mites based on their carbohydrase activities. Soil Biol Biochem 25:1491–1497

Simpson GG (1953) The major features of evolution. Columbia University Press, New York

Solounias N, Dawson-Saunders B (1988) Dietary adaptations and palaecology of the late Miocene ruminants from Pikermi and Samos in Greece. Palaeogeogr Palaeoecl 65:149–172

Solounias N, Moelleken S (1993) Dietary adaptations of some extinct ruminants determined by premaxillary shape. J Mammal 74:1059–1071

Solounias N, Moelleken SMC (1999) Dietary determination of extinct bovids through cranial foraminal analysis, with radiographic applications. Ann Mus Goulandris 10:267–290

Solounias N, Semprebon G (2002) Advances in the reconstruction of ungulate ecomporphology with application to early fossil equids. Am Mus Novit 3366:1–49

Solounias N, Teaford M, Walker A (1988) Interpreting the diet of extinct ruminants: the case of a non-browsing giraffid. Paleobiology 14:287–300

Solounias N, Fortelius M, Freeman P (1994) Molar wear rates in ruminants: a new approach. Ann Zool Fenn 31:219–227

Solounias N, Moelleken S, Plavcan J (1995) Predicting the diet of extinct bovids using masseteric morphology. J Vertebr Paleontol 15:795–805

Spalinger DE, Robbins CT, Hanley TA (1986) The assessment of handling time in ruminants: the effect of plant chemical and physical structure on the rate of breakdown of plant particles in the rumen of mule deer and elk. Can J Zool 64:312–321

Spencer LM (1995) Morphological correlates of dietary resource partitioning in the African bovidae. J Mammal 76:448–471

Sponheimer M, Reed KE, Lee-Thorp JA (1999) Combining isotopic and ecomorphological data to refine bovid paleodietary reconstruction: a case study from the Makapansgat Limeworks hominin locality. J Hum Evol 36:705–718

Sponheimer M, Lee-Thorp JA, DeRuiter D, Smith JM, Van der Merwe NJ, Reed K, Grant CC, Ayliffe LK, Robinson TF, Heidelberger C, Marcus W (2003) Diets of Southern African bovidae: stable isotope evidence. J Mammal 84:471–479

Sprent JA, McArthur C (2002) Diet and diet selection of two species in the macropodid browser–grazer continuum: do they eat what they 'should'? Aust J Zool 50:183–192

Stöckmann W (1979) Differences in the shape of the mandibles of African bovidae in relation to food composition. Zool Jahrb Syst 106:344–373

Thenius E (1992) The okapi from Zaire - a "living fossil" or a secondary rainforest inhabitant? Z Zool Syst Evol 30:163–179

Troelsen JE, Campbell JB (1968) Voluntary consumption of forage by sheep and its relation to the size and shape of particles in the digestive tract. Anim Prod 10:289–296

Turnbull WD (1970) Mammalian masticatory apparatus. Fieldiana Geol 18:147–356

Van Hoven W (1991) Mortalities in kudu populations related to chemical defence in trees. J Afr Zool 105:141–145

Van Soest PJ (1975) Physico-chemical aspects of fiber digestion. In: McDonald IW, Warner ACI (eds) Digestion and metabolism of the ruminant. University of New England Publishing, Armidale, NSW, pp 352–365

Van Soest PJ (1994) Nutritional ecology of the ruminant, 2nd edn. Cornell Univ Press, Ithaca, New York

Van Soest PJ (1996) Allometry and ecology of feeding behavior and digestive capacity in herbivores: a review. Zoo Biol 15:455–479

Van Wieren SE (1996a) Nutrient extraction from mixed grass-browse diets by goats and sheep. In: Van Wieren SE (ed) Digestive strategies in ruminants and nonruminants. Thesis University of Wageningen, pp 67–79

Van Wieren SE (1996b) Browsers and grazers: foraging strategies in ruminants. In: Van Wieren SE (ed) Digestive strategies in ruminants and nonruminants. Thesis University of Wageningen, pp 119–146

Vrba ES (1978) The significance of bovid remains as indicators of environment and predation patterns. University of Chicago Press, Chicago

Waldo DR, Smith LW, Cox EL (1972) Model of cellulose disappearance from the rumen. J Dairy Sci 55:125–129

Walker A, Hoeck HN, Perez L (1978) Microwear of mammalian teeth as indicators of diet. Science 201:908–910

Warner RG, Flatt WP, Loosli JK (1956) Dietary factors influencing the development of the ruminant stomach. Agr Food Chem 4:788–792

Wattiaux MA, Satter LD, Mertens DR (1992) Effect of microbial fermentation on functional specific gravity of small forage particles. J Anim Sci 70:1262–1270

Westoby M, Leishman MR, Lord JM (1995) On misinterpreting the "phylogenetic correction". J Ecol 83:531–534

Williams SH, Kay RF (2001) A comparative test of adaptive explanations for hypsodonty in ungulates and rodents. J Mammal Evol 8:207–229

Williams J, Ostrowski S, Bedin E, Ismail K (2001) Seasonal variation in energy expenditure, water flux and food consumption of Arabian oryx (*Oryx leucoryx*). J Exp Biol 204:2301–2311

Wilman D, Riley JA (1993) Potential nutritive value of a wide range of grassland species. J Agr Sci 120:43–49

Wilson JR (1993) Organization of forage plant tissues. In: Jung HG, Buxton DR, Hatfield RD, Ralph J (eds) Forage cell wall structure and digestibility. American Society of Agronomy, Madison, pp 1–32

Wilson JR, McLeod NM, Minson DJ (1989) Particle-size reduction of the leaves of a tropical and a temperate grass by cattle. I. Effect of chewing during eating and varying times of digestion. Grass Forage Sci 44:55–63

Witzel U, Preuschoft H (1999) The bony roof of the nose in humans and other primates. Zool Anz 238:103–115

Witzel U, Preuschoft H (2002) The functional shape of the human skull, as documented by three-dimensional FEM studies. Anthropol Anz 60:113–135

Witzel U, Preuschoft H (2005) Finite-element model construction for the virtual synthesis of the skulls in vertebrates: case study of Diplodocus. Anat Rec Part A 283:391–401

Wofford H, Holechek JL (1982) Influence of grind size on four- and forty-eight hour in vitro digestibility. Proc W Sect Am Soc Anim Sci 33:261–263

Woodall PF (1992) An evaluation of a rapid method for estimating digestibility. Afr J Ecol 30:181–185

Chapter 4
Nutritional Ecology of Grazing and Browsing Ruminants

Alan J. Duncan and Dennis P. Poppi

4.1 Introduction

Ruminants are, without exception, obligate herbivores subsisting as they do on a diet composed entirely of plant material. However, plant material is a diverse resource and within the Ruminantia there is a range of feeding niches with different herbivore classes focussing their foraging effort on different vegetation types (Hofmann 1989). The plant material available to herbivores comes in a range of morphological types with the major types being grasses, forbs, and browse. Grasses (including the morphologically similar sedges) are monocotyledonous plants characterised by a basal meristem, a low growth form and a relative lack of lignified support structures (except for some of the tall tropical grasses which could be considered to be morphologically classed as browse). Browse plants, comprising shrubs and trees are largely dicotyledonous plants characterised by an apical meristem, a low or high growth form and a well developed system of lignified support structures. Forbs are generally intermediate in character being dicotyledonous, with apical meristems but generally showing low growth forms and a relative lack of lignified growth structures. This variation in growth form, degree of lignification and location of the meristem has implications for resource allocation within the plant, e.g., relative investment in structural tissue and defensive chemicals. This in turn affects the nutritional value of plants for potential herbivores. The purpose of this chapter is to consider the way in which plants are digested by ruminants and how this varies with feeding habit. The chapter will first provide some simple background on digestion of plants by ruminants and how this is influenced by the presence of plant secondary metabolites in plant material. The discussion will then turn to differences between browsers and grazers and review the evidence for differences in the way in which they process the food they consume. Finally, the extent to which browsers and grazers deal differently with plant secondary metabolites will be explored.

4.2 Nutritive Value of Plant Material for Ruminant Herbivores

4.2.1 Digestion of plant material by Herbivores

Plant material can be considered to be composed of cell wall material and cell contents. The structure of plants is provided by cell walls composed of cellulose and hemicellulose with varying amounts of lignin to increase the strength of the cell walls. Cellulose cannot be digested by enzymes of mammalian origin so herbivores rely on symbiotic micro-organisms within their digestive tracts capable of cellulolytic activity (Hungate 1966). Bacteria are presumed to exist in the digestive tract of all herbivores because cellulase is not in the suite of mammalian enzymes. In ruminants, as well as bacteria, protozoa and fungi are present in the digestive tract. They are contained largely within a pre-gastric chamber known as the rumen but also in hind-gut organs such as the caecum and colon. Cellulolytic microorganisms convert cellulose to volatile fatty acids (VFAs), predominantly acetate, butyrate and propionate, mainly by anaerobic metabolic pathways. It is these VFAs which are used as an energy source by ruminants. The degree to which cell walls are digested by micro-organisms depends on a number of factors, but notably on the extent to which the cell wall is lignified and thus protected from colonisation and utilisation by microbes, on the length of time the plant material resides in the rumen prior to further passage via the omasum to the abomasum, and on the availability of nitrogen for microbial growth within the rumen. The length of time required for degradation of cell walls by micro-organisms has led to the evolution of a mechanism for selective retention of cell-wall particles while allowing soluble components of plant material and small particles to pass quickly on (Van Soest 1994). In the ruminant this selective retention occurs at the outflow of the reticulo-rumen into the omasum. Cell contents consisting mainly of sugars, starch, and protein are all rapidly and completely digestible by ruminant animals.

4.2.2 How do Browse and Grass Differ in Nutritive Value?

Grasses and browse represent very different food resources for ruminant herbivores. These differences in food characteristics led to the ideas of niche separation amongst ungulates (Bell/Jarman principle; Bell 1971; Jarman 1974) which were adopted in Hoffman's classic papers that classified the spectrum of ruminant herbivores into the grazers, intermediate feeders, and browsers ('concentrate selectors', to use Hoffman's terminology; Hofmann and Stewart 1972; Hofmann 1973, 1989). Hofmann presents a view of the nutritive characteristics of grass and browse in which grasses are rich in relatively unlignified cell wall material and hence available for cellulolysis by rumen micro-organisms. Browse, on the other hand, has a lower proportion of cell-wall material but is rich in cell contents. Furthermore, the cell-wall component of browse in heavily lignified and hence refractory to microbial cellulolysis (Hofmann 1989). These characteristics, according to Hofmann, led to distinct feeding styles among herbivores specialising on either grass or browse. Grazers would tend to retain material in their rumen for long periods to allow maximum opportunity for micro-organisms to digest cell-wall material. Browsers, on the other hand, with most of their nutrition derived from rapidly fermentable cell contents, should be expected to adopt a strategy of passing material rapidly through the rumen.

Does Hofmann's view of the nutritive characteristics of grasses and browse still hold with the advent of more biologically realistic methods of assessing nutritive value? It has been suggested that Hofmann's view of nutritive value came from an era during which chemical assessments of nutritive value predominated and the role of plant secondary metabolites in limiting the nutritive value of browse plants was not fully realised (Robbins et al. 1995). Grasses certainly show higher neutral detergent fibre (NDF) percentages than the leaves of browse and forbs. Typical values for grasses are in the region of 50–70% NDF while browse values tend to be in the region of 30–50% (Van Soest 1982). However, numerous recent papers point to the inadequacy of detergent extraction methods for characterising the nutritive value of browse (reviewed in Mould 2003). Polyethylene glycol (PEG) interacts with tannins, effectively neutralising their protein-binding properties (Jones and Mangan 1977). Addition of PEG to in vitro systems for assessing nutritive value has repeatedly demonstrated the often large depressive effects of tannins on fermentation rates in the rumen. As an example, in a study of the nutritive value of a range of plant species found on the Tibetan Plateau, Long et al. (1999) found that grasses and sedges had higher NDF values and, by implication, lower digestibility than browse and forbs. These differences in digestibility were less marked when in sacco techniques were used to assess digestibility, and when gas production methods were used the differences disappeared altogether. Addition of PEG to gas production syringes demonstrated the large effect of tannins on rates of fermentation in forb and browse samples (Fig 4.1).

Poppi et al. (1999) reviewed the chemical composition of a large number of grasses and dicotyledonous plants including tropical and temperate species. They concluded that whilst crude protein (CP) content varied widely among plant species and stages of maturity, the CP/neutral detergent-soluble ratio was relatively constant, indicating that within the cell content and metabolic fraction of the plant there was a constancy in the mix of photosynthetic enzymes and enzymes involved in other metabolic pathways. The proportion of plant cell wall was the feature which separated plants in terms of composition per unit of dry matter.

As well as the chemical differences in the properties of grass and browse, large structural differences exist. The grass material consumed by ruminants tends to be long and fibrous in structure whereas the leaves of browse plants break down into small polygonous particles during mastication (Spalinger et al. 1993). It has recently been suggested that the physical characteristics of grass and browse are a major determinant of the nutritional strategies of grazers and browsers. Clauss et al. (2003) argue that the tendency of grass-based diets to stratify in the rumen into a fibrous raft floating on a liquid phase has led to morphological adaptation

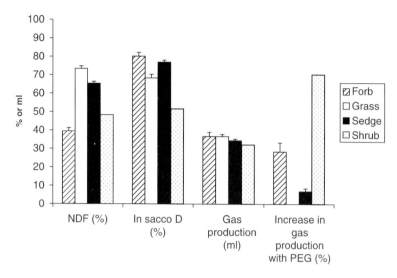

Fig. 4.1 Mean values of NDF, in sacco digestibility, gas production, and increase in gas produced in presence of PEG for a range of plant species of different morphological types found on the Tibetan Plateau (based on the data of Long et al. 1999) Error bars represent SEM. (Only one shrub representative appeared in the original dataset)

of the rumen among grass specialists. The stratification of rumen contents, they argue, facilitates longer retention of solids in the rumen and hence more thorough fermentation. Browse material, on the other hand, is distributed more uniformly within the rumen leaving less opportunity for selective retention of large particles, and browsers therefore have to rely on alternative strategies for deriving nutrients from their food (see below). Thornton and Minson (1973) observed a higher packing density in the rumen of sheep of legume forage compared to grass forage and used that concept to explain the higher intake of legumes. This suggests that it is the different morphology of the herbage types per se that is important and that the way in which the herbage material behaves in the rumen is not controlled by the animal.

4.2.3 Plant Secondary Metabolites

As well as cell walls and cell contents, plant material contains the so-called plant secondary metabolites (PSM), a diverse array of biologically active compounds some of which may have an adaptive role in protecting plant material from herbivory (Rosenthal and Janzen 1979). Plant secondary metabolites are largely absent from graminaceous species but are generally abundant in forbs and especially browse species (Harborne 1988). As well as performing a range of other ecological functions, some plant secondary metabolites limit nutritive value of plant material in a number of ways. Condensed tannins, which are polymers of flavonoid sub-units, form complexes with proteins and, to some extent carbohydrates, (Zucker 1983). In this way they limit the extent of microbial degradation of cell wall material in the rumen. A number of other classes of PSMs act at the level of the rumen by inhibiting microbial fermentation directly. These include phenolic acids (Theodorou et al. 1987) and terpenes (Oh et al. 1967) as well as a range of other compounds (Wallace 2004). Still other PSMs exert their effects following absorption and distribution to the tissues. These include the alkaloids such as the pyrrolizidine alkaloids found in *Senecio* species which form stable pyrroles in the liver and thus act as cumulative hepatotoxins (Stegelmeier et al. 1999). The range of PSMs is large and their physiological action is known in only sketchy detail for a few well-studied plants, usually those of agronomic importance.

4.3 Do Browsers and Grazers Differ in the Way They Process Their Food?

It was Hofmann who laid the groundwork for the debate over the different digestive strategies of browsers vs. grazers. Hofmann's arguments are summarised in Hofmann (1989). In summary, Hofmann argues that the different feeding niches of browsers and grazers have led to morpho-physiological adaptations connected to the chemical characteristics of their diets. Thus, browsers, which consume a diet rich in cell contents but whose cell walls are refractory to digestion adopt a strategy involving rapid passage rates, small reticulo-rumen volume, copious saliva production to hasten passage and buffer high fermentation rates in the rumen, rapid absorption of VFAs, less selective particle retention in the rumen, and greater reliance on hind-gut fermentation. Grazers on the other hand have adopted a different strategy involving slow passage rate, large reticulo-rumen volumes to allow sufficient time for cellulose digestion, slower saliva flow and less need to absorb VFAs quickly. Grazers also retain small particles more effectively to allow extensive energy extraction from cellulose in the rumen. These ideas have sparked a long-running debate in the literature; the main sceptics of the Hofmann hypothesis have argued that variation in strategies for dealing with different diets can be explained wholly by body size effects without the need to invoke classifications based on morpho-physiological adaptation (Gordon and Illius 1994, 1996; Robbins et al. 1995). More recent contributions to the debate have suggested that the polarity of positions is unnecessary with both body size and morpho-physiological specialisation contributing to observed variation in digestive strategy (Iason and Van Wieren 1999; Perez-Barberia et al. 2004). Some of the evidence for these varying positions is reviewed in the following sections (summarised in Table 1).

Table 4.1 Summary of evidence for and against physiological adaptations of browsers (BR) and grazers (GR) to their feeding habit

Adaptation	Browsers	Grazers	Supporting evidence	Equivocal evidence	Contrary evidence	Reference
Reticulo-rumen (RR) volume relative to body size	Low	High	Meta-analysis of wet RR contents indicated that GR had higher values than BR			Clauss et al. 2003
					Van Wieren conducted statistical analysis on Hofmann's data and found that RR volume did not differ between BR and GR	Van Wieren 1996
Passage rate	High	Low	Indirect evidence of faster passage rate in BR comes from their body fat composition which is richer in unsaturated lipids, pointing to less biohydrogenation in the rumen		Meta-analysis shows that for a given body mass, MRT of BR and GR do not differ However, GR and BR body masses did not overlap to any great extent in the dataset analysed. Also, the quality of the MRT data may have been limited since Foose data relied on bulked daily faecal samples. It was not clear what foods were used in the MRT trials; if BR were fed on grass hay, one would not expect shorter retention times	Gordon and Illius 1994

Mean particle size escaping rumen	High	Low	Larger faecal particle size in BR	Clauss 2002
			Meta-analysis indicates that GR retain particles more efficiently in the rumen than BR	Clauss 2001
Efficiency of digestion of fibre in the rumen	Low	High	Meta-analysis of digestibility trials shows that digestibility (D) is negatively related to lignin and positively related to body mass but that GR show more efficient NDF digestion than BR	Iason and van Wieren 1999
			NDF (cell wall) digestibility values from published studies were compared across a range of herbivores and found not to differ between BR and GR	Robbins et al. 1995
Salivary gland size to buffer highly fermentable feeds	High	Low	BR have higher parotid salivary gland weights than GR for a given body weight	Robbins et al. 1995
			Resting saliva flow does not differ between GR and BR, but, saliva flow during browsing was not measured. This is the parameter that would be expected to differ if an oesophageal groove mechanism operated in BR	Robbins et al. 1995

(continued)

Table 4.1 (continued)

Adaptation	Browsers	Grazers	Supporting evidence	Equivocal evidence	Contrary evidence	Reference
Absorptive surface area of rumen to rapidly absorb VFAs to prevent acidosis	High	Low	Van Wieren (1996) reanalysed Hofmann's 1973 data and found that density of rumen papillae on the ventral wall was higher in BR than GR			Van Wieren 1996
			Measurements of absorptive area of rumen between BR and GR show that, for a given body size, BR have higher surface area than GR			Demment and Longhurst 1987
Rumen musculature to deal with fibrous diets	poorly developed	well developed	Meta-analysis of published data shows that RR pillar thickness greater in GR than BR			Clauss et al. 2003
Fermentation rate in rumen	High	Low			Robbins argues that the nutritive value and hence rumen fermentation rates of browse diets have been over-emphasised without considering the suppressing effects of PSMs/tannins. However, this does not tally with the dataset of Gordon and Illius (1994) which found fermentation rates in small ruminants (mainly BR) to be higher than in large ruminants	Robbins et al. 1995

			Meta-analysis of fermentation rates shows that the same relationship applies to GR and BR when body size effects are taken into account	Gordon and Illius 1994
Tannin-binding salivary proteins	Occurs	Does not occur	Salivary tannin-binding proteins found in saliva of mule deer (*Odocoileus hemionus*), a browser, but not in the saliva of sheep and cattle	Austin et al. 1989
			When sheep and deer were fed quebracho tannin-containing diets, sheep showed a reduction in digestibility while deer did not	Robbins 1991
			STBPs not found in tannin-adapted cattle (grazer)	
			Rhinoceros saliva shows tannin-binding capacity that conforms to feeding niche except for intermediate feeders	Makkar and Becker 1998 Clauss et al. 1995

(continued)

Table 4.1 (continued)

Adaptation	Browsers	Grazers	Supporting evidence	Equivocal evidence	Contrary evidence	Reference
Plant secondary metabolite tolerance or detoxifying ability of rumen microbes	High	Low	Bacterium (*Streptococcus caprinus*) capable of degrading tannin-protein complexes described; found in feral goats grazing Acacia but not domestic goats. Presumably its presence in the rumen was a function of the diet consumed			Brooker et al. 1994
					Found strains of tannin-tolerant bacteria in the rumen of sheep, goats and antelope adapted to Acacia diets. Bacteria were not cellulolytic. Since they were found in the rumen of sheep (a grazer/intermediate feeder), this suggests that the presence of tannin-tolerant bacteria in the rumen is a function of diet	Odenyo and Osuji 1998
				Gas production (GP) from rumen fluid from a range of ruminants incubated with a range of browse species was tested. Dik-dik showed particularly high GP but other results were less clear-cut. As argued by Jones		Odenyo et al.1999

et al. (2001), the high dik dik GP could be related to the high DM of the rumen fluid which was not corrected for in this study. Difficult to separate out the effects of diet and animal species

Rumen fluid from African game species and sheep was screened for its ability to degrade forages rich in tannins. Game-species-derived rumen fluid was found to be no better at degrading tannin-rich forages than sheep rumen fluid. Addition of PEG consistently improved protein degradation. Authors argue that if browser rumen fluid contains tannin-resistant bacteria, then one would expect to see similar in vitro nitrogen digestibility (IVND) values in the presence and absence of PEG. The fact that this was not observed is taken as evidence that tannin-tolerant bacteria are not present in BR rumen fluid

Jones et al. 2001

4.3.1 Reticulo-Rumen Size

According to Hofmann, browsers would be expected to have a relatively smaller reticulo-rumen than grazers since their strategy is to retain food in the rumen for a relatively short period of time, relying less as they do on the energy derived from cellulolysis of cell wall material, focussing instead on deriving nutrients from cell contents. Van Wieren (1996) conducted a statistical analysis of Hofmann's original data (Hofmann 1973) and concluded that there was no statistically significant difference in reticulo-rumen size between grazers and browsers as classified by Hofmann. Perez-Barberia et al. (2001) also carried out a multivariate analysis of stomach morphology traits of Hofmann, controlling for the effects of body size and phylogeny. Their study also supported the conclusion that browsers and grazers cannot be separated morphologically once the effects of body size and phylogeny have been taken into account. Interestingly, in a follow-up study Perez-Barberia et al. (2004) found that even after accounting for phylogeny and body size, differences in NDF digestibility between browsers and grazers were evident, suggesting that the anatomical basis of digestive function has not yet been properly understood. Clauss et al. (2003) conducted a meta-analysis of wet rumen contents using data from a range of published sources. Rumen volume was regressed against body weight, and a significantly larger intercept was found for grazers than browsers. These authors assumed that gut volume was linearly related to body weight (Parra 1978), but, interestingly, their regression of wet rumen contents as a percentage of body weight against body weight showed a positive slope, thus raising questions about the quality of the data used to investigate these relationships.

4.3.2 Retention Time and Passage Rate

Browsers would be expected to pass material through their rumens relatively quickly since soluble cell components are given up rapidly and cell wall digestion is not a priority for them. There is no good quality data on passage rate in the existing literature. Gordon and Illius (1994) used published data to show that variation in mean retention time of a range of ruminants could be wholly explained by body size effects; values for browsers and grazers lay along a common regression line. However, much of the base data was derived from a study in which retention times were estimated using daily bulked samples (Foose 1982), which would yield rather crude estimates of mean retention time. Furthermore, the diet types used in the published studies were unclear. As pointed out by Clauss and Lechner-Doll (2001), indirect anecdotal evidence of shorter retention time in browsing ruminants comes from the composition of their body fat. The unsaturated lipid found in plant material is saturated during passage through the rumen. Meat from browsing animals has a higher unsaturated lipid content and some would speculate that this indicates shorter rumen residence times.

Most wildlife studies are compromised because they do not strictly compare animals on the same diets under the same feeding conditions, and there is an inevitable difference in the feed type and physiological state of the animals. The most controlled conditions arise in comparison of domestic ruminants such as sheep, cattle, goats, and deer. The largest difference is between cattle and sheep; under controlled conditions cattle have a longer retention time and slower passage rate of material out of the rumen, with which is associated a higher digestibility (Hendricksen et al. 1981; Poppi et al. 1981). Despite the large difference in live weight between the two species, the rumen fill of dry matter (DM), expressed relative to metabolic live weight, was similar and could not explain these differences. These studies showed that the difference in digestibility was not related to any inherent difference in rate of digestion but rather to the rate of passage from the rumen, which affected the time available for digestion and hence the final extent of digestion. Most of the rumen particles were of a size that had little resistance to escape, and yet rumen conditions, presumably aspects of rumen volume and raft characteristics, affected the rate of passage. deVega and Poppi (1997) showed by inserting common particles (either digested or undigested of one size <1mm) in a range of rumens that the rumen conditions, as set by the diet, was the most important factor affecting rate of passage rather than the extent of digestion of the particle. It seems likely that for retention time, passage rate, and possibly other digestion parameters, apparent grazer/browser differences may be less to do with morpho-physiological differences than with the nature of the diet consumed.

4.3.3 Mean Particle Size Escaping Rumen

According to the Hofmann view, browsers would be expected to be less efficient at retaining material in their reticulo-rumen since they rely less on cellulolytic digestion of fibre in the rumen. They would therefore be expected to allow larger particles to escape the rumen than the grazers. This hypothesis is supported by the work of Clauss and co-authors who propose selective retention of large particles in the rumen of grazers as being a driving force (Clauss et al. 2003) which allowed the exploitation of graminaceous vegetation and radiation of ruminant forms following the appearance of grasslands during the late Miocene. A meta-analysis of ratios of solid to liquid phase markers in a range of ruminants indicated browsers to have a lower 'selectivity index' than grazers, i.e., they were less selective in their retention of solid material (Clauss and Lechner-Doll 2001). In further work, comparison of mean particle size in the faeces of around 90 captive ruminants showed higher mean particle size in the faeces of browsers than grazers after allowing for the effects of body size (Clauss et al. 2002). The effects of diet were not controlled for and, as acknowledged by the authors, most of the samples analysed would have been derived from animals offered alfalfa hay diets.

Clauss et al. (2003) go on to argue that the physico-mechanical characteristics of browse and grass herbage have been key to the diversification of ruminants. As previously described, grass material tends to form long, fibrous particles in the rumen which form a fibrous raft floating on a liquid phase, while browse material is more friable and breaks down into smaller, polygonal particles which do not stratify to the same extent (see section 4.2.2, above). According to this view, an important adaptation among grazing ruminants is therefore a stronger rumen musculature to deal with the stratified rumen contents typical of animals consuming grass-based diets. Browsing ruminants' inability to subsist on predominantly grass diets is therefore attributed to their weak rumen musculature which cannot deal with the rumen load resulting from grass diets (Clauss et al. 2002). In support of this idea Clauss et al. (2003) included data on rumen muscle dimensions in their meta-analysis and showed that grazers have thicker ruminal pillars for a given body weight than browsers.

4.3.4 Absorptive Surface Area of Rumen

Faster passage rates and smaller rumen volumes in browsers compared to grazers would reduce the time available for absorption of nutrients and VFAs. Hofmann asserts that browsing ruminants have adapted to this constraint by increasing the surface area of the rumen (surface enlargement in Hofmann's terminology) to facilitate the faster absorption of VFAs that would be necessary given the rapid fermentation rates found in browsers (Hofmann 1989). Van Wieren (1996) reanalysed Hofmann's data and showed that browsers do appear to show more extensive papillation of the rumen wall than grazers of similar body size. A similar conclusion was arrived at by Demment and Longhurst (1987) using their own data derived from samples of East African ruminants collected from shot animals. Extent of papillation varies within a species depending on metabolisable energy intake or VFA production (Van Soest 1994) and so experiments need to be carefully designed to avoid this confounding factor. The biggest difference is seen between cattle on concentrate diets compared to forage-based diets where the intakes are quite different and the papillae density and size markedly different. The weight of rumen contents (as distinct from surface area) appears to be directly proportional to liveweight ($W^{1.0}$; Van Soest 1994). Rumen volume varies with physiological state and as energy demand is increased the rumen can increase in size up to a point to overcome the physico-chemical limitations to increasing intake to meet this increased energy demand. For example, in an experiment where rumen volume was compared in barren cows and lactating cows suckling twins and fed the same forage, the volume of digesta and size of the rumen were both found to be higher in the lactating cows (Tulloh 1966a, 1966b).

4.3.5 Saliva Flow

Saliva flow would be expected to be more rapid in browsers than grazers for two reasons. Firstly, the strategy of passing material rapidly through the rumen to maximise utilisation of nutrients derived from cell contents would be facilitated by high rates of liquid flow through the rumen. Secondly, an important function of saliva in the ruminant is the buffering of rumen contents. The rapid fermentation rates envisaged for browsers would require rapid rates of saliva flow to buffer the rapid production rates of VFAs. The larger salivary glands found in browsers than in grazers by Robbins et al. (1995) provided indirect evidence for more copious production of saliva in browsing compared with grazing ruminants. However, Robbins et al. (1995) also used a meta-analysis of resting saliva flow rates in a range of ruminants to show that saliva flow rates did not differ between grazers and browsers once body size effects had been accounted for. The study was subsequently criticized for only measuring resting saliva flow rates when saliva flow rates during food ingestion would have been more appropriate (Ditchkoff 2000). Ditchkoff (2000) also suggested that critics of Hofmann had misunderstood his hypotheses related to salivary flow. Hofmann suggests that browsers have a mechanism for passing food past the rumen, directly into the abomasum via the ventricular groove thus avoiding the inefficiencies of rumen fermentation for readily available substrates (Hofmann 1989). A similar mechanism is used by juvenile ruminants to allow ruminal bypass of milk (Orskov et al. 1970), although in juvenile ruminants it is a conditioned reflex. Ditchkoff contends that conventional measurements of passage rates do not take account of this mechanism and cannot therefore be used to refute Hofmann's ideas. However, since the passage rates used in meta-analyses were whole tract values, the basis for Ditchkoff's criticism is unclear.

4.3.6 Fermentation Rate and Fibre Digestibility in the Rumen

The supposed browser strategy of focussing on cell contents (neutral detergent solubles; NDS) to derive nutrients from plants would be accompanied by faster fermentation rates since cell contents are more-or-less immediately available sources of energy and protein, and would not require the slow cellulolysis for extraction of energy from cell wall material. On the other hand, grazers would be expected to be more efficient at extracting nutrients from cell wall material and hence would be expected to show higher rates and extents of NDF digestion. The meta-analysis of Gordon & Illius (1994) indicates that once body size effects are taken into account, fermentation rates in the rumen of browsers and grazers do not differ. Similarly, Robbins et al. (1995) did not detect differences in digestibility of NDF between browsers and grazers. However, the opposite conclusion was reached in a similar meta-analysis but using a different data set by Iason Van Wieren (1999), and a similar study which accounted for the effects of phylogeny reached a similar conclusion (Perez-Barberia et al. 2004). There is, however, no strong empirical evidence to suggest substantial differences in extent of NDF digestion or digestion of cell contents among species. Thus, all rumens irrespective of size or classification of animal (concentrate feeder, grazer, etc) essentially digest NDF at the same rate; any differences in digestion can be explained simply by differences in rate of passage and retention time allowing greater or lesser time for digestion when species are compared. Van Soest (1994) highlighted this by stating that the claims for differences in digestibility are not supported by 'rational biochemistry and kinetics' and that 'there are no magic enzymes and no magic rumen bacteria that can exceed physico-chemical limitations'. The differences observed in rate or extent of DM digestion are largely influenced by the diet selected by the animal. If the proportion of NDF (and hence cell contents) varies because of diet selection, it follows that rate and extent of digestion will also differ. Strict comparisons on the same diet are not common and Van Soest's conclusions would appear most appropriate.

There is one circumstance relating to low CP diets in which differences between species may appear. In these diets, rumen ammonia levels are low and rate of fibre digestion may be reduced. This is the classical explanation for the response of ruminants to urea (or non-protein N supplements) where rate of fibre digestion and microbial protein production are increased in response to the limiting nutrient, N. Several species comparisons have been done on the same low CP diets whereby differences between species have been explained by differences in N recycling to the rumen (Norton et al. 1979; Watson & Norton 1982; Alam et al. 1984). Species which have a higher salivary flow rate or maintain higher plasma urea levels place themselves at a physiological advantage to enhance N recycling to the rumen. Increasing N recycling to the rumen will be of advantage to the species in times of low CP diets. The evidence is variable but it does indicate that in times of acute shortage of N for the microbes in the rumen that any species which can increase N recycling to the rumen will have an advantage (but still within the classical physico-chemical response of N requirement by microbes for digestion and growth).

For example, goats have been compared with sheep in many studies and they are categorised differently by the Hofmann criteria. With high digestibility and CP diets, there were no differences between the species in site of digestion, digestibility, and use of nutrients (Alam et al. 1983, 1987a, 1987b). With low CP diets there was an advantage to goats in many digestion parameters, one of which was maintenance of a higher rumen ammonia N level (Watson and Norton 1982; Alam et al. 1984). This appeared to occur in goats because of their lower water intake, higher plasma urea concentration, and hence higher recycling of urea to the rumen (Alam et al. 1984, 1987b). The water conservation mechanism of goats enhanced the N conservation mechanism leading to the maintenance of better N conditions within the rumens of N deficient animals.

4.3.7 Rumen Microbial Genetic Profiles and Digestion End Products

The end products of digestion in the rumen are volatile fatty acids (VFA), carbon dioxide, methane, and microbial cells. VFA and microbial cells provide the bulk of the nutrients absorbed by the host. In general, there are three main VFAs produced by all anaerobic rumen fermentation: acetic acid, propionic acid, and butyric acid, with acetic acid predominant under most forage feeding regimes. Smaller amounts of short chain branch chain fatty acids are also found when there are large amounts of protein degraded within the rumen. The changes in VFA proportions have been extensively studied in ruminants, with propionic acid increasing in concentration as starch is fermented and when legume forages are digested (Annison et al. 2002). Assuming that browsers and concentrate feeders select diets high in temperate plant species and/or high in cell content, then it may be implied that there would also be a shift in VFA proportion from acetate to propionate. Clemens et al. (1983) examined a wide range of East African wild ruminants ranging in weight from 5 or 6 kg (dik-dik) to 850 kg (African buffalo) and found that acetate as a proportion of total VFA did not vary greatly; what variation occurred was related more to body size than diet type. Domestic ruminants show a much greater range in VFA proportion than occurred in this study, probably because the dietary intake of starch and other soluble carbohydrates vary more widely for domestic ruminants than can be achieved by animals that only graze. Clemens and Maloiy (1984) showed with the same wide grouping of animals a wide variation in colonic function (absorption of fluids, VFA concentration, etc.) but did not report VFA proportions. It is unlikely that VFA proportions in the large intestine would vary unless large amounts of starch escaped digestion, which is unlikely in animals consuming herbage.

Ishaque et al. (1971) in a little known study showed that a group of sheep on a common diet could be separated into two groups based on level of propionate and microbial protein production, with both being positively associated. The high propionate, high microbial protein production group was suggested to have a different rumen microbial population. The methods to determine this were not available at that time but now the use of the 16S rDNA sequence has allowed the microbial diversity within the rumen to be explored (Mackie et al. 2002). Some studies have already shown that only a small percentage of the diversity of microbial species has been described or cultured to date (Mackie et al. 2002). Larue (2005) have outlined novel lines of Clostridium species associated with the microbial sub-population tightly adhered to the fibre, and Tolosa et al. (2004) have identified species which increase under high molasses-based diets. This work is rapidly expanding and we will see more information on microbial ecology under different species and diets, especially with respect to wild ruminants in natural situations. For example, McEwan et al. (2005) have shown in, Soay sheep a day-length-sensitive sheep species that, despite being fed the same diet, the rumen microbial diversity varied with the day length under which the sheep were housed. Mackie et al. (2003) and Sundset et al. (2004), studying reindeer in

Norway, showed that there was greater variation in the diversity of Oscillospira sp. in wild reindeer than in cattle and sheep. They identified novel rumen bacteria and showed differences in rumen microbial diversity between reindeer consuming natural diets and formulated diets. The extent to which these differences relate to diet, to genotype of the host animal, or simply to the interaction of individuals with sources of microbial inoculums has yet to be determined.

The same approach has been used to study kangaroos (herbivores with a rumen-like forestomach) in Australia. Ouwerkerk et al. (2005a) identified a range of new bacterial species in kangaroos using denaturing gradient gel electrophoresis (DGGE), PCR, and the phylogenetic-tree-construction approach. Most interesting is that they were able to explain the lack of methane production under anaerobic fermentation in kangaroos by the discovery of a group of reductive acetogens in these animals (Ouwerkerk et al. 2005b). Kangaroos had reductive acetogens and no methanogens, while the reverse was true for sheep. Thus two different microbial populations had evolved under anaerobic foregut fermentation in two distinct species despite co-existence for around 200 years within Australia; no cross transfer of these microbes appears to have occurred. It is, therefore, possible that browsers and grazers may differ in their microbial populations, and this could have implications for food digestion.

4.4 Problems with the Meta-Analysis Approach

From the foregoing it is clear that consensus has yet to be reached on whether browsers and grazers show morpho-physiological adaptation to diet type or whether variation in digestive parameters can be explained by body size effects. The reality is probably somewhere between these two extremes (Iason and Van Wieren 1999). Given the very different physical and chemical characteristics of vegetation consumed by browsers and grazers, it would be surprising if different ruminant classes had not developed physiological adaptations to their diet type. Unfortunately, most of the evidence used in arguments for one or other position is derived from meta-analyses which often rely on confounded data of questionable quality. Browsers tend to be smaller than grazers, and so defining relationships between body size and digestive parameters can be problematic. Browsers tend to eat browse while grazers tend to eat grass. It is, therefore, difficult to separate the effects of diet type from the effects of morpho-physiological variation in influencing the way in which ruminants handle their food. For example, faecal particle size was found to be greater in the faeces of browsing compared with grazing ruminants (Clauss et al. 2002) but these samples came from captive zoo animals during winter when all animals would be consuming a predominantly alfalfa hay diet. Furthermore, it is unclear what influence diet type has on the fermentation characteristics used in the various meta-analyses used to support or refute the Hofmann classification.

4.5 Detoxification of Plant Secondary Metabolites: Do Browsers and Grazers Differ?

One of the major differences between the chemical characteristics of grass- and browse-dominated diets is the presence of plant secondary compounds which are prominent in the diets of browsers but largely absent from the diet of grazers. This marked distinction in chemistry is thought to result from the different growth forms of monocotylenous and dicotyledonous plants: monocots have a basal meristem so that growth originates low in the plant. This makes them more resilient to offtake of material by foraging herbivores. Dicots have an apical meristem and loss of tissue to herbivory has more serious resource implications than for monocots. This leads to strong selection pressure to defend tissue by chemical means and has led to the diverse array of plant secondary compounds found in forbs, shrubs and trees (Harborne 1988). Plant secondary compounds are diverse in both chemical structure and physiological action on herbivores. Many attempts have been made to classify plant secondary compounds according to their mode of action, and a useful and broad division is into those compounds which reduce digestive efficiency in the gut and those which act at the tissue level following absorption and circulation in the blood. The most important digestibility reducers are the tannins which bind dietary and endogenous protein leading to reduced protein digestibility via a direct effect on dietary protein and an indirect effect through suppression of digestive enzyme activity (Robbins et al. 1987a, 1987b). A range of other secondary compounds exert negative effects on digestion through anti-microbial effects in the rumen and hind-gut (Wallace 2004). Plant secondary compounds which exert their effects post-absorptively can act as either acute or chronic toxins. For example, cyanogenic glycosides yield free cyanide following ingestion and this is an acute mammalian toxin (Majak 1992). Pyrrolizidine alkaloids, on the other hand, cause the deposition of toxic pyrroles in the liver leading to chronic toxicity (Mattocks 1986).

A range of physiological adaptations to the presence of plant secondary metabolites in the diet have been proposed and some research has reported on possible differences between browsers and grazers. This will now be reviewed.

4.5.1 Salivary Tannin-Binding Proteins

Salivary tannin-binding proteins were originally demonstrated in rats fed tannin-containing sorghum diets (Mehansho et al. 1983). Sorghum-fed rats showed a dramatic increase in salivary gland weight which was accompanied by increased synthesis of a group of salivary glycoproteins with a strong affinity for tannins. The high proportion of the amino acid proline in these salivary glycoproteins led to them being described as 'proline-rich proteins' but the more functionally descriptive term 'salivary tannin-binding protein' (STBP) is now widely used. Their hypothesised mode of action is as sacrifice proteins to limit the digestion-reducing effects of plant tannins. Because of their high affinity for tannin, STBPs are hypothesised to spare dietary protein for utilisation further down the gut. The presence of similar proteins in ruminant saliva has subsequently been reported although the proportion of proline in the salivary mucoproteins of ruminants is much lower than in rats (Austin et al. 1989). Furthermore, in ruminant herbivores, STBPs appear to be constitutive and not to be inducible following exposure to dietary protein.

The role of salivary tannin-binding proteins has been reviewed by McArthur et al. (1995) who suggest that their ancestral role was in the maintenance of oral homeostasis but that in mammals with a high concentration of dietary tannins they have evolved to protect animals against the detrimental effects of plant tannin. There is certainly evidence for the occurrence of STBPs in the saliva of browsers and their absence in grazing herbivores; in one study salivary tannin-binding proteins were found in saliva of mule deer (*Odocoileus hemionus hemionus*), a browser, but not in the saliva of sheep and cattle, grazers (Austin et al. 1989). Further indirect evidence for a browser/grazer difference came from a study in which sheep and deer were fed quebracho tannin-containing diets; sheep showed a reduction in digestibility while deer did not (Robbins et al. 1991). STBPs were not found in the saliva of cattle even following a period of adaptation to a tannin-rich diet (Makkar and Becker 1998). Salivary tannin-binding proteins are also found in non-ruminant herbivores such as rhinoceroses. Salivary tannin-binding capacity in the saliva of captive rhinoceros species appears to conform to dietary feeding niche, at least to some extent; Black rhinoceroses (*Diceros bicornis*, browser) have higher salivary tannin binding capacity than White rhinoceroses (*Ceratotherum simum*, grazer) although Indian rhinoceroses (*Rhinoceros unicornis*, intermediate feeder) appear to have even higher tannin binding capacity than the Black species in saliva (Clauss et al. 2005). Although evidence supports the idea that STBPs are an important physiological adaptation to browse diets, tannins are only one group of plant secondary compounds and there is no evidence that the saliva of browsing animals can neutralise the many other secondary compounds present in browse diets. Histatins have also been shown to precipitate tannins, and to a greater extent than STBPs. They are a group of peptides found in human saliva but not studied in ruminants (Naurato et al. 1999).

The binding of tannins with protein can have both detrimental and beneficial effects. At low levels of condensed tannin, protein from plant material and saliva is bound most probably during mastication but also within the rumen contents. This protein is protected from degradation and increases the supply of protein to the small intestine where it is released and absorbed. However, high levels of condensed tannin may lead to over-protection and detrimental effects on fibre and protein digestion within the animal. In these cases the infusion of PEG experimentally is able to determine the effect of condensed tannins on the host by binding to and reversing the effects of condensed tannin. The role of STBPs in these circumstances of high condensed tannin will most likely be beneficial. These issues have been extensively reviewed in Foley et al. (1999). They identified a range of herbivore species where STBPs have a role in binding the condensed tannin, enabling species to deal with variable and high levels of condensed tannin in their diet. Perez-Maldonado and Norton (1996a, 1996b) clearly showed that no difference exists between goats and sheep in binding and metabolising condensed tannins in tropical legumes with high levels of condensed tannin. They also could find no evidence for STBPs in goats (B.W. Norton, pers. comm.).

4.5.2 Ruminal Detoxification

The anterior position of the microbial fermentation chamber in the ruminant digestive tract increases the potential for microbial biotransformation of plant secondary compounds in ruminants—as compared to herbivores occupying similar feeding niches but relying on hind-gut fermentation. The rumen is a rich source of enzymatic activity and many of the biotransformations that occur under the action of mammalian detoxification enzymes can occur readily within the rumen environment. The role of the rumen in microbial detoxification of plant secondary metabolites has been well reviewed (Carlson and Breeze 1984). Classic examples of the role of the rumen micro-flora in reducing toxicity of plants consumed by ruminant herbivores include the degradation of oxalates found in *Halogeton glomeratus* (Allison and Reddy 1984), mimosine found in *Leucaena leucocephala* (Jones 1981), glucosinolate breakdown products found in Brassica species (Duncan and Milne 1992) and pyrrolizidine alkaloids found in *Senecio* species (Lanigan 1970). Mammalian metabolism of plant secondary compounds can result in activation rather than detoxification and a similar phenomenon can occur in the rumen. For example, S-methyl cysteine sulphoxide found in cruciferous species is transformed by rumen bacteria into the toxic principle dimethyl disulphide, which is responsible for the haemolytic anaemia observed in farm livestock fed on brassica material (Smith 1980).

Although the role of the rumen in metabolism of plant secondary compounds is not in doubt, there is some question over whether the rumen microbial populations of different ruminant species have markedly different capacities to degrade plant secondary compounds. This uncertainty also applies to the question of whether browsers and grazers differ at the rumen level in their capacity to degrade plant secondary compounds. The extent to which the host animal can manipulate the composition of the rumen microbial population is unclear and this would be a pre-requisite for browser/grazer differences. The classic experiments of Jones and colleagues with mimosine-degrading bacteria demonstrated that transfer of bacteria from one ruminant population to another could also transfer tolerance to an otherwise detrimental plant toxin (Jones 1981). This work was influential and suggested many possibilities for using rumen microbial communities for protection of host animals from toxicity in the agricultural context. However, manipulating rumen microbial communities for particular ends has proved more challenging than first supposed, and attempts to transfer tolerance to plant toxins using rumen inoculae have not been convincing (Brooker et al. 1994; Miller et al. 1995, 1996). Since rumen microbial populations are largely controlled by substrate availability, mechanisms by which host animals could vary their resident rumen microbial population, other than by altering the diet selected, are unclear. Genotypic differences in digestive tract function in ruminants have been recently reviewed (Hegarty 2004) with the conclusion that the main digestive function trait under genetic control is mean retention time. Factors such as rate of saliva flow, the morphology of the reticulo-omasal orifice, and the degree of urea recycling to the rumen (Hungate et al. 1960) could all contribute to variation in mean retention time, and are all factors which could vary genetically.

There has been particular interest, in the last decade, in isolating bacterial strains which are capable of degrading tannin, or are at least tolerant of tannin. Such strains had been isolated (Brooker et al. 1994; Odenyo and Osuji 1998) but they have not been found to be necessarily host-animal specific. In a study by Odenyo et al. (1999) gas production from rumen fluid from a range of ruminants incubated with a range of browse species was measured. Rumen fluid from dik-dik showed particularly high gas production values but results for other species were less clear-cut. The study design made it difficult to distinguish the effects of diet and animal species. Jones et al. (2001) screened rumen fluid from various African game species for its ability to degrade test forages rich in tannins. Results were compared with values found with sheep. It was shown that game-species-derived rumen fluid was no better at degrading tannin-rich forages than sheep rumen fluid. Addition of PEG consistently improved protein degradation. The authors argue that if browser rumen fluid contains tannin-resistant bacteria then one would expect to see similar in vitro nitrogen digestibility values in the presence and absence of PEG. The fact that this was not observed was taken as evidence that tannin-tolerant bacteria are not present in browser rumen fluid.

Mackie et al. (2002) collated a list of known secondary compounds which were metabolised by specific microbial species. The compounds were non-protein amino acids, phenolics, trihydrobenzenoids, ferulic and p-coumaric acid, condensed and hydrolysable tannins, phyto-oestrogens, oxalate, pyrrolizidine alkaloids, and mycotoxins. Whilst these compounds may be metabolised, there is no guarantee that the toxic effects are completely removed as the action depends on level and extent of exposure. Whether the populations of these secondary compound-degrading bacteria differ between grazers and browsers has not been investigated. In conclusion, the question of whether the rumen microbial populations of browsers and grazers show different capacities to degrade plant secondary compounds, thus conferring differential tolerance to toxicity on their hosts, has not yet been resolved. Intuitively, one would expect such genotypic differences to be more readily expressed at the level of the detoxification mechanisms of the host itself, rather than indirectly through the rumen microbial population.

4.5.3 *Post Absorptive Metabolism*

Following ingestion and digestion of plant material, nutrients and toxins are absorbed from the digestive tract. Biotransformation in the rumen may lead to detoxification of plant secondary metabolites but this is not always the case; some

PSMs may escape detoxification while others may increase in toxicity as a result of microbial action as described above. PSMs absorbed from the digestive tract may be detoxified in the tissues, primarily in the liver. Most of our understanding of post-absorptive metabolism of xenobiotics comes from the discipline of pharmacology, and most research in the area is focussed on the metabolism of artificial drugs in the context of human and veterinary medicine. As a result, most of the animal models that have been used in this area of research are either laboratory rodents or domestic livestock. Very little information on tissue-level detoxification is available for wildlife species. Watkins and Klaassen (1986) used probe substrates to quantify cytochrome P450 activity in a range of species including fish and birds as well as domestic livestock. Vast differences in apparent enzyme activity were reported which did not follow obvious phylogenetic lines. The study highlighted the complexity of hepatic biotransformation of xenobiotics, the multiple enzymes involved, the overlap in substrate specificity, and the difficulties in using probe substrates developed in laboratory rats for quantifying activity of enzymes in other species. Our understanding of the P450 superfamily of enzymes has increased considerably since the Watkins and Klaassen (1986) study with the development of molecular and immunological methods of quantifying hepatic enzyme activity. There is a need to use some of this information for cross-species comparisons in the context of browser-grazer differences, but this has not yet been reported.

Comparisons to date have been at the level of gross anatomy. A general observation is that concentrate feeders have a larger liver than grazers (Van Soest 1994; Foley et al. 1999) and that this confers an advantage to them in the further detoxification of plant secondary compounds or their derivatives after gastro-intestinal tract degradation. They are also more likely to be exposed to such compounds given their feeding habit. Goats metabolise anthelmintic drugs faster than sheep, indicating some differences in metabolic activity (Hennessy et al. 1993). In the case of the latter study there was no difference between goats and sheep in metabolism of condensed tannins and their degradation products and appearance in urine.

4.6 Conclusions

Plant material is variable in structural characteristics and hence in nutritive value. Plant secondary compounds also show marked diversity across the plant kingdom. Browsers and grazers thus subsist on very different diets. Despite this variation, our discussion has demonstrated that there is still much that we do not understand about how herbivores have adapted nutritionally and physiologically to this variation in diet. Much of our understanding of ruminant nutrition derives from the agricultural context and has focussed on a very few species which do not adequately represent the browser-grazer continuum. What is now needed is to apply some of the methodology derived from several decades of agricultural research to conduct controlled experiments aimed at understanding the ways in which a range of wild herbivore species deal with the varied plant sources that form their diets in natural situations.

Such research will allow a better understanding of the impact of wild herbivores on vegetation in natural habitats and will facilitate the more effective management of vegetation using herbivores as tools for land management.

In this chapter we have considered the various levels at which food processing by browsers and grazers might differ. As we have seen, the research community has, so far, given most attention to possible differences in digestive anatomy and physiology and how this might differ between grazers and browsers. There has been much debate in the literature but no clear consensus has emerged about whether grazers and browsers show marked morpho-physiological differences which might influence their digestive efficiency. Considerable recent attention has also been given to the question of whether rumen microbial degradation of plant secondary metabolites is more effective in browsers than grazers. The balance of evidence suggests no great differences at the rumen level. In contrast, very scant attention has been paid to possible browser-grazer differences in post-absorptive metabolism of plant secondary metabolites. It is perhaps at this level that we might expect physiological differences between browsers and grazers to be most strongly expressed, since species-level genetic variation in Phase 1 and Phase 2 enzyme activity has been shown to be extensive in biomedical studies. Despite the difficulties inherent in making measurements of tissue enzyme activity in wildlife, we suggest that this would be an interesting avenue for future work.

References

Alam MR, Poppi DP, Sykes AR (1983) Intake, digestibility and retention time of two forages by kids and lambs. Proc New Zeal Soc An 43:119–121

Alam MR, Borens F, Poppi DP, Sykes AR (1984) Comparative digestion in sheep and goats. In: Barker SK, Gawthorne JB, Mackintosh JB, Purser DB (eds) Ruminant physiology – concepts and consequences. Proceedings Symposium, University of Western Australia, Perth, p 184

Alam MR, Lawson GD, Poppi DP, Sykes AR (1987a) Comparison of the site and extent of digestion of nutrients of a forage in kids and lambs. J Agr Sci 109:583–589

Alam MR, Poppi DP, Sykes AR (1987b) Comparative aspects of water-intake and its flow through the gastrointestinal-tract of kids and lambs. J Agr Sci 108:253–256

Allison MJ, Reddy CA (1984) Adaptations of gastrointestinal bacteria in response to changes in dietary oxalate and nitrate. In: Klug MJ, Reddy CA (eds) Current perspectives in microbial ecology. American Society of Microbiology, Washington, DC, pp 248–256

Annison, EF, Lindsay, DB and Nolan, JV (2002). Digestion and metabolism, In:Freer M, Dove H (eds) Sheep nutrition. CABI/CSIRO, Wallingford New York, pp 95–118

Austin PJ, Suchar LA, Robbins CT, Hagerman AE (1989) Tannin-binding proteins in saliva of deer and their absence in saliva of sheep and cattle. J Chem Ecol 15:1335–1347

Bell RHV (1971) Grazing ecosystem in Serengeti. Sci Am 225:86–93

Brooker JD, O'Donovan LA, Skene I, Clarke K, Blackall L, Muslera P (1994) *Streptococcus caprinus* sp.nov., a tannin-resistant ruminal bacterium from feral goats. Lett Appl Microbiol 18:313–318

Carlson JR, Breeze RG (1984) Ruminal metabolism of plant toxins with emphasis on indolic compounds. J Anim Sci 58:1040–1049

Clauss M, Lechner-Doll M (2001) Differences in selective reticulo-ruminal particle retention as a key factor in ruminant diversification. Oecologia 129:321–327

Clauss M, Lechner-Doll M, Streich WJ (2002) Faecal particle size distribution in captive wild ruminants: an approach to the browser/grazer dichotomy from the other end. Oecologia 131:343–349

Clauss M, Lechner-Doll M, Streich WJ (2003) Ruminant diversification as an adaptation to the physicomechanical characteristics of forage. A reevaluation of an old debate and a new hypothesis. Oikos 102:253–262

Clauss M, Gehrke J, Hatt JM, Dierenfeld ES, Flach EJ, Hermes R, Castell J, Streich WJ, Fickel J (2005) Tannin-binding salivary proteins in three captive rhinoceros species. Comp Biochem Phys A 140:67–72

Clemens ET, Maloiy GMO (1984) Colonic absorption and secretion of fluids, electrolytes and organic-acids in East-African wild ruminants. Comp Biochem Phys A 77:51–56

Clemens ET, Maloiy GMO, Sutton JD (1983) Molar proportions of volatile fatty-acids in the gastrointestinal-tract of East-African wild ruminants. Comp Biochem Phys A 76:217–224

Demment MW, Longhurst WM (1987) Browsers and grazers: constraints on feeding ecology imposed by gut morphology and body size. In: Santana OP, Da Silva AG, Foote WC (eds) Proceedings of the IVth International Conference on Goats, Symposia, vol 2, March 8–13, Brasilia, Brazil, pp 989–1004

deVega A, Poppi DP (1997) Extent of digestion and rumen condition as factors affecting passage of liquid and digesta particles in sheep. J Agr Sci 128:207–215

Ditchkoff SS (2000) A decade since "diversification of ruminants": has our knowledge improved? Oecologia 125:82–84

Duncan AJ, Milne JA (1992) Rumen microbial degradation of allyl cyanide as a possible explanation for the tolerance of sheep to brassica-derived glucosinolates. J Sci Food Agr 58:15–19

Foley WJ, Iason GR, McArthur C (1999) Role of plant secondary metabolites in the nutritional ecology of mammalian herbivores: how far have we come in 25 years? Pages In: Jung H-JG, Fahey GC (eds) Nutritional ecology of herbivores. Proceedings Vth International Symposium Nutrition of Herbivores, Am Soc Anim Sci, Savoy, IL, pp 130–209

Foose TM (1982) Trophic strategies of ruminant versus non-ruminant herbivores. Thesis. University of Chicago

Gordon IJ, Illius AW (1994) The functional significance of the browser-grazer dichotomy in African ruminants. Oecologia 98:167–175

Gordon IJ, Illius AW (1996) The nutritional ecology of African ruminants - a reinterpretation. J Anim Ecol 65:18–28

Harborne JB (1988) Introduction to ecological biochemistry. Academic Press, London

Hegarty RS (2004) Genotype differences and their impact on digestive tract function of ruminants: a review. Aust J Exp Agr 44:458–467

Hennessy, DR (1993) Pharmokinetic disposition of benzimidazole drugs in the ruminant gastrointestinal tract. Parasitol Today 9:329–333

Hendricksen RE, Poppi DP, Minson DJ (1981) The voluntary intake, digestibility and retention time by cattle and sheep of stem and leaf fractions of a tropical legume (*Lablab purpureus*). Aust J Agr Res 32:389–398

Hofmann RR (1973) The ruminant stomach: stomach structure and feeding habits of East African game ruminants. East African Literature Bureau, Nairobi

Hofmann RR (1989) Evolutionary steps of ecophysiological adaptation and diversification of ruminants: a comparative view of their digestive system. Oecologia 78:443

Hofmann RR, Stewart DRM (1972) Grazer or browser: a classification based on the stomach structure and feeding habits of East African ruminants. Mammalia 36:226–240

Hungate RE (1966) The rumen and its microbes. Academic Press, New York

Hungate RE, Phillips GD, Hungate DP, MacGregor A (1960) A comparison of rumen fermentation in European and Zebu cattle. J Agr Sci 54:196–201

Iason GR, Van Wieren SE (1999) Digestive and ingestive adaptations of mammalian herbivores to low-quality forage. In: Olff H, Brown VK, Drent RH (eds) Herbivores: between plants and predators. Blackwell, Oxford, pp 337–369

Ishaque M, Thomas PC, Rook JAF (1971) Consequences to host of changes in rumen microbial activity. Nature–New Biol 231:253–256

Jarman PJ (1974) The social organisation of antelope in relation to their ecology. Behaviour 48:215–266

Jones RJ (1981) Does ruminal metabolism of mimosine explain the absence of Leucaena toxicity in Hawaii? Aust Vet J 57:55–56

Jones RJ, Meyer JHF, Bechaz FM, Stoltz MA, Palmer B, Van der Merwe G (2001) Comparison of rumen fluid from South African game species and from sheep to digest tanniniferous browse. Aust J Agr Res 52:453–460

Jones WT, Mangan JL (1977) Complexes of condensed tannins of sainfoin (*Onobrychis viccifolia* Scop.) with fraction 1 leaf protein and submaxillary mucoprotein, and their reversal by polyethylene glycol and pH. J Sci Food Agr 28:126–136

Lanigan GW (1970) Metabolism of pyrrolizidine alkaloids in the ovine rumen II. Some factors affecting rate of alkaloid breakdown by rumen fluid in vitro. Aust J Agr Res 21:633–639

Larue R, Yu ZT, Parisi VA, Egan AR, Morrison M (2005) Novel microbial diversity adherent to plant biomass in the herbivore gastrointestinal tract, as revealed by ribosomal intergenic spacer analysis and rrs gene sequencing. Environ Microbiol 7:530–543

Long RJ, Apori SO, Castro FB, Orskov ER (1999) Feed value of native forages of the Tibetan Plateau of China. Anim Feed Sci Tech 80:101–113

Mackie RI, McSweeney CS, Klieve AV (2002) Microbial ecology of the ovine rumen. In: Freer M, Dove H (eds) Sheep nutrition. CABI/CSIRO, Wallingford/New York, pp 71–94

Mackie RI, Aminov RI, Hu WP, Klieve AV, Ouwerkerk D, Sundset MA, Kamagata Y (2003) Ecology of uncultivated Oscillospira species in the rumen of cattle, sheep, and reindeer as assessed by microscopy and molecular approaches. Appl Environ Microbiol 69:6808–6815

Majak W (1992) Metabolism and absorption of toxic glycosides by ruminants J Range Manage 45:67–71

Makkar HPS, Becker K (1998) Adaptation of cattle to tannins: role of proline-rich proteins in oak-fed cattle. Anim Sci 67:277–281

Mattocks AR (1986) Chemistry and toxicology of pyrrolizidine alkaloids. Academic Press, San Diego

McArthur C, Sanson GD, Beal AM (1995) Salivary proline-rich proteins in mammals: roles in oral homeostasis and counteracting dietary tannin. J Chem Ecol 21:663–691

McEwan NR, Abecia L, Regensbogenova M, Adam CL, Findlay PA, Newbold CJ (2005) Rumen microbial population dynamics in response to photoperiod. Lett Appl Microbiol 41:97–101

Mehansho H, Hagerman A, Clements S, Butler L, Rogler J, Carlson DM (1983) Modulation of proline-rich protein biosynthesis in rat parotid glands by sorghums with high tannin levels. Proc Nat Acad Sci 80:3948–3952

Miller SM, Brooker JD, Blackall LL (1995) A feral goat rumen fluid inoculum improves nitrogen-retention in sheep consuming a mulga (*Acacia aneura*) diet. Aust J Agr Res 46:1545–1553

Miller SM, Brooker JD, Phillips A, Blackall LL (1996) *Streptococcus caprinus* is ineffective as a rumen inoculum to improve digestion of mulga (*Acacia aneura*) by sheep. Aust J Agr Res 47:1323–1331

Mould FL (2003) Predicting feed quality - chemical analysis and in vitro evaluation. Field Crop Res 84:31–44

Naurato N, Wong P, Lu Y, Wroblewski K, Bennick A (1999) Interaction of tannin with human salivary histatins. J Agr Food Chem 47:2229–2234

Norton BW, Moran JB, Nolan JV (1979) Nitrogen metabolism in Brahman cross, buffalo, Banteng and shorthorn steers fed on low-quality roughage. Aust J Agr Res 30:341–351

Odenyo AA, Osuji PO (1998) Tannin-tolerant ruminal bacteria from East African ruminants. Can J Microbiol 44:905–909

Odenyo AA, McSweeney CS, Palmer B, Negassa D, Osuji PO (1999) In vitro screening of rumen fluid samples from indigenous African ruminants provides evidence for rumen fluid with superior capacities to digest tannin-rich fodders Aust J Agr Res 50:1147–1157

Oh HK, Sakai T, Jones MB, Longhurst WM (1967) Effect of various essential oils isolated from Douglas Fir needles upon sheep and deer rumen microbial activity. Appl Microbiol 15:777–784

Orskov ER, Benzie D, Kay RNB (1970) The effect of feeding procedure on closure of the oesophageal groove in sheep. Brit J Nutr 24:785–795

Ouwerkerk D, Klieve AV, Forster RJ, Templeton JM, Maguire AJ (2005a Characterization of culturable anaerobic bacteria from the forestomach of an eastern grey kangaroo, *Macropus giganteus*. Lett Appl Microbiol 41:327–333

Ouwerkerk D, Maguire AJ, Klieve AV (2005b) Reductive acetogenesis in the foregut of macropod marsupials in Australia. In: Soliva CR, Takahashi J, Kreuzer M (eds) Publication Series, vol 27. Institute of Animal Science, EFTH, Zurich, pp 98–101

Parra, R (1978) Comparison of foregut and hindgut fermentation in herbivores. In: Montgomery GG (ed) The ecology of arboreal folivores. Smithsonian Institute, Washington, DC, pp 205–230

Perez-Barberia FJ, Gordon IJ, Illius AW (2001) Phylogenetic analysis of stomach adaptation in digestive strategies in African ruminants. Oecologia 129:498–508

Perez-Barberia FJ, Elston DA, Gordon IJ, Illius AW (2004) The evolution of phylogenetic differences in the efficiency of digestion in ruminants. Proc R Soc Lond B Bio 271:1081–1090

Perez-Maldonado RA, Norton BW (1996a) The effects of condensed tannins from *Desmodium intortum* and *Calliandra calothyrsus* on protein and carbohydrate digestion in sheep and goats. Brit J Nutr 76:515–533

Perez-Maldonado RA, Norton BW (1996b) Digestion of c-14-labeled condensed tannins from *Desmodium intortum* in sheep and goats. Brit J Nutr 76:501–513

Poppi DP, Minson DJ, Ternouth JH (1981) Studies of cattle and sheep eating leaf and stem fractions of grasses. 1. The voluntary intake, digestibility and retention time in the reticulo-rumen. Aust J Agr Res 32:99–108

Robbins CT, Hagerman AE, Hjelijord O, Baker DL (1987a) Role of tannins in defending plants against ruminants: reduction in protein availability. Ecology 68:98–107

Robbins CT, Mole S, Hagerman AE, Hanley TA (1987b) Role of tannins in defending plants against ruminants: reduction in dry matter digestion. Ecology 68:1606–1615

Robbins CT, Hagerman AE, Austin PJ, McArthur C, Hanley TA (1991) Variation in mammalian physiological responses to a condensed tannin and its ecological implications J Mammal 72:480–486

Robbins CT, Spalinger DE, Vanhoven W (1995) Adaptation of ruminants to browse and grass diets - are anatomical-based browser-grazer interpretations valid? Oecologia 103:208–213

Rosenthal GA, Janzen DH (1979) Herbivores: their interaction with secondary plant metabolites. Academic Press, New York

Smith RH (1980) Kale poisoning: the brassica anaemia factor. Vet Rec 107:12–15

Spalinger DE, Robbins CT, Hanley TA (1993) Adaptive rumen function in elk (*Cervus elaphus nelsoni*) and mule deer (*Odocoileus hemionus hemionus*). Can J Zool 71:601–610

Stegelmeier BL, Edgar JA, Colegate SM, Gardner DR, Schoch TK, Coulombe RA, Molyneux RJ (1999) Pyrrolizidine alkaloid plants, metabolism and toxicity. J Nat Toxins 8:95–116

Sundset MA, Cann IOK, Mathiesen SD, Mackie RI (2004) Rumen microbial ecology in reindeer - adaptations to a unique diet. J Anim Feed Sci 13:717–720

Theodorou MK, Gascoyne DJ, Akin DE, Hartley RD (1987) Effect of phenolic acids and phenolics from plant cell walls on rumen-like fermentation in consecutive batch culture. Appl Environ Microbiol 53:1046–1050

Thornton RF, Minson DJ (1973) Relationship between apparent retention time in rumen, voluntary intake, and apparent digestibility of legume and grass diets in sheep. Aust J Agr Res 24:889–898

Tolosa MX, Dinh TV, Klieve AV, Ouwerkerk D, Poppi DP, McLennan SR (2004) Molecular characterisation of rumen bacterial populations in cattle fed molasses diets. Anim Prod Aust 25:328

Tulloh NM (1966a) Physical studies of the alimentary tract of grazing cattle. IV. Dimensions of the tract in lactating and non-lactating cows. New Zeal J Agr Res 9:999–1008

Tulloh NM (1966b) Physical studies of the alimentary tract of grazing cattle. III. Seasonal changes in capacity of the reticulo-rumen of dairy cattle. New Zeal J Agr Res 9:252–260

Van Soest PJ (1982) Nutritional ecology of the ruminant. O&B Books, Corvallis, OR

Van Soest PJ (1994) Nutritional ecology of the ruminant. Cornell University Press, Ithaca, NY

Van Wieren SE (1996) Digestive strategies in ruminants and non-ruminants. University of Wageningen, Netherlands

Wallace RJ (2004) Antimicrobial properties of plant secondary metabolites. P Nutr Soc 63:621–629

Watkins JBI, Klaassen CD (1986) Xenobiotic biotransformation in livestock: comparison to other species commonly used in toxicity testing. J Anim Sci 63:933–942

Watson C, Norton BW (1982) The utilisation of pangola grass hay by sheep and Angora goats. P Aust Soc Anim Prod 41:467–470

Zucker WV (1983) Tannins: does structure determine function? An ecological perspective. Am Nat 121:335–365

Chapter 5
The Comparative Feeding Behaviour of Large Browsing and Grazing Herbivores

Kate R. Searle and Lisa A. Shipley

5.1 Introduction

Herbivores exploit a food resource that is fundamentally different from that of most other trophic levels. Their food exists in an apparent surplus, food items are rarely eaten in entirety, and the concepts of pursuit and catchability are irrelevant (Owen-Smith and Novellie 1982; Spalinger & Hobbs 1992a). Plant foliage is generally of low nutritive value so foraging time and digestive time are important constraints. Ungulate herbivores can be broadly classified into two groups—those that feed mainly on grass, and those than feed mainly on browse. This division is important in relation to the both the way in which an animal forages, and the definition of what constitutes a patch, because grass and browse material present themselves in very different ways to the foraging herbivore, and because grasses differ from browses in architecture and physical structure (Table 5.1).

5.2 The Functional Response

An herbivore's foraging behaviour derives from the interaction between characteristics of the herbivore and characteristics of the plants on which it feeds. Intake rate is the currency through which an herbivore measures the outcomes of different foraging strategies (Stephens and Krebs 1986), with an aim to maximising energy intake rate over short time scales (i.e., minutes, hours). At longer time scales (days) other currencies and constraints come to influence the foraging process, such as digestive constraints, gut capacity (Provenza 1995; Hirakawa 1997; Iason et al. 1999; Villalba and Provenza 1999, 2000, 2005; Launchbaugh et al. 2001), and predation risk (Brown 1988, 1999; Beecham and Farnsworth 1999; Abrams and Schmitz 1999; Frair et al. 2005). Hence, over the course of a day an herbivore may simultaneously choose short-term strategies that maximise instantaneous intake rate (e.g., selection of bites within a feeding

Table 5.1 A comparison of general chemical and structural characteristics of grasses (monocots) and browses (herbaceous and woody dicots)

Characteristic	Grasses	Browses
Cell wall	Thick	Thin
	Greater proportion cellulose/hemicellulose	Greater proportion lignin
Plant defences	Silica	Thorns and spines, secondary chemicals
Plant architecture	Fine-scaled heterogeneity within plant	Coarse-scaled heterogeneity within plant
	Meristem at base	Meristem at tips
	Low growth form	Low to high growth from
	Compact	Complex, diffuse, branching architecture
Dispersion	Uniform	Dispersed

station), whilst employing more long-term strategies that seek to minimise time spent foraging (e.g., choosing patches within a habitat; Bergman et al. 2001). These decisions occur very frequently and form the grain of the foraging hierarchy, representing an important underlying base to higher level behaviours.

First, we will examine important components that control the relationship between plant characteristics and intake rate, and examine the short-term decisions herbivores make in relation to maximising instantaneous intake rate. Because herbivores can be classified into two general groups, browsers and grazers, by the main components of their diets, we will examine how these foraging behaviours vary between these groups.

An animal's intake rate is typically expressed using the functional response, which describes the relationship between intake rate and some measure of plant abundance (e.g., biomass, sward height, density, plant mass, bite mass). The concept of the functional response was first developed for predators (disk equation; Holling 1959, 1965), though it has since been adapted for herbivores (Spalinger Hobbs 1992a, 1992b; Gross et al. 1993; Beckerman 2005). As such, it is a key concept in herbivore foraging behaviour, linking individual behaviour to higher level processes such as plant–herbivore interactions and population dynamics. Dynamics of the functional response provide the impetus for herbivore movement—changes to intake rate motivate movements in search of new food sources. The functional response is therefore a key component in the reciprocal interactions of herbivores and the spatio-temporal patterns of their food resources, which ultimately determine important ecosystem functions such as nutrient cycling, fire regimes, and sediment and water transfer (Fig. 5.1).

The seminal paper of Spalinger and Hobbs (1992a) provided a mechanistic interpretation of the way several key processes shape and control the functional response of herbivores foraging within patches (Eq. 1). For herbivores feeding on spatially concentrated foods in which foods are apparent and spaced such that animals can reach the next bite before completely chewing the previous (Process

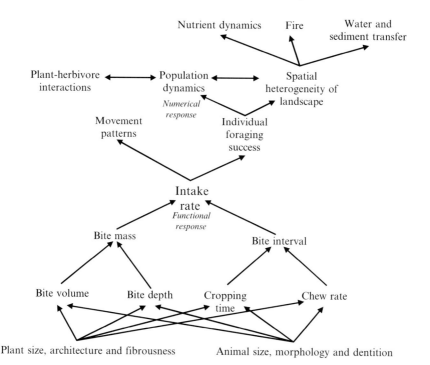

Fig. 5.1 Conceptual model demonstrating linkages between plant characteristics, intake rate and behaviour, and higher order population dynamics and ecosystem function

III; Spalinger and Hobbs 1992a), such as a grass sward or browse patch, instantaneous intake rate (I, g min^{-1}) is represented as an asymptotic function of bite size (S, g):

$$I = \frac{R_{max} S}{R_{max} h + S} \qquad \text{(Eq. 1)}$$

where R_{max} (g min^{-1}) is the maximum rate of processing of plant tissue in the mouth that would occur in the absence of cropping, and h (min) is the average time required to crop a single bite in the absence of chewing. This model rests upon the assumption that intake rate is controlled by the competition between cropping and chewing leading to a Type II functional response, just as competition between searching for and handling prey items creates a Type II functional response in Holling's (1959) original disk equation.

When foods are apparent and dispersed in space such that previous bites are fully consumed before encountering the next (Process II; Spalinger and Hobbs 1992a), variables describing handling time (h and R_{max}) drop out of the model and variables describing rate of encounter with plants (V_{max}, m/min, travel velocity in the absence of cropping, D, bites/m^2, bite density, and δ, m/bite, decrease in velocity caused by cropping bites) are included in the model:

$$I = \frac{V_{max}\sqrt{D}}{1+\delta\sqrt{D}} \cdot S$$

(Eq. 2)

Finally, when bites are dispersed, but are cryptic, animals must search to encounter bites (Process I; Spalinger and Hobbs 1992a), thus W (m), the width of the search path or the detection limits imposed by vision or smell are included in the model:

$$I = \frac{DV_{max}W}{1+\delta WD} \cdot S$$

(Eq. 3)

5.2.1 Components of the Functional Response

Because herbivores experience selective pressures for obtaining food rapidly while foraging (Stephens and Krebs 1986), coevolutionary processes between plants and herbivores have shaped different mouth morphology for grazing and browsing herbivores (Table 5.2; Hofmann and Stewart 1972; Jarman 1974; Janis and Ehrhardt 1988; Pérez-Barbería and Gordon 2001). Therefore, plant architecture and animal anatomy influence each of the components of functional response, including finding, cropping, and chewing bites of vegetation.

5.2.1.1 Cropping Time

The time necessary to crop bites of vegetation (h, min, Eq. 1) is influenced by fibre composition, dispersion of plant parts, and structural defences of plants, all of which differ between grasses and browses. First, grasses tend to have thicker

Table 5.2 A comparison of mouth and tooth anatomy of grazing and browsing herbivores

Characteristic	Grazer	Browser
Mouth	Wide muzzle	Narrow muzzle
	Smaller mouth opening	Larger mouth opening
	Stiffer lips	Flexible, often muscular lips, long tongue
Teeth	Lower incisors of similar size, wider contact area	Central incisors broader than outside ones, less contact area for manipulation
	Wide incisor row	Narrow incisor row
	Incisors project forward	Incisors more upright
	High-crowned teeth	Low-crowned teeth
	Less surface area, more shearing action, more ridges	Large surface area of contact between lower and upper teeth for crushing action

cell walls than do browses. Within the cell wall, grasses contain a larger proportion of cellulose, whereas browses, especially the woody parts, contain more lignin. Many grasses contain silica, causing leaf blades to have sharp and rigid edges (Demment and Van Soest 1985; Bodmer 1990; Gordon and Illius 1994; Owen-Smith 1997). Cellulose, lignin, and silica increase the tensile strength, or 'toughness' of the plant, thus requiring an herbivore to exert greater force and expend more time to sever the bite (Reid 2000). Tensile strength, or 'toughness', increases as a function of age and diameter of tillers and stems cropped (Van Soest 1981; Nelson and Moser 1994; Wright and Illius 1995; Shipley et al. 1999). Tensile strength can also limit the size of bites cropped by both grazers and browsers, thus further reducing intake rates (Vivås et al. 1991). For example, large bites taken by grazers necessarily require cropping more tillers and a greater proportion of mature tillers, both of which increase tensile strength. When grazing swards of high tensile strength, grazers sometimes fail to sever the bites of grasses apprehended, and must do so in order to release some of the forage before completing the biting motion. In grazing macropods, for example, this requires longer occlusal contact between upper and lower incisors (Hume 1999). When browsers take larger bites, they often must include more woody material and crop larger-diameter or older stems. As twig diameter increases, browsers must switch from using incisors to molars to crop bites, which slows intake (Cooper and Owen-Smith 1986). Eventually, stem diameters become too large to crop at all.

The dispersion of plant parts and physical plant defences can also influence cropping time. When leaves and tillers are small and dispersed, herbivores must spend more time rounding up multiple leaves and stems with the tongue and lips before cropping the bite, than they would when consuming more uniform, compact bites (Illius and Gordon 1987). Because browses are typically more spatially diffuse than grasses, cropping bites is often more time-consuming for browsers. Furthermore, structural defences of woody browses, such as spines and thorns, also slow cropping by impeding stripping motions and by separating leaves (Dunham 1980; Cooper and Owen-Smith 1986; Belovsky et al. 1991; Gowda 1996; Illius et al. 2002). Animals must manipulate thorny plants more slowly and carefully in their mouths to avoid pain and injury (Dunham 1980; Cooper & Owen-Smith 1986; Belovsky et al. 1991). Grazers and browsers typically have evolved different mouth anatomy that may allow them to reduce cropping time on the plants for which they are adapted. For example, the wide incisor row of grazers may allow them to spread out the force of severing many tillers in one bite (Janis and Ehrhardt 1988). In contrast the pointed muzzles and mobile lips of browsers are likely adaptations for quickly obtaining small bites separated in space or defended by thorns, and for gathering or stripping leaves from branches in one motion (Myers and Bazeley 1992). In browsing macropods, the first upper incisor is larger than the second and third incisor, minimising the amount of upper and lower incisor contact, which allows manipulation of food items by the incisors (Sanson 1989). Unfortunately, few studies have directly compared the ability of browsers and grazers to crop bites in order to evaluate quantitatively whether their different mouth anatomy influences cropping efficiency on similar plants. When comparing a group of 13 herbivores cropping bites of fresh alfalfa, however, Shipley et al. (1994) found that relative to their size, grazers generally had a shorter cropping time than the intermediate and browsing species, especially those with long muzzles.

5.2.1.2 Bite Size

Bite size cropped (S, g, Eq. 1–3) strongly influences intake and the functional response of browsing and grazing herbivores. In many herbivores, intake rate varies as much as 10-fold with increasing bite size (Black and Kenney 1984; Spalinger and Hobbs 1992b; Gross et al. 1993; Shipley et al. 1994). Animals achieve this greater intake rate because large bites require relatively fewer interruptions in chewing and swallowing caused by cropping new bites than do small bites. The size of bites animals obtain depends on both the architecture of the plant and the size and morphology of its mouth.

Relative to many browses, grasses are often uniformly distributed in a 3-dimensional layer of food. The bite mass a grazer can obtain from this layer depends on the height and bulk density of grass, and the width of the grazer's incisor row (Black and Kenney 1984; Gordon and Illius 1988; Ungar et al. 1991; Illius and Gordon 1992). Therefore, the wider the grazer's incisor width, the larger the bite it can obtain from a grass sward (Illius and Gordon 1987; Janis and Ehrhardt 1988). When grasses are particularly short, intake rates can become so low, that feeding on them becomes unprofitable and they are avoided (Arnold 1987; Laca et al. 1993). Even within grazers, Murray and Illius (2000) found that when feeding on short swards the wildebeest (*Connochaetes taurinus*) obtained twice the bite size and intake rate of narrower-muzzled topi (*Damaliscus lunatus*). On mid-length, differentiated grass swards, however, topi were able to select the higher quality leaves and thus have a higher intake than wildebeest.

In contrast to grasses, browses often grow with a fractal, branching geometry which separates plant tissue in space and makes predicting bite mass cropped by browsers difficult to model and predict from plant volume (Shipley et al. 1994). For example, browsers often avoid cropping small thin stems branching at wide angles from which they can obtain little mass (Vivås et al. 1991; Myers & Bazeley 1992). Instead, large browsers tend to prefer woody plants that provide larger leaves, thicker or longer annual growth stems, and non-thorny plants like willows that allow them to strip many leaves in one bite (Danell et al. 1994; Shipley et al. 1998; Stapley 1998). Browses also tend to vary in nutritional quality on a small scale, thus browsers must often select specific plant parts in each bite (Jarman 1974; de Reffye and Houllier 1997). Furthermore, structural defences, such as thorns, separate leaves and reduce the size of bite a browser can take (Pellew 1984; Milewski et al. 1991; Stapley 1998; Illius et al. 2002). The browser's narrower muzzle, prehensile lips (e.g., elephant's trunk) and long tongue allow greater selection of smaller, more nutritious plant parts, and allow it to obtain bites interspersed with thorns (Janis and Ehrhardt 1988; Hofmann 1989; Belovsky et al. 1991). These features also permit large browsers, like moose, kudus, and giraffes, to obtain large bites by stripping many leaves from one stem in a sideways motion (Pellew 1984; Stapley 1998).

Although few direct comparisons of bite size of grazing and browsing herbivores have been made, Shipley et al. (1994) compiled data on maximum bite sizes reported in the literature for 16 grazing and 17 browsing herbivores feeding on their natural forages. Although maximum bite mass increased with body mass with the same slope, browsers of the same size took larger maximum bites than grazers. This result suggests that grazers may be more limited by their mouth size when cropping bites of grasses than are browsers when cropping browses. This result has implications for the form of gain functions observed in grazers and browsers. Linear gain functions have often been reported for browsing herbivores (Åström et al. 1990; Rominger 1995; Rominger et al. 1996; Illius et al. 2002). Illius et al. (2002) observed linear gain functions in roe deer feeding in browse patches, and demonstrated that decelerating gain functions were not observed in browse patches because increased bite rate compensated for declines in bite mass as patches were depleted. In contrast, grazing herbivores are often observed to exhibit decelerating gain functions, especially on tall grasses (Laca et al. 1994; Ginnett et al. 1999). Searle et al. (2005) showed that decelerating gain curves in herbaceous patches are fundamentally a function of declining bite size. However, this relationship is mediated through changes in the time required to form and process those bites. Their results showed that bite mass declined during the course of patch depletion, but that cropping time and processing time increased due to accumulation of plant material in the mouth interfering with bite formation (increased cropping time) and bite processing (increased chewing rate).

5.2.1.3 Chewing Investment and Chewing Rates

How efficiently food can be processed in the mouth (R_{max}, g/min, Eq. 1) depends on both the rate of chewing (chews/min) and how many chews must be invested per g of food comminuted (chews/g) The fibre content of plants and the size and quality of the teeth play a role in these processes (Balch 1971; McLeod et al. 1990; Shipley and Spalinger 1992; Shipley et al. 1994). Because cell wall thickness and fibre content is often relatively greater in grasses, herbivores must spend more time chewing a unit mass of grass than of browse, either while ingesting it, or later during rumination (Robbins 1983; Choong et al. 1992; Wright and Illius 1995). In addition, grazers are expected to have a smaller reticulo-omasal orifice (Hofmann & Stewart 1972; Clauss & Lechner-Doll 2001, but see also Pérez-Barbería and Gordon 2001), thus grazers may need to comminute food to a smaller particle size to escape the rumen than do browsers, further increasing their chewing investment. Fibre content, and thus chewing time, increases as both grasses and browses mature (Hacker and Minson 1981). Spines on leaves of dicots may also reduce chewing efficiency for browsers (Illius et al. 2002).

The efficiency with which an herbivore can chew, and thus ingest, its food also depends on the size and wear of its molars. Silica in grasses can promote rapid tooth wear and thus reduce chewing efficiency (McNaughton and Georgiadis

1986; Riet-Correa et al. 1986). As a response, grazers of some taxonomic groups have developed high-crowned, or hypsodont, molars (Vaughan et al. 2000). These high-crowned molars allow for longer wear when chewing higher fibre and silica. For example, molars and premolars of rock hyraxes (*Procavia johnsoni*), which feed preferentially on grasses, are hypsodont, having high crowns and short roots (Hoeck 1975, 1989). In contrast, yellow-spotted hyraxes (*Heterohyrax brucei*), a browsing hyrax, have brachydont tooth structure with shorter crowns and longer roots. In herbivorous marsupials, grazing macropod marsupials have extended the wear of molars through sequential molar progression anteriorly along a curved tooth row (Sanson 1980, 1989). This morphology allows only two molars to occlude at a time, concentrating shearing force in a smaller area of occlusion to maximise the effectiveness of mastication. As the anterior molars wear out and are lost, they are replaced by posterior molars. In contrast, browsing macropod marsupials have a flat tooth row in which all molars occlude at once. The larger occlusal surface area allows a crushing action and precise molar occlusion. Furthermore, the tooth enamel of the grazing red kangaroo (*Macropus giganteus*) is harder than that of the browsing, swamp wallaby (*Wallabia bicolour*; Palamara et al. 1984). However, similar patterns in tooth structure have not been found in ruminants. Larger molars provide more area for breaking down plant tissue while chewing, and thus R_{max} increases with body mass at the same rate as does the surface area of the molars (Shipley et al. 1994). However, relative to body size, grazers and browsers had the same molar surface area, chewing investment, and chewing rates (Shipley et al. 1994).

5.2.1.4 Encountering

The rate of encounter with bites of vegetation (λ, bites/min) may also differ in habitats occupied by grazing and browsing herbivores. However, modelling encounter rate is complex because the animal's mechanics of travel, perceptive abilities, spatial memory, and volition play a strong and interconnected role. Navigating to food items requires an animal to detect, travel to, and choose to eat food items at more than one spatial scale. Detection depends on characteristics of the animal (e.g., sight and olfaction, spatial memory, and travel speed (Bailey et al. 1989; Roese et al. 1991; Speakman and Bryant 1993), and characteristics of the plant (e.g., crypticity, size and contrast, visual and olfactory cues, predictability, (Blough 1989; Howery et al. 2000), and occasionally characteristics of the animal's conspecifics (e.g., social facilitation; Howery et al. 1998; Bailey et al. 2000). Efficient encounter rate requires that an animal avoid unfilled or low quality patches and move directly among food items. Although we have no reason to believe that grazers and browsers differ fundamentally in their perceptual abilities or memory, grasses and browses tend to differ in their spatial distribution and contrast.

Homogeneous food sources that are relatively fixed and predictable in time and space, as may be true of grasses, provide fewer visual cues for easily locating foods, but animals are more able to depend on their spatial memory to find food resources in these habitats (Howery et al. 2000). Spatial memory allows herbivores to revisit nutrient rich sites and avoid low quality sites, but often requires animals to store an enormous amount of information (Bailey et al. 1989; Bailey and Rittenhouse 1989; Laca 1998). On the other hand, heterogeneous food sources, as may often be true of browses, may provide more visual cues that herbivores can use to predict and locate forage resources from a distance. Visual cues allow animals to speed up and travel directly between food items or patches. Because patches are easier to detect than individual food items, herbivores tend to have lower intake when food is distributed randomly or uniformly than when it is distributed in patches (Laca and Ortega 1996).

Animals that forage within a social group may more easily detect patches of forages through social facilitation (Howery et al. 1998; Bailey et al. 2000). In general, grazers that occupy open habitat are more likely to live and forage in groups than are browsers that more typically occupy areas with taller and denser vegetation, possibly because of the added advantage in detecting and avoiding predators (Krebs and Davies 1993). In feral cattle, for example, the integrity of the social group is high (Lazo 1994). Therefore, social facilitation beyond that provided by the mother (Howery et al. 1998) may enhance encounter rates at larger scale for grazers.

After food items have been detected, encounter rate depends on the distance between food items and how fast an animal can travel while foraging. Maximum foraging speed is likely similar between browsers and grazers, even if they differ in body mass. Shipley et al. (1996) found no relationship in maximum travel velocity while foraging for animals ranging from 5 g to 500 kg. However, average travel speed was slower for large animals when plants were spaced closer together because it took larger animals longer to accelerate and decelerate between bites. If food items are difficult to detect, animals must slow their travel rate while foraging to increase detection ability (Getty and Pulliam 1991; Speakman and Bryant 1993; Shipley et al. 1996). Animals may also purposely choose to increase travel speed when plants are apparent, but spaced widely apart (Shipley and Spalinger 1995). Therefore, herbivores often show rapid coarse-scaled movements among patches within a matrix and slower, fine-scaled movements when searching for bites within a patch.

Finally, diet selectivity can also influence encounter rate. When food is more heterogeneous in quality and dispersion, animals must make decisions about maximizing nutrient intake more frequently than do animals encountering more homogenous resources. The more selective an animal is, the slower it may have to travel to detect and make a decision about a plant item, and likely the further apart and more cryptic the items may be (Murray 1991; Laca & Demment 1996). Browsers are expected to be more selective when foraging because browses tend to vary greatly in nutritional content and browsers may not be able to digest plant fibre as well as grazers. Therefore, encounter rate may be more influenced by diet selection in browsers than grazers.

5.3 Foraging in Patches

Spatially subdividing natural environments into units that have functional relevance for an ecological process remains a fundamental challenge in many areas of ecological research. For foraging herbivores, the question of what constitutes a patch has been at the core of behavioural research for several decades. The patch concept forms a framework on which most research into foraging behaviour for herbivores rests, and it underpins many of the widely used models in foraging behaviour research. However, it is a concept that continues to defy rigorous definition, and applications that rely on it often fail to produce predictions that match with observed behaviour. One of the greatest challenges in defining patches for foraging herbivores arises because herbivores exploit food resources that are generally continuous, making it very difficult to break up the foraging environment into discrete units that are robust and repeatable.

Early attempts to define patches centred on reference to their appearance or differences from their surroundings (Wiens 1976; Kotliar and Wiens 1990). However, a more functional definition has since been introduced where patches are delineated by reference to a change in the rate of a process or behavior that relates specifically to the foraging animal (Sih 1980; Senft et al. 1987; Bailey et al. 1996). Hierarchy theory (O'Neill et al. 1986) has been used to provide a framework for the delineation of heterogeneous foraging landscapes for large herbivores (Senft et al. 1987; Kotliar and Wiens 1990; Milne 1991; Laca and Ortega 1995; Bailey et al. 1996). This framework captures heterogeneity by subdividing a landscape into a set of nested spatial scales, or hierarchical levels, each having functional relevance to the foraging animal.

Use of this hierarchical framework requires us to define some terms; for a large herbivore these hierarchical levels may consist of individual plants, nested into feeding stations (an aggregation of bites, where all bites can be removed without movement of the animal's forelegs, Goddard 1968), nested into small patches, then plant communities, ultimately aggregating up to home ranges. We can treat each of these hierarchical levels as a patch in the traditional sense (sensu MacArthur and Pianka 1966; Charnov 1976) because at each level the foraging herbivore must make a decision as to which individual units it chooses to eat from, and how much effort to devote to exploiting an individual unit before moving on. For instance, at lower hierarchical levels an herbivore encounters feeding stations as it moves through its foraging environment, and it must first decide which individual feeding stations to forage within, and second, how much forage to consume from each feeding station before moving on.

Natural landscapes present the foraging herbivore with considerable variation within all these hierarchical levels (i.e., different plant species co-occurring within a feeding station). Any attempt to define patches for large herbivores must, therefore, acknowledge this spatial heterogeneity, and act to incorporate the multi-layered decision process that foraging herbivores follow in complex landscapes. This section examines the selection of forage patches by ungulates over a range of spatial scales; specifically asking whether patches are functionally different for grazers versus browsers, and what consequences these differences have for the way these animals perceive and interact with their foraging environment.

5.3.1 Definition of Patches

The growth form of grasses differs from that of browse, and these differences have important consequences for the ways in which herbivores interact with these two food resources. Jarman (1974) initiated much of our current understanding of these interactions in his seminal paper discussing the ecology of antelope in Africa. Through describing significant differences in the growth forms and dispersion of grasses and browse (differences in meristem position, proportion of the plant composed of actively growing tissue, chemical and physical defences), Jarman (1974) was able to relate behavioural aspects of antelope ecology to specific differences in forage resources. Of particular relevance to any discussion of patch definition for herbivores is Jarman's (1974) conclusion that, on average, a grass plant is likely to be more homogeneous in food quality than a browse plant, and that grasses typically present a more contiguous distribution than browse plants. Woody browse plants tend to be composed of 'an assembly of parts of heterogeneous value' (Jarman 1974); younger shoots and leaves will offer relatively high nutritional content, whilst more mature leaves and stems or trunks will offer only negligible food value. This underlying heterogeneity derived from differences in plant growth form and plant dispersion interacts with the selective capabilities of the herbivore to determine the ultimate variation in food quality experienced by grazers and browsers. It is this variation that ultimately drives the perception of patches of functional significance to the forager.

The distribution of grasses and browse across a landscape can differ substantially. Grasses often occur in extensive contiguous swards that may be dominated by relatively few grazing-tolerant species under intermediate to high levels of grazing (Jarman 1974; McNaughton 1979; Walker 1987; Stuart-Hill and Tainton 1989), whereas browse species often occur in more isolated clusters within a landscape (Jarman 1974). Both grasses and browse employ physical and chemical defences against herbivory, but these are generally more effective in browse species (Cooper and Owen-Smith 1986; Milewski et al. 1991; Cooper and Ginnett 1998). Browse species tend to offer the forager a greater opportunity for higher nutritional value because on average grasses have a higher dietary fibre content, though for browsing animals ultimate fibre intake is determined by how much woody stem is eaten (Van Soest 1981). Intra-plant variation in quality is a general feature of both grasses and browse; individual plants may contain considerable morphological and phenological differences in quality. These differences vary widely through a number of mechanisms, and have profound consequences for the pattern and scale of spatial variation in the quality of grass and browse forage. It is this spatial variation that browsers and grazers are able to exploit to increase foraging efficiency and diet quality. We will now examine the patterns and mechanisms that underlie spatial variation in the quality of browse and grasses for foraging herbivores.

5.4 Spatial Variation in the Quality of Grass

Grasslands can appear at first glance as monotonous stretches of physiologically and structurally similar grasses, uniformly distributed within the landscape. This perception has encouraged the view that the foraging behaviour of large grazers is relatively simple, with animals using large, unselective mouths to uniformly mow their way through grass swards. In reality, grasslands are a 'highly dynamic, spatially and temporally heterogeneous complex of ecological relationships' (McNaughton 1989), and grazing herbivores interact in a reciprocal manner with this variation, both responding to, and creating heterogeneity.

For grazing herbivores the quality of their food resources is determined not so much by plant defence mechanisms (secondary compounds and physical structures), but more by the concentrations of nutrients and minerals—namely nitrogen, phosphorous, potassium, sodium, and calcium. Whilst grasses can contain secondary compounds, they are generally lower in defensive chemicals than are other forages (Cates and Rhoades 1977; McNaughton 1983; Malachek and Balph 1987). Physical defences, such as morphological adaptations and silicification appear to be the predominant evolutionary response of grasses to grazers (McNaughton 1989). However, it is the spatial variation in the concentrations of nutrients and minerals in natural grasslands that is the key driver underlying the diets and distributions of grazing mammals (Owen-Smith and Novellie 1982; McNaughton 1988; McNaughton 1990; Owen-Smith 1994; Olff and Ritchie 1998). Explaining the causes of this spatial variation in grass quality is difficult because the patterns evident in natural grasslands are the outcome of many interacting biotic and abiotic processes.

Additional complexity is added because, as with most ecological patterns, the relative importance of these constituent processes changes with spatial scale (Roberts 1987; Mutanga et al. 2004). However, spatial patterns in grass quality have been quantified in several grassland systems. The presence of coherent plant communities has been demonstrated in the Serengeti and Masai Mara, arguably the most extensive mammalian grazing systems in the world. McNaughton (1983) demonstrated that almost a hundred stands of grasses within these systems could be classified into just seventeen objectively identifiable plant communities, each having a characteristic, coherent species composition. Spatial variation in grass quality has been documented at most of the functionally relevant spatial scales for grazing herbivores—from individual plants and stands of plants, up to landscape and regional scale variation. McNaughton (1989) presented a foraging hierarchy for the Serengeti ecosystem and used forage nutrient analyses to indicate that grazing herbivores confront substantial heterogeneity in forage quality during the growing season at each hierarchical level. Clearly then, there is ample opportunity for grazers to increase their foraging efficiency by choosing to forage in those communities that best suit their dietary requirements.

The majority of research into spatial variation in grass quality has focused on regional and landscape scale variation, and has tended to examine the influence of individual factors in isolation. Traditional approaches have linked patterns in grass quality at regional scales with temperature and rainfall (Robbins 1983; McNaughton 1983; Roberts 1987; Skarpe 1992; Kumar et al. 2002). At landscape scales, topography becomes an important determinant of grass quality through mechanisms expressed differentially with slope, aspect, and altitude (McNaughton 1983; Roberts 1987; Seagle and McNaughton 1992; Chaneton et al. 2005), and soil characteristics greatly influence grass quality through spatial variation in soil nutrients, structure, and texture (Anderson and Talbot 1965; Bell 1982; McNaughton 1983; Bakker et al. 1983; Kumar et al. 2002; Mutanga et al. 2004). For instance, McNaughton (1989) found evidence for considerable variation in mineral contents within a grass species distributed between local patches of forage in the Serengeti; phosphorous content varied by as much as 52% between swards, and sodium by as much as 486% among swards.

At a finer scale, spatial variation in grass quality is influenced by biotic factors. Studies have emphasized the importance of species type and genotype, phenology, aboveground biomass, and tree/shrub canopy cover (Bakker et al. 1983; Wilson 1984; McNaughton 1988; McNaughton 1990; Olff and Ritchie 1998; Ludwig et al. 2001, 2004). Edaphic factors play a significant role in manipulating grass quality on local scales. A study in the Kruger National Park in South Africa demonstrated that grass nutrient distribution was significantly correlated with slope, aspect, aboveground biomass, and percentage grass cover (Mutanga et al. 2004). A strong general relationship was found between grass quality and micro-variations in slope, altitude and aspect, factors which influence grass quality through their effects on soil temperature and water run-off. The interplay between grass quality, topography, and hydrology is important in many semi-arid grazing systems. Fine-scale redistribution of sediments and nutrients through run-off and run-on is well documented, particularly in systems dominated by tussock-forming grasses (Greene et al. 1994; Mcivor et al. 1995, 2005; Turner 1998; Eldridge 1998; Northup et al. 1999, 2005; Ludwig et al. 1999, 2000, 2005; HilleRisLambers et al. 2001; Rietkerk et al. 2002, 2004; Augustine 2003). The nitrogen concentration of individual grasses within a species was found to vary with slope in Kruger National Park: some species had greater nitrogen concentrations on steep slopes, whereas others had higher concentrations on flat to gentle slopes (Mutanga et al. 2004). The same study found an inverse relationship between percentage cover and biomass of grass, and grass quality. It is thought that this can be explained by the dilution of nutrients in high biomass regions (Wilson 1984; Mutanga et al. 2004).

Trees in savanna ecosystems can cause significant changes in the quality of grasses under, and nearby, their canopies. Trees alter nutrient availability for grasses, generally increasing soil nutrient availability and plant productivity in the understorey (Kellman 1979; Bernhard-Reversat 1982; Belsky et al. 1989, 1993;

Belsky 1994), though some negative net effects of savanna trees on understorey productivity have been documented (Stuart-Hill and Tainton 1989; Mordelet and Menaut 1995). A study in East Africa found that soil nutrient availability around *Acacia tortilis* trees increased with tree age and size, being highest under dead trees and lowest in open grassland patches (Ludwig et al. 2004). The presence of trees seemed to cause a shift from nitrogen limitation in open grassland to phosphorous limitation under tree canopies; soil moisture was lower under tree canopies, and the species composition was very different under tree canopies in comparison to open areas (Ludwig et al. 2004). Whilst this study found no difference in aboveground biomass under versus outside live tree canopies, likely due to increased competition for soil moisture, the quality of grasses was different. A previous study reported similar results, finding greater mid-wet season nutrient concentrations in grasses under than outside tree canopies (Ludwig et al. 2001). This variation represents an important opportunity for herbivores to select for better quality forage beneath tree canopies in this system.

Within-species level variation is also present in grass species. Lignification is under genetic control and considerable differences in lignin concentration and composition have been found among individuals growing in different temperatures, soil moisture and fertility, and light levels, and even among genotypes within a species (see Moore and Jung 2001 for a review). Within-plant variation in quality increases as grass plants mature and the proportion of tissues with large amounts of structural components increases, but some plant parts such as leaves remain of relatively high quality. McNaughton (1989) demonstrated significant within-plant variation in minerals such as aluminium, calcium, iron, potassium, and magnesium for two grass species within two swards in the Serengeti during the wet season. This presents grazers with a great potential for exercising selectivity (Fryxell 1991), and provides opportunities for selective exploitation of plant parts of high nutrient values within the sward. Grass blades can, therefore, present the grazing animal with considerable nutrient and mineral heterogeneity.

5.5 Spatial Variation in the Quality of Browse

Species of browse plants vary dramatically in their digestibility to both mammalian and insect herbivores (Kamupingene et al. 2004; Scogings et al. 2004). This variation is of undoubted importance to diet selection of browsing herbivores, but perhaps more relevant to their perception of patches is the considerable variation *among individuals* of the same browse species in a suite of compounds known to influence herbivore digestion and energy gain.

Several mechanisms can cause variation in plant defences within a population. First, variance in defence against browsing can have a strong genetic component (Dimock et al. 1976; Silen et al. 1986; Laitinen et al. 2002), and genetic variation in plant resistance to browsing has been documented in several browse species (Danell et al. 1990; Rousi et al. 1997; Roche and Fritz 1997; Mutikainen et al. 2000; Vourc'h et al. 2002; Laitinen et al. 2002). Lignification is also under genetic control, and considerable differences in lignin concentration and composition, and hence digestibility, can be found amongst browse species, and even amongst genotypes within species (Moore and Jung 2001). Second, palatability of plants can be altered as environmental variation in plant resource availability affects defence chemical expression (Coley et al. 1985; Moore and Jung 2001; Vourc'h et al. 2002; Vila et al. 2002; Laitinen et al. 2002). Third, ontogeny causes variations in plant defences as stage-specific constraints and selections influence a plant at different stages in its life cycle (Kearsley and Whitham 1989; Karban and Thaler 1999; Laitinen et al. 2000; Swihart and Bryant 2001; Moore and Jung 2001; Scogings et al. 2004). Finally, the browsing history of a plant can influence its defence levels; both increases (Chapin et al. 1985; Stolter et al. 2005), decreases (Karban and Baldwin 1997; Agrawal and Karban 1999; Vourc'h et al. 2002), and no difference (Vila et al. 2002) in nutritional quality have been documented in species subjected to browsing.

All of these mechanisms contribute to variation in browse quality for herbivores across multiple hierarchical levels. Species-level variation in important compounds has been well documented (Ricklefs and Matthew 1982; Hatcher 1990; Kamupingene et al. 2004; Scogings et al. 2004), but additional important variation exists at lower hierarchical levels. Individual trees within a species can vary greatly in concentrations of nutrients and secondary metabolites related to palatability and deterrency of grazing (Suomela and Ayres 1994). Strong variability in concentrations of terpenes and phenolic compounds within individual eucalyptus trees has been demonstrated (Lawler et al. 1998), and variation in leaf quality (total phenolics) was best explained at the individual tree level for mountain birch (*Betula pubescens tortuosa*; Suomela et al. 1995). Individuals of two tropical deciduous forest tree species (*Spondias mombin* and *Bursera simaruba*) varied significantly in palatability and leaf characteristics (Howard 1990). Variation in phenolic and monoterpenoid compounds amongst individuals of several *Artemesia* spp was found to be large (Welch and McArthur 1981; White et al. 1982; Cluff et al. 1982; Zhang and States 1991; Gershenzon and Croteau 1991; Wilt et al. 1992).

At still lower levels, extensive within-tree variation is thought to arise from the modular organisation of trees into hierarchical levels that are more or less physiologically autonomous and can even differ genetically due to somatic mutations (Suomela and Ayres 1994). In mountain birch up to 44% of the variance in water content, specific weight, and toughness was explained at the ramet, branch, and shoot level (Suomela and Ayres 1994). Leaf quality (secondary chemistry, nutrient availability, and moisture content) within tropical tree species has been found to vary within individuals (Howard 1990). Leaf nitrogen content has been found to be relatively invariant within individual trees, suggesting that the processes of nitrogen allocation are relatively well integrated on a whole plant basis (Suomela and Ayres 1994). Woody shrubs show similar trends of within-plant variation. Significant variation at the sub-individual level is not uncommon in browse species; variation in leaf quality, measured as the content of carbohydrates and amino acids, was generally larger amongst branches within a tree, than variation between trees in mountain birch (Suomela et al. 1995). Indeed, different genotypes have highly individualistic responses to environmental variation in terms of defence against herbivores (Rousi et al. 1993; Mutikainen et al. 2000; Laitinen et al. 2000, 2002). Up to 70% greater resistance to mammalian herbivores was found at the intra-clonal level of *B pendula* from a restricted area in south-central Finland (Mutikainen et al. 2000). As a result, the resistance to herbivory of an individual tree in relation to other trees in the population may be dependent on the specific conditions of a particular year (Laitinen et al. 2000); this presents the browsing herbivore with an incredible array of variation in its feeding choices. However, at least one study has shown that the degree of resistance to mammalian herbivores in one tree species (*Betula pubescens*) was influenced more by genotype than environmental variation (artificial fertilisation) (Mutikainen et al. 2000). Not only does the chemical composition vary dramatically within individuals of a browse species, leaf biomass can also vary significantly at different heights within a tree canopy (Woolnough and du Toit 2001; Nordengren et al. 2003). Indeed the vertical distribution of nitrogen, fibre, and defensive compounds varied substantially within trees of birch (*B pubescens*) and willow (*Salix* spp; Nordengren et al. 2003).

Several studies have suggested that the temporal component of variation in quality within browse plants may be less than the spatial component. Studies on mountain birch in Finland have found qualitative variation within trees, ramets, and branches to be quite constant from year to year (Suomela and Ayres 1994; Suomela et al. 1995). This constancy may reflect the presence of sources (e.g., mature leaves) and sinks (e.g., growing meristems and reproductive structures) that remain relatively stationary throughout the tree, or persistent microclimatic effects such as shading by neighbouring trees. This temporal qualitative consistency may provide the browsing herbivore with an opportunity to learn about its foraging environment, and exploit variable resources more efficiently by returning to previously browsed patches in successive years. It has been suggested that browsing can increase the quality of trees and shrubs in subsequent years for browsing herbivores. Repetitive browsing from year to year has been demonstrated in several studies; moose foraging on Scots Pine in Sweden and Finland (Löyttyniemi 1985; Faber and Lavsund 1999), moose browsing on birch in Sweden (Danell et al. 1985), moose browsing on willow in Alaska (Bowyer and Bowyer 1997), and red deer and roe deer browsing on Sitka spruce trees in Scotland (Welch et al. 1991). These studies suggest that initial browsing resulted in browse of a higher quality making them more palatable increasing the likelihood for re-browsing. Few studies have examined quantitative changes in browse quality after browsing, though one study on willow leaves found an increase in leaf nitrogen concentrations in those leaves severely browsed by moose in the previous winter (Stolter et al. 2005). This change in quality may well promote the perception of individual trees, or portions of individual trees as patches, representing a functionally relevant unit that can be manipulated through browsing to maximise foraging efficiency.

5.6 Perception of Patches

Hierarchical models of patchiness in environments provide us with a framework for functionally linking the physiology and behaviour of foraging animals to the spatial patterns of their food resources (Senft et al. 1987; Kotliar and Wiens 1990;

Bailey et al. 1996; Hobbs 1999). The smallest scale at which an animal can respond to patchiness is called the grain (Kotliar and Wiens 1990). This effectively represents the lower limit at which selection by the foraging animal can occur; below this limit the animal is simply not able to respond to any patchiness in the environment. The differing growth forms and spatial distributions of grasses and browse promote important differences in the most effective ways a foraging animal can exploit these resources. This is because the grain for an animal grazing a grass sward will not be the same as that of an animal browsing a clump of willow. The lower spatial limit at which selection by the animal remains profitable will differ, and as a consequence the perception of patches by grazers and browsers will likely to be quite different.

Because the architecture and dispersion of forage resources of grazers exhibits a more coarse-grained pattern in natural landscapes, grazers are more likely to make food choices on a larger spatial scale than do browsers. In grasslands, bite mass, and thus intake rate, is predominantly constrained by the height and bulk density of the grass canopy. The same is true for browsing herbivores, where the arrangement of leaves on a branch determines the potential range of bite sizes available to the animal. Because grasses tend to occur in contiguous swards, it may not pay a grazing herbivore to invest in behaviours that allow it to discriminate on very fine scales, such as the bite. However, for the browsing herbivore, the structural arrangement of leaves within a single tree can present considerable complexity. For instance, some species such as willow allow animals to strip many leaves in one bite, but in species where stems branch at wide angles bites are effectively separated in space, and so individual groups of leaves effectively represent separate decisions for the foraging animal. As such it may pay the browsing herbivore to invest in behaviours that allow for discrimination on fine scales that are more appropriate for its fine-grained foraging environment.

In addition, grazers focus more on nutrient and mineral concentrations as the main currency for forage selection (McNaughton 1988; Seagle and McNaughton 1992; Papachristou and Nastis 1993), which vary most among plants and swards, whereas browsers focus more on secondary chemicals (Provenza and Malecheck 1984;; Provenza et al. 1990; Palo et al. 1992; Papachristou and Nastis 1993; Jia et al. 1995; Lawler et al. 1998; Stolter et al. 2005; but see Shipley et al. 1998), which vary most within plants. Thus browsers and grazers may respond to fundamentally different currencies while foraging, each of which has a different spatial scale of variation in the environment. As a consequence, we expect that grazers perceive their environment on a larger spatial scale than browsers, and that this will be reflected in the evidence for their patch perceptions. Defining patches for grazers and browsers requires considering these differences in the spatial heterogeneity of food resources; a practice rendered difficult because there is no innate way to know the scale at which an animal perceives its environment. However, several approaches have been developed whereby likely patch perceptions can be identified using analysis of observed intake rates and movement patterns of foraging herbivores.

5.6.1 Evidence for Patch Perceptions of Grazers and Browsers

Attempts to functionally define patches for herbivores consider how changes in plant characteristics and spatial distribution interact with intake rate. Competing models describing how intake rate might be influenced by plant characteristics, such as biomass and density, can be confronted with data to identify the mechanics most likely responsible for observed behaviour. Hobbs et al. (2003) used this approach to analyse the short-term foraging behaviour of a range of herbivores feeding in artificial patches of alfalfa. They found that intake rates were best described using a threshold model that utilises a distance d^*, which identifies the plant spacing at which intake rate switches from being regulated by food processing (bite mass) to regulation by food encounter (plant density). This distance d^* thus represents a threshold for the functional expression of heterogeneity (Hobbs et al. 2003). At distances below d^* the foraging herbivore will be insensitive to changes to heterogeneity in the spatial arrangement of plants because its intake rate is regulated by bite mass, not encounter rate of plants. Hobbs et al. (2003) predicted that d^* would scale positively with increasing body size in herbivores. However, Fortin (2006) examined the allometry of distance threshold d^* in several species of herbivores foraging in artificial patches of alfalfa, and found overall greater evidence for a negative relationship between d^* and body size. This suggests that larger herbivores should generally experience a decrease in intake rate at closer spacing of plants than smaller herbivores. The consequences of this finding are that small herbivores should be able to maintain close to maximal intake rates when foraging in environments where their food resource is patchily distributed (Fortin 2006). This outcome has considerable significance for patch perceptions of the two feeding guilds, given that grazing herbivores are generally of larger body size than browsing herbivores.

5.6.2 Patch Perceptions by Grazers

Grazing can create areas of relative homogeneity within swards, or larger grazing lawns of grasses of similar age, structure and physiognomy (Bakker et al. 1983; McNaughton 1984). Studies have demonstrated that grazing sustains grasses in an actively growing, highly nutritious state (Seagle et al. 1992). Grassland systems with a long evolutionary history of mammalian grazing tend to contain grasses with a high capacity for compensatory growth after defoliation (McNaughton 1979, 1983; McNaughton and Chapin 1985; Milchunas et al. 1988). Consequently, grazing herbivores can maintain small, highly productive areas within which grass is kept in a state of active growth with higher overall nutrient content (particularly nitrogen), digestibility, and biomass than surrounding areas (Misleavy et al. 1982; McNaughton 1983; 1984, 1988; Bakker et al. 1983). However, grazing is very rarely uniform and as such both natural and commercial grasslands are often a mosaic of short, heavily grazed patches and tall, lightly grazed patches (McNaughton 1984). For instance, ten years of sheep grazing in an initially uniform *Holcus lanatus*-dominated sward created a micro-pattern of heavily grazed patches alternating with taller, lightly grazed patches on a scale of 5–30 m (Bakker et al. 1983). This micro-pattern was more or less stable over time, probably maintained by a higher protein content (greater proportion of young leaves) in the patches subjected to repetitive sheep grazing (Bakker et al. 1983). The creation of patches within grasslands, and observations on the movement patterns of herbivores can be used to quantify the scale at which grazing herbivores perceive their environment.

Foraging patterns of elk (*Cervus elaphus*) have been examined using first passage times that describe the time an animal spends in an area of a given size. This approach identified movement scales of approximately 270 m while elk were foraging; elk tended to forage in areas having intermediate biomass and low movement costs (Frair et al. 2005). Similarly, elk in Yellowstone National Park were found to allocate more time to foraging in areas of greater biomass of grasses and forbs, and searched more intensively (creating larger craters in the snow) where food biomass was higher, consistent with patch use models (Fortin et al. 2005). Elk readily use all areas within small patches but tend to restrict their movements to smaller areas within larger patches when foraging (McIntyre and Wiens 1999). Cattle have been shown to respond to patches of forage. In one of the rare occasions of quantitative support for the Marginal Value Theorem (Charnov 1976), Laca et al. (1993) demonstrated that cattle grazing from grassland patches adjusted their patch residence times in accordance with predictions for different patch types and inter-patch distances. On a smaller spatial scale, cattle have further been shown to exhibit selectivity within patches for feeding stations (WallisDeVries et al. 1999). However, the number of bites per feeding station was not affected by patch size, suggesting that selection between and within feeding stations are essentially different processes (WallisDeVries et al. 1999). Once a feeding station had been chosen, no further selectivity was shown; the number of bites per feeding station was similar in different patch types (WallisDeVries et al. 1999). Selection therefore operated mainly in directing the foraging path and in selecting feeding stations (WallisDeVries et al. 1999).

More detailed studies suggest that the behaviour of large grazers at the level of feeding stations appears to be consistent with patch use models: an alternating pattern of a series of steps between feeding stations followed by grazing (a series of bites) within each feeding station (Laca et al. 1994). Sheep have been shown to exhibit no discrimination between individual plant species within feeding stations unless there were very strong preference differences between the composite plant species (Gordon et al., unpubl). Sheep simply consumed the different grass species in proportion to their abundance within each feeding station. Cattle have been shown to graze grass swards systematically, removing grass in upper horizons before biting from the lower horizon (Ungar et al. 1992; Laca et al. 1994; Ginnett et al. 1999). Both cattle (Wade et al. 1989; Ungar et al. 1991; Laca et al. 1992) and sheep (Milne et al. 1982; Burlison et al. 1991) have been shown to remove a constant proportion of the available sward height when grazing. However, unlike cattle and sheep, reindeer—a mixed feeder (*Rangifer tarandus*

tarandus)—took constant bite sizes irrespective of vascular plant biomass on tundra (Trudell and White 1981), and wood bison (*Bison bison athabascae*) took constant bite sizes irrespective of sward height in an experimentally manipulated *Carex* grassland (Bergman et al. 2000). The authors suggest this is because, unlike other Bovinae, bison rarely use their tongues to increase bite area on tall swards, so bite area tends to remain constant. This evidence suggests that large grazing herbivores are relatively unselective within feeding stations in grass swards, and that perception of heterogeneity in their environment likely occurs at the feeding station scale.

Selectivity at the sub-feeding station level seems, for grazers, to be limited to manipulating the level of depletion before moving on, especially in large-bodied herbivores that have a limited capacity to benefit from exploitation of fine-scale heterogeneity (WallisDeVries et al. 1999). WallisDeVries et al. (1999) demonstrated that the selectivity of cattle was significantly impaired in a fine-grained environment. They suggest that cattle were unable to adjust their movement patterns to a sufficient degree as to match the scale of patchiness in the fine-grained environment once grain size becomes sufficiently small—in this study approximating the body length of the animals—the cost of frequently turning to seek out the most profitable food becomes too high (Jiang and Hudson 1993; WallisDeVries et al. 1999). Elk grazing grassland patches have been shown to tend to leave feeding stations when their lateral neck angle reaches a critical point, suggesting a biokinetic mechanism for feeding station departure rules (Jiang and Hudson 1993).

5.6.3 Patch Perceptions by Browsers

Browsers may treat a portion of an individual tree or shrub as a patch, the entire tree or shrub as a patch, or alternatively, whole stands of trees could be considered as a patch. The clustering of leaves on browse species may prompt browsing herbivores to perceive individual trees as patches of food (Gordon 2003). Several studies have indicated that moose (*Alces alces*) perceive individual trees as patches. Åström et al. (1990) demonstrated that handling time per tree increased with increasing tree size for moose feeding on deciduous trees. Danell et al. (1991) found that moose did not consume from stands of trees in proportion to their availability, rather they disproportionately directed foraging towards more profitable tree types, indicating that selection again occurred on the scale of the tree. Furthermore, Edenius et al. (2002) were able to show that moose did not perceive aspen stands as discrete patches; random sites and aspen stands were utilized equally by moose in terms of overall use of forage. Instead, moose seemed to use individual aspen ramets in accordance with diet theory. Owen-Smith and Novellie (1982) found that kudus (*Tragelaphus strepsiceros*) foraging in South Africa tended to end step sequences in less than four steps, which approximately corresponds to the size of individual bushes; they observed that kudus commonly took 2 or 3 paces without leaving the bush upon which they had been feeding. This suggests that kudus were exploiting their foraging environment treating individual browse bushes as a patch unit. A further patch size of approximately 18 paces was identified, corresponding to a patch radius of about 75m (Owen-Smith and Novellie 1982). The authors suggest that this area may correspond to the 'field of view' over which a kudo can see ahead and choose its direction of movement to the next feeding station. Movements over greater distances probably involved search for the next group of woody plants or cluster of forbs, rather than for individual food items (Owen-Smith and Novellie 1982).

Browsers also make diet choices at a smaller scale, the bite or the grain of the patch hierarchy. Browsers respond to both food quality and availability by eating a higher quality diet and taking less from each plant at high food densities (Vivås and Sæther 1987; Andersen and Sæther 1992; Shipley and Spalinger 1995). This behaviour seems to be primarily controlled at the bite scale. Browsing roe deer (*Capreolus capreolus*) have been shown to exhibit a high degree of selectivity at the bite scale whilst feeding from patches of browse species (Illius et al. 2002). Animals took larger bites from larger patches (branches)—similar to grazers on taller swards—but bite mass declined continuously as patch exploitation progressed, implying that animals were selecting larger items to eat first, prompting the authors to reject optimal patch use as the mechanism for exploitation of food in favour of diet optimization—a trade-off between diet quality and quantity (Illius et al. 2002). Moose (*Alces alces*) have also been shown to select larger bites on their first visit to Scots pine (*Pinus sylvestris*) and aspen (*Populus tremula*; Edenius 1991). Other studies have shown that moose tend to consume birch twigs of greater diameter as food abundance declines, with a higher proportion of the intake taken from nutritious 'top twigs' in high density plots (Vivås and Sæther 1987; Shipley and Spalinger 1995). If browsers were responding to variation in their food resources at the patch scale, the near linearity of gain functions reported for several browsers (Åström et al. 1990; Shipley and Spalinger 1995; Illius et al. 2002) would predict that optimal patch use should involve nearly complete defoliation of browse patches. Yet many studies have proven otherwise, demonstrating low amounts of biomass removal by browsing ungulates. In contrast great success in predicting diet optimization at the bite scale has been demonstrated. Shipley et al. (1999) developed a general model that predicted optimal bite size for browsing herbivores, defined as the bite size that results in the greatest daily net energy intake, based on constraints in harvesting and digesting foods. Tests of this model with moose, red deer (*Cervus elaphus*) and roe deer provided good evidence that browsing herbivores selected bite sizes in relation to the chemistry and morphology of plants, animal body size, and digestive strategy (Shipley et al. 1999). An additional model predicting optimal bite size for moose browsing on birch in Norway based on maximization of net energetic gain, incorporating intake rate and digestibility, also resulted in accurate predictions (Vivås et al. 1991). This suggests that browsers exploit variation in their food resource, primarily through selectivity at the bite scale.

5.7 Summary

All herbivores are faced with meeting their energy requirements from a food source that provides relatively little digestible energy and protein, requires extensive time to harvest and digest, and is patchily-distributed over landscapes. As a consequence, the foraging behaviours of grazers and browsers have many elements in common. The shape of the functional response for both grazers and browsers responds to plant and animal characteristics, and determines how much time animals have to spend to acquire nutrients, and how much time remains for other life-sustaining and fitness-enhancing activities, such as predator evasion and rearing of young. There are commonalities in some of the mechanistic components of the functional response for grazers and browsers; cropping time and chewing investment increase with bite size for both grazers and browsers because of the increase in structural tissues associated with larger bites. Grasses and browses each have structural components that reduce bite size and chewing rates, and grazers and browsers both must search for and travel to food resources. The different architecture of grasses and browses, especially the more spatially diffuse organisation of browse plants, and how that architecture influences bite size forms the most important difference in functional response between grazers and browsers.

Herbivores interact reciprocally with the distribution of their forage resources, perceiving and responding to variation in plant quality and quantity across multiple spatial scales. However, structural and chemical differences in grasses and browses have required grazers and browsers to adapt morphologies and behaviours that allow them to best acquire nutrients from these different plant types. In response, grasses and browses have co-evolved mechanisms for reducing the effects of herbivory on their fitness. These differences can have striking consequences for the organisation of herbivores across landscapes. For instance, most species of herbivores that aggregate into large herds are grazers rather than browsers (Fryxell and Sinclair 1988), and this may be because grasses tend to be of particularly poor quality at maturation (Owen-Smith 1982) so the benefits of aggregating and maintaining swards of forage in an early maturational stage are high (Fryxell 1991). The combination of a relatively contiguous distribution of grasses, and a tendency for variation in quality to occur between swards of grasses, encourages grazing herbivores to interact with their forage resource on a comparable spatial scale, utilising their environment to make use of small patches and feeding stations as the basic unit of exploitation. In contrast, the more diffuse distribution of browse plants, and the tendency for variation in quality to occur within individual trees and shrubs, promotes a more fine-scale interaction between browsers and their forage resources such that browsers utilise their environment to make use of portions of individual plants and bites as the basic unit of exploitation. There is mounting evidence that the application of patch optimisation models to browsers may be misplaced. Gain functions for browsing herbivores are often linear or piecewise-linear, and the majority of studies have failed to support predictions based on patch optimisation models. Standard applications of patch models for browsing herbivores are in need of revision; energy intake, time constraints, predation avoidance, social behaviours, and diet optimisation incorporating digestive constraints should all be included in future research directions.

The differences in foraging strategies we have outlined here are manifested across the life histories of these two ungulate guilds, having important consequences for social and mating systems, production and care of young, and distribution and migration patterns. As such, seemingly minor differences in forage and foraging tactics between grazers and browsers can lead to profound contrasts at higher ecological levels.

References

Abrams PA, Schmitz OJ (1999) The effect of risk of mortality on the foraging behaviour of animals faced with time and digestive capacity constraints. Evol Ecol Res 1:285–301

Agrawal A, Karban R (1999) Why induced defenses may be favoured over constitutive strategies in plants. In: Tollrian R, Harvell CD (eds) The ecology and evolution of inducible defenses. Princeton University Press, Princeton, NJ, pp 45–61

Andersen R, Sæther BE (1992) Functional response during winter of a herbivore, the moose, in relation to age and size. Ecology 73:542–550

Anderson GD, Talbot LM (1965) Soil factors affecting the distribution of the grassland types and their utilization by wild animals on the Serengeti plains, Tanganyika. J Ecol 53:33–56

Arnold GW (1987) Influence of the biomass, botanical composition and sward height of annual pastures on foraging behaviour by sheep. J Appl Ecol 24:759–772

Åström M, Lundberg P, Danell K (1990) Partial prey consumption by browsers - trees as patches. J Anim Ecol 59:287–300

Augustine DJ (2003) Spatial heterogeneity in the herbaceous layer of a semi-arid savanna ecosystem. Plant Ecol 167:319–332

Bailey DW, Rittenhouse LR, (1989) Management of cattle distribution. Rangelands 11:159–161

Bailey DW, Rittenhouse LR, Hart RH, Richards RW (1989) Characteristics of spatial memory in cattle. Appl Anim Behav Sci 23:331–340

Bailey DW, Gross JE, Laca EA, Rittenhouse LR, Coughenour MB, Swift DM, Sims PL (1996) Mechanisms that result in large herbivore grazing distribution patterns. J Range Manage 49:386–400

Bailey DW, Howery LD, Boss DL (2000) Effects of social facilitation for locating feeding sites by cattle in an eight-arm radial maze. Appl Anim Behav Sci 68:93–105

Bakker JP, de Leeuw J, van Wieren S (1983) Micropatterns in grassland vegetation created and sustained by sheep grazing. Vegetatio 55:153–161

Balch CC (1971) Proposal to use time spent chewing as an index of the extent to which diets for ruminants possess the physical property of fibrousness characteristic of roughages. Brit J Nutr 26:383–392

Beckerman AP (2005) The shape of things eaten: the functional response of herbivores foraging adaptively. Oikos 110:591–601

Beecham JA, Farnsworth KD (1999) Animal group forces resulting from predator avoidance and competition minimization. J Theor Biol 198:533–548

Bell RHV (1982) The effect of soil nutrient availability on community structure in African ecosystems. In: Huntley BJ, Bakker BH (eds) Ecology of tropical savannas. Springer, Berlin Heidelberg New York, pp 193–216

Belovsky GE, Schmitz OJ, Slade JB, Dawson TJ (1991) Effects of spines and thorns on Australian arid zone herbivores of different body masses. Oecologia 88:521–528

Belsky AJ (1994) Influences of trees on savanna productivity - tests of shade, nutrients, and tree–grass competition. Ecology 75:922–932

Belsky AJ, Amundson RG, Duxbury JM, Riha SJ, Ali AR, Mwonga SM (1989) The effects of trees on their physical, chemical, and biological environments in a semi-arid savanna in Kenya. J Appl Ecol 26:1005–1024

Belsky AJ, Mwonga SM, Amundson RG, Duxbury JM, Ali AR (1993) Comparative effects of isolated trees on their undercanopy environments in high-rainfall and low-rainfall savannas. J Appl Ecol 30:143–155

Bergman CM, Fryxell JM, Gates CG (2000) The effect of tissue complexity and sward height on the functional response of Wood Bison. Funct Ecol 14:61–69

Bergman CM, Fryxell JM, Gates CG, Fortin D (2001) Ungulate foraging strategies: energy maximizing or time minimizing? J Anim Ecol 70:289–300

Bernhard-Reversat F (1982) Biogeochemical cycle of nitrogen in a semi-arid savanna Oikos 38:321–332

Black JL, Kenney PA (1984) Factors affecting diet selection by sheep: II. Height and density of pasture. Aust J Agr Res 35:551–563

Blough DS (1989) Contrast as seen in visual search reaction times. J Exp Anal Behav 52:199–211

Bodmer RE (1990) Ungulate frugivores and the browser–grazer continuum. Oikos 57:319–325

Bowyer JW, Bowyer RT (1997) Effects of previous browsing on the selection of willow stems by Alaskan moose. Alces 33:11–18

Brown JS (1988) Patch use as an indicator of habitat preference, predation risk, and competition. Behav Ecol Sociobiol 22:37–47

Brown JS (1999) Vigilance, patch use and habitat selection: foraging under predation risk Evol Ecol Res 1:49–71

Burlison AJ, Hodgson J, Illius AW (1991) Sward canopy structure and the bite dimensions and bite weight of grazing sheep. Grass Forage Sci 46:29–38

Cates RG, Rhoades CJ (1977) Patterns in the production of anti-herbivore chemical defences in plant communities. Biochem Syst Ecol 5:185–194

Chaneton EJ, Perelman SB, León RJC (2005) Floristic heterogeneity of flooding pampas grasslands: a multi-scale analysis. Plant Biosyst 139:245–254

Chapin III FS, Bryant JP, Fox JF (1985) Lack of induced chemical defense in juvenile Alaskan woody plants in response to simulated browsing. Oecologia 67:457–459

Charnov EL (1976) Optimal foraging, the Marginal Value Theorem. Theor Popul Biol 9:129–136

Choong MF, Lucas PW, Ong JSY, Pereira B, Tan HTW, Turner IM (1992) Leaf fracture toughness and sclerophyll: their correlations and ecological implications. New Phytol 121:597–610

Clauss M, Lechner-Doll M (2001) Difference in selective reticulo-omasal particle retention as a key factor in ruminant diversification Oecologia 129:321–327

Cluff LK, Welch BL, Pederson JC, Brotherson JD (1982) Concentration of monoterpenoids in the rumen ingesta of wild mule deer. J Range Manage 35:192–194

Coley PD, Bryant JP, Chapin III FS (1985) Resource availability and plant antiherbivore defense. Science 230:895–899

Cooper SM, Owen-Smith N (1986) Effects of plant spinescence on large mammalian herbivores. Oecologia 68:446–455

Cooper SM, Ginnett TF (1998) Spines protect plants against browsing by small climbing mammals. Oecologia 113:219–221

Danell K, Huss-Danell K, Bergström R (1985) Interactions between browsing moose and two species of birch in Sweden. Ecology 66:1867–1878

Danell K, Gref R, Yazdani R (1990) Effects of mono- and diterpenes in Scots Pine needles on moose browsing. Scand J Forest Res 5:535–539

Danell K, Edenius L, Lundberg P (1991) Herbivory and tree stand composition - moose patch use in winter. Ecology 72:1350–1357

Danell K, Bergstrom R, Iedenius L (1994) Effects of large mammalian browsers on architecture, biomass, and nutrients of woody-plants. J Mammal 75:833–844 de Reffye P, Houllier F (1997) Modelling plant growth and architecture: some recent advances and applications to agronomy and forestry. Curr Sci India 73:984–992

Demment MW, Van Soest PJ (1985) A nutritional explanation for body-size patterns of ruminant and nonruminant herbivores. Am Nat 125:641–672

Dimock II EJ, Silen RR, Allen VE (1976) Genetic resistance to Douglas-fir damage by snowshoe hare and black-tailed deer. Forest Sci 22:106–121

Dunham KM (1980) The feeding behaviour of a tame impala *Aepyros melampus*. Afr J Ecol 18:253–257

Edenius L (1991) The effect of resource depletion on the feeding behavior of a browser winter foraging by moose on Scots pine. J Appl Ecol 28:318–328

Edenius L, Ericsson G, Naslund P (2002) Selectivity by moose versus the spatial distribution of aspen: a natural experiment. Ecography 25:289–294

Eldridge DJ (1998) Trampling of microphytic crusts on calcareous soils, and its impact on erosion under rain-impacted flow. Catena 33:221–239

Faber WE, Lavsund S (1999) Summer foraging on Scots pine *Pinus sylvestris* by moose *Alces alces* in Sweden - patterns and mechanisms. Wildl Biol 5:93–106

Fortin D, Morales JM, Boyce MS (2005) Elk winter foraging at fine scale in Yellowstone National Park. Oecologia 145:335–343

Fortin, D (2006) The allometry of plant spacing that regulates food intake rate in mammalian herbivores. Ecology 87:1861–1866

Frair JL, Merrill EH, Visscher DR, Fortin E, Beyer HL, Morales JM (2005) Scales of movement by elk (*Cervus elaphus*) in response to heterogeneity in forage resources and predation risk. Landscape Ecol 20:273–287

Fryxell JM (1991) Forage quality and aggregation by large herbivores Am Nat 138:478–498

Fryxell J M, Sinclair ARE (1988) Causes and consequences of migration by large herbivores. Trends Ecol Evol 3:237–241

Gershenzon J, Croteau R (1991) Herbivores: their interactions with secondary plant metabolites. Academic Press, San Diego

Getty T, Pulliam HR (1991) Random prey detection with pause-travel search. Am Nat 138:1459–1477

Ginnett TF, Dankosky JA, Deo G, Demment MW (1999) Patch depression in grazers: the roles of biomass distribution and residual stems. Funct Ecol 13:37–44

Goddard J (1968) Food preferences of two black rhinoceros populations. E Afr Wildl J 6:1–18

Gordon IJ (2003) Browsing and grazing ruminants: are they different beasts? Forest Ecol Manage 181:13–21

Gordon IJ, Illius AW (1988) Incisor arcade structure and diet selection in ruminants. Funct Ecol 2:15–22

Gordon IJ, Illius AW (1994) The functional significance of the browser–grazer dichotomy in African ruminants. Oecologia 98:167–175

Gowda JH (1996) Spines of *Acacia tortilis*: what do they defend and how? Oikos 77:279–284

Greene RSB, Kinnell PIA, Wood JT (1994) Role of plant cover and stock trampling on runoff and soil-erosion from semiarid wooded rangelands. Aust J Soil Res 32:953–973

Gross JE, Shipley LA, Hobbs NT, Spalinger DE, Wunder BA (1993) Functional response of herbivores in food-concentrated patches: tests of a mechanistic model. Ecology 74:778–791

Hacker JB, Minson DJ (1981) The digestibility of plant parts. Herbage Abstracts 51:459–482

Hatcher PE (1990) Seasonal and age-related variation in the needle quality of five conifer species. Oecologia 85:200–212

HilleRisLambers R, Rietkerk M, van den Bosch F, Prins HHT, de Kroon H (2001) Vegetation pattern formation in semi-arid grazing systems. Ecology 82:50–61

Hirakawa H (1997) Digestion-constrained optimal foraging in generalist mammalian herbivores. Oikos 78:37–47

Hobbs NT (1999) Responses of large herbivores to spatial heterogeneity in ecosystems. In: Jung HG, Fahey GC (eds) Nutritional ecology of herbivores: Proceedings of the Vth International Symposium on the nutrition of herbivores. Am Soc Anim Sci, Savoy IL, pp 97–129

Hobbs NT, Gross JE, Shipley LA, Spalinger DE, Wunder BA (2003) Herbivore functional response in heterogeneous environments: a contest among models. Ecology 84:666–681

Hoeck HN (1975) Differential feeding behaviour of sympatric *Hyrax procavia-johnstoni* and *Heterohyrax brucei*. Oecologia 22:15–47

Hoeck, H N (1989) Demography and competition in Hyrax - a 17 year study Oecologia 79:353–360

Hofmann RA (1989) Evolutionary steps of ecophysiological adaptation and diversification of ruminants: a comparative view of their digestive system. Oecologia 78:443–457

Hofmann RA, Stewart DRM (1972) Grazer or browser: a classification based on the stomach-structure and feeding habits of East African ruminants. Mammalia 36:226–240

Holling CS (1959) Some characteristics of simple types of predation and parasitism. Can Entomol 41:385–398

Holling CS (1965) The functional response of predators to prey density and its role in mimicry and population regulation. Mem Entomol Soc Can 48:1–46

Howard JJ (1990) Infidelity of leaf-cutting ants to host plants: resource heterogeneity or defense induction? Oecologia 82:394–401

Howery LD, Provenza FD, Banner PE, Scott CB (1998) Social and environmental factors influence cattle distribution on rangeland. Appl Anim Behav Sci 55:231–244

Howery LD, Bailey DW, Ryle CB, Renken WJ (2000) Cattle use visual cues to track food locations. Appl Anim Behav Sci 67:1–14

Hume ID (1999) Marsupial nutrition. Cambridge University Press, Cambridge

Iason GR, Mantecon AR, Sim DA, Gonzalez J, Foreman E, Bermudez FF, Elston DA (1999) Can grazing sheep compensate for a daily foraging time constraint? J Anim Ecol 68:87–93

Illius AW, Gordon IJ (1987) The allometry of food intake in grazing ruminants. J Anim Ecol 56:989–1000

Illius AW, Gordon IJ (1992) Modelling the nutritional ecology of ungulate herbivores evolution of body size and competitive interactions. Oecologia 89:428–434

Illius AW, Duncan P, Richard C, Mesochina P (2002) Mechanisms of functional response and resource exploitation in browsing roe deer. J Anim Ecol 71:723–734

Janis CM, Ehrhardt D (1988) Correlation of relative muzzle width and relative incisor width with dietary preference in ungulates. Zool J Linn Soc–Lond 92:267–284

Jarman PJ (1974) The social organization of antelope in relation to their ecology. Behaviour 48:215–266

Jia J, Niemela P, Danell K (1995) Moose *Alces alces* bite diameter selection in relation to twig quality on four phenotypes of Scots pine *Pinus sylvestris*. Wildlife Biol 1:47–55

Jiang Z, Hudson RJ (1993) Optimal grazing of wapiti *cervus elaphus* on grassland patch and feeding station departure rules. Evol Ecol 7:488–498

Kamupingene GT, Abate AL, Kimambo AE (2004) Crude protein degradability, fibre and tannin levels of browse forages in an extensive farming system. J Anim Feed Sci 13:111–114

Karban R, Baldwin IT (1997) Induced responses to herbivory. University of Chicago Press, Chicago

Karban R, Thaler JS (1999) Plant phase change and resistance to herbivory. Ecology 80:510–517

Kearsley MJC, Whitham TG (1989) Developmental changes in resistance to herbivory: implications for individuals and populations. Ecology 70:442–444

Kellman M (1979) Soil enrichment by neotropical savanna trees. J Ecol 67:565–577

Kotliar NB, Wiens JA (1990) Multiple scales of patchiness and patch structure - a hierarchical framework for the study of heterogeneity. Oikos 59:253–260

Krebs JR, Davies MJ (1993) An introduction to behavioural ecology. Blackwell, Oxford

Kumar L, Rietkerk M, van Langevelde F, van de Koppel J, van Andel J, Hearne J, de Ridder N, Stroosnijder L, Skidmore AK, Prins HHT (2002) Relationship between vegetation growth rates at the onset of the wet season and soil type in the Sahel of Burkina Faso. Ecol Model 149:143–152

Laca EA (1998) Spatial memory and food searching mechanisms of cattle. J Range Manage 51:370–378

Laca EA, Ortega IM (1995) Integrating foraging mechanisms across spatial and temporal scales. In: West NE (ed) Rangelands in a sustainable biosphere. Society for Range Management, Denver, pp 129–132

Laca EA, Demment MW (1996) Foraging strategies of grazing animals. In: Hodgson J, Illius AW (eds) The ecology and management of grazing systems. CABI, Wallingford, UK, pp 137–158

Laca EA, Ortega IM (1996) Integrating foraging mechanisms across spatial and temporal scales. In: West NE (ed) Vth International Rangeland Congress Society of Range Management, Salt Lake City, pp 129–132

Laca EA, Ungar ED, Seligman N, Demment MW (1992) Effects of sward height and bulk density on bite dimensions of cattle grazing homogeneous swards. Grass Forage Sci 47:91–102

Laca EA, Distel RA, Griggs TC, Deo GP, Demment MW (1993) Field test of optimal foraging with cattle: the marginal value theorem successfully predicts patch selection and utilisation. In: Proceedings of XVII International Grassland Congress, New Zealand and Queensland, February, pp 709–701

Laca EA, Distel RA, Griggs TC, Demment MW (1994) Effects of canopy structure on patch depletion by grazers. Ecology 75:706–716

Laitinen M, Julkunen-Tiitto R, Rousi M (2000) Variation in phenolic compounds within a birch (*Betula pendula*) population. J Chem Ecol 26:1609–1622

Laitinen J, Rousi M, Tahvanainen J (2002) Growth and hare, *Lepus timidus*, resistance of white birch, *Betula pendula*, clones grown in different soil types. Oikos 99:37–46

Launchbaugh KL, Provenza FD, Pfister JA (2001) Herbivore response to anti-quality factors in forages. J Range Manage 54:431–440

Lawler IR, Foley WJ, Eschler BM, Pass DM, Handasyde K (1998) Intraspecific variation in Eucalyptus secondary metabolites determines food intake by folivorous marsupials. Oecologia 116:160–169

Lazo A (1994) Social segregation and the maintenance of social stability in a feral cattle population. Anim Behav 48:1133–1141

Löyttyniemi K (1985) On repeated browsing of Scots Pine saplings by moose (*Alces alces*). Silva Fenn 19:387–391

Ludwig JA, Eager RW, Williams RJ, Lowe LM (1999) Declines in vegetation patches, plant diversity, and grasshopper diversity near cattle watering-points in Victoria River District, northern Australia. Rangeland J 21:135–149

Ludwig JA, Wiens JA, Tongway DJ (2000) A scaling rule for landscape patches and how it applies to conserving soil resources in savannas. Ecosystems 3:84–97

Ludwig F, de Kroon H, Prins HHT, Berendse F (2001) Effects of nutrients and shade on tree–grass interactions in an East African savanna. J Veg Sci 12:579–588

Ludwig F, de Kroon H, Berendse F, Prins HHT (2004) The influence of savanna trees on nutrient, water and light availability and the understorey vegetation. Plant Ecol 170:93–105

Ludwig JA, Wilcox BP, Breshears DD, Tongway D, Imeson AC (2005) Vegetation patches and runoff–erosion as interacting ecohydrological processes in semiarid landscapes. Ecology 86:288–297

MacArthur RH, Pianka ER (1966) On optimal use of a patchy environment. Am Nat 100:603–609

Malachek JC, Balph DF (1987) Diet selection by grazing and browsing livestock. In: Hacker JB, Ternouth JH (eds) The nutrition of herbivores. Academic Press, Sydney, pp 121–132

McIntyre NE, Wiens JA (1999) Interactions between landscape structure and animal behavior: the roles of heterogeneously distributed resources and food deprivation on movement patterns. Landscape Ecol 14:437–447

Mcivor JG, Williams J, Gardener CJ (1995) Pasture management influences runoff and soil movement in the semi-arid tropics. Aust J Exp Agr 35:55–65

Mcivor JG, McIntyre S, Saeli I, Hodgkinson JJ (2005) Patch dynamics in grazed subtropical native pastures in south-east Queensland. Austral Ecol 30:445–464

McLeod MN, Kennedy PM, Minson DJ (1990) Resistance of leaf and stem fractions of tropical forage to chewing and passage in cattle. Brit J Nutr 63:105–119

McNaughton SJ (1979) Grazing as an optimization process: grass–ungulate relationships in the Serengeti. Am Nat 113:691–703

McNaughton SJ (1983) Serengeti grassland ecology: the role of composite environmental factors and contingency in community structure. Ecol Monogr 53:291–320

McNaughton SJ (1984) Grazing lawns: animals in herds, plant form, and coevolution. Am Nat 124:863–866

McNaughton SJ (1988) Mineral nutrition and spatial concentrations of African ungulates. Nature 334:343–345

McNaughton SJ (1989) Interactions of plants of the field layer with large herbivores. Symposium Zool Soc Lond 61:15–29

McNaughton SJ (1990) Mineral nutrition and seasonal movements of African migratory ungulates. Nature 345:613–615

McNaughton SJ, Chapin IFS (1985) Effects of phosphorus nutrition and defoliation on C4 graminoids from the Serengeti Plains. Ecology 66:1617–1629

McNaughton SJ, Georgiadis NJ (1986) Ecology of African grazing and browsing mammals. Annu Rev Ecol Syst 17:39–66

Milchunas DG, Sala OE, Lauenroth WK (1988) A generalized model of the effects of grazing by large herbivores on grassland community structure. Am Nat 132:87–106

Milewski AV, Young TP, Madden D (1991) Thorns as induced defenses - experimental evidence. Oecologia 86:70–75

Milne BT (1991) Heterogeneity as a multiscale characteristic of landscapes. In: Kolasa J, Pickett STA (eds) Ecological heterogeneity. Springer, Berlin Heidelberg New York, pp 68–84

Milne JA, Hodgson J, Thompson R, Souter WG (1982) The diet ingested by sheep grazing swards differing in white clover and perennial ryegrass content. Grass Forage Sci 37:209–218

Misleavy P, Mott GO, Martin FG (1982) Effect of grazing frequency on forage quality and stolon characteristics of tropical perennial grasses. Soil Crop Sci Soc Fl 41:77–83

Moore KJ, Jung HG (2001) Lignin and fiber digestion. J Range Manage 54:420–430

Mordelet P, Menaut JC (1995) Influence of trees on aboveground production dynamics of grasses in a humid savanna. J Veg Sci 6:223–228

Murray MG (1991) Maximizing energy retention in grazing ruminants. J Anim Ecol 60:1029–1045

Murray MG, Illius AW (2000) Vegetation modification and resource competition in grazing ungulates. Oikos 89:501–508

Mutanga O, Prins HHT, Skidmore AK, van Wieren S, Huizing H, Grant R, Peel M, Biggs H (2004) Explaining grass–nutrient patterns in a savanna rangeland of southern Africa. J Biogeogr 31:819–829

Mutikainen P, Walls M, Ovaska J, Keinänen M, Julkunen-Tiitto R, Vapaavuori E (2000) Herbivore resistance in *Betula pendula*: effect of fertilization, defoliation and plant genotype. Ecology 81:49–65

Myers JH, Bazeley DR (1992) Thorns, spines, prickles and hairs: are they stimulated by herbivory and do they deter herbivores? In: Tallamy DW, Raupp NJ (eds) Phytochemical induction by herbivores. Wiley, New York, pp 325–344

Nelson CJ, Moser LE (1994) Plant factors affecting forage quality. in Fahey Jr GC (ed) Forage quality, evaluation and utilization. American Society of Agronomy, Crop Science Society of America, and Soil Science Society, Madison, WI, pp 115–154

Nordengren C, Hofgaard A, Ball JP (2003) Availability and quality of herbivore winter browse in relation to tree height and snow depth. Ann Zool Fenn 40:305–314

Northup BK, Brown JR, Holt JA (1999) Grazing impacts on the spatial distribution of soil microbial biomass around tussock grasses in a tropical grassland. Appl Soil Ecol 13:259–270

Northup BK, Dias CD, Brown JR, Skelly WC (2005) Micro-patch and community scale spatial distribution of herbaceous cover in a grazed eucalypt woodland. J Arid Environ 60:509–530

O'Neill RV, Deangelis DL, Waide JB, Allen TFH (1986) A hierarchical concept of ecosystems. Princeton University Press, Princeton, NJ

Olff H, Ritchie ME (1998) Effects of herbivores on grassland plant diversity Trends Ecol Evol 13:261–265
Owen-Smith N (1982) Factors influencing the consumption of plant products by large herbivores. In: Huntley BJ, Walker BH (eds) Ecology of tropical savannas. Springer, Berlin Heidelberg New York, pp 359–404
Owen-Smith N (1994) Foraging responses of kudus to seasonal changes in food resources: elasticity in constraints. Ecology 75:1050–1062
Owen-Smith N (1997) Control of energy balance by a wild ungulate, the kudu (*Tragelopahus strepsiceros*) through adaptive foraging behaviour. P Nutr Soc 56:15–24
Owen-Smith N, Novellie P (1982) What should a clever ungulate eat? Am Nat 119:151–178
Palamara J, Phakey PP, Rachinger WA, Sanson GD, Oranus HJ (1984) On the nature of the opaque and translucent enamel regions of some Macropodinae (*Macropus giganteus*, *Wallabia bicolor* and *Peradorcus concinna*). Cell Tissue Res 238:329–337
Palo RT, Bergstrom R, Danell K (1992) Digestibility distribution of phenols and fiber at different twig diameters of birch in winter: implications for browsers Oikos 65:450–454
Papachristou TG, Nastis AS (1993) Factors affecting forage preference by goats grazing kermes oak shrubland in northern Greece. In: Papanastasis V, Nikolaidis A (eds) Management of Mediterranean shrublands and related forage resources. Seventh meeting, FAO European sub-network on Mediterranean pastures and fodder crops. Chania, Greece, pp 167–170
Pellew RA (1984) Food consumption and energy budgets of the giraffe. J Appl Ecol 21:141–159
Pérez-Barbería FJ, Gordon IJ (2001) Relationships between oral morphology and feeding style in the Ungulata: a phylogenetically controlled evaluation. P Roy Soc Lond B Bio 268:1023–1032
Provenza FD (1995) Postingestive feedback as an elementary determinant of food preference and intake in ruminants. J Range Manage 48:2–17
Provenza FD, Malecheck JC (1984) Diet selection by domestic goats in relation to blackbrush twig chemistry. J Appl Ecol 21:831–841
Provenza FD, Burrit EA, Clausen TP, Bryant JP, Reichardt PB, Distel RA (1990) Conditioned flavour aversion: a mechanism for goats to avoid tannins in blackbrush. Am Nat 136:810–828
Reid ED (2000) Differential expression of herbivore resistance in late- and mid-seral grasses of the tall grass prairie. Dissertation, University of Idaho, Moscow, ID
Ricklefs RE, Matthew KK (1982) Chemical characteristics of the foliage of some deciduous trees in southeastern Ontario. Can J Botany 60:2037–2045
Riet-Correa F, Mendez MC, Schild AL, Oliveira JA, Zenebon O(1986) Dental lesions in cattle and sheep due to industrial pollution caused by coal combustion. Pesquisa Vet Brasil 6:23–31
Rietkerk M, Ouedraogo T, Kumar L, Sanou S, van Langevelde F, Kiema A, van de Koppel J, van Andel J, Hearne J, Skidmore AK, de Ridder N, Stroosnijder L, Prins HHT (2002) Fine-scale spatial distribution of plants and resources on a sandy soil in the Sahel. Plant Soil 239:69–77
Rietkerk M, Dekker SC, de Ruiter PC, van de Koppel J (2004) Self-organized patchiness and catastrophic shifts in ecosystems. Science 305:1926–1929
Robbins CT (1983) Wildlife feeding and nutrition. Academic Press, London
Roberts BR (1987) The availability of herbage. In: Hacker JB, Ternouth JH (eds) The nutrition of herbivores. Academic Press, London, pp 47–63
Roche MS, Fritz RS (1997) Genetics of resistance of *Salix sericea* to a diverse community of herbivores. Evolution 51:1490–1498
Roese JH, Risenhoover KL, Folse LF (1991) Habitat heterogeneity and foraging efficiency: an individual-based foraging model. Ecol Monogr 57:133–143
Rominger EM (1995) Late winter foraging ecology of woodland caribou. Dissertation, Washington State University, Pullman
Rominger EM, Robbins CT, Evans MA (1996) Winter foraging ecology of woodland caribou in northeastern Washington. J Wildlife Manage 60:719–728
Rousi M, Tahvanainen J, Henttonen H, Uotila I (1993) Effects of shading and fertilization on resistance of winter-dormant birch (*Betula pendula*) to voles. Ecology 74:30–38

Rousi M, Tahvanainen J, Henttonen H (1997) Clonal variation in susceptibility of white birches (*Betula* spp.) to mammalian and insect herbivores. Forest Sci 43:396–402

Sanson GD (1980) The morphology and occlusion of the malariform cheek teeth in some Macropodinae (Marsupialia:Macropodidae). Aust J Ecol 28:341–365

Sanson GD (1989) Morphological adaptations of teeth to diets and feeding in the Macropodoidea. In: Grigg G, Jarman PJ, Hume ID (eds) Kangaroos, wallabies and rat-kangaroos. Surrey Beatty, Sydney, pp 151–168

Scogings PF, Dziba LE, Gordon IJ (2004) Leaf chemistry of woody plants in relation to season, canopy retention and goat browsing in a semiarid subtropical savanna. Austral Ecol 29:278–286

Seagle SW, McNaughton SJ (1992) Spatial variation in forage nutrient concentrations and the distribution of Serengeti grazing ungulates. Landscape Ecol 7:229–241

Seagle SW, McNaughton SJ, Ruess RW (1992) Simulated effects of grazing on soil nitrogen and mineralization in contrasting Serengeti grassslands. Ecology 73:1105–1123

Searle KR, Vandervelde T, Hobbs NT, Shipley LA (2005) Gain functions for large herbivores: tests of alternative models. J Anim Ecol 74:181–189

Senft RL, Coughenour MB, Bailey DW, Rittenhouse LR, Sala OE, Swift DM (1987) Large herbivore foraging and ecological hierarchies. Bioscience 37:789–799

Shipley LA, Spalinger DE (1992) Mechanics of browsing in dense food patches effects of plant and animal morphology on intake rate. Can J Zool 70:1743–1752

Shipley LA, Spalinger DE (1995) Influence of size and density of browse patches on intake rates and foraging decisions of young moose and white-tailed deer. Oecologia 104:112–121

Shipley LA, Gross JE, Spalinger DE, Hobbs NT, Wunder BA (1994) The scaling of intake rate in mammalian herbivores. Am Nat 143:1055–1082

Shipley LA, Blomquist S, Danell K (1998) Diet choices made by free-ranging moose in northern Sweden in relation to plant distribution, chemistry, and morphology. Can J Zool 76:1722–1733

Shipley LA, Illius AW, Danell K, Hobbs NT, Spalinger DE (1999) Predicting bite size selection of mammalian herbivores: a test of a general model of diet optimization Oikos 84:55–68

Shipley LA, Spalinger DE, Gross JE, Hobbs NT, Wunder BA (1996) The dynamics and scaling of foraging velocity and encounter rate in mammalian herbivores. Funct Ecol 10:234–244

Sih A (1980) Optimal foraging: partial consumption of prey. Am Nat 116:281–290

Silen RR, Randall WL, Mandel NL (1986) Estimates of genetic parameters for deer browsing of Douglas fir. Forest Sci 32:178–184

Skarpe C (1992) Dynamics of savanna ecosystems. J Veg Sci 3:293–300

Spalinger DE, Hobbs NT (1992a) Mechanisms of foraging in mammalian herbivores: new models of functional response. Am Nat 140:325–348

Spalinger DE, Hobbs NT (1992b) Herbivore functional response: a mechanistic model of food intake rate in patchy environments. B Ecol Soc Am 73:209

Speakman JR, Bryant DM (1993) The searching speeds of foraging shorebirds: redshank (*Tinga tetanus*) and oystercatcher (*Haematopus ostralegus*). Am Nat 142:296–319

Stapley L (1998) The interaction of thorns and symbiotic ants as an effective defense mechanism of swollen-thorn acacias. Oecologia 115:401–405

Stephens DW, Krebs JR (1986) Foraging theory. Princeton University Press, Princeton, NJ

Stolter C, Ball JP, Julkunen-Tiitto R, Lieberei R, Ganzhorn JU (2005) Winter browsing of moose on two different willow species: food selection in relation to plant chemistry and plant response. Can J Zool 83:807–819

Stuart-Hill GC, Tainton NM (1989) The competitive interaction between *Acacia karroo* and the herbaceous layer and how this is influenced by defoliation. J Appl Ecol 26:285–298

Suomela J, Ayres MP (1994) Within-tree and among-tree variation in leaf characteristics of mountain birch and its implications for herbivory. Oikos 70:212–222

Suomela J, Ossipov J, Haukioja E (1995) Variation among and within mountain birch trees in foliage phenols, carbohydrates, and amino acids, and in growth of *Epirrata autumnata* larvae. J Chem Ecol 21:1421–1446

Swihart RK, Bryant JP (2001) Importance of biogeography and ontogeny of woody plants in winter herbivory by mammals. J Mammal 81:1–21

Trudell J, White RG (1981) The effect of forage structure and availability on food intake, biting rate, bite size and daily eating time of reindeer. J Appl Ecol 18:63–81

Turner MD (1998) Long-term effects of daily grazing orbits on nutrient availability in Sahelian West Africa: I. Gradients in the chemical composition of rangeland soils and vegetation. J Biogeogr 25:669–682

Ungar ED, Genizi A, Demment MW (1991) Bite dimensions and herbage intake by cattle grazing short hand-constructed swards. Agron J 83:973–978

Ungar ED, Seligman NG, Demment MW (1992) Graphical analysis of sward depletion by grazing. J Appl Ecol 29:427–435

Van Soest PJ (1981) Nutritional ecology of the ruminant. Cornell University Press, Ithaca, NY

Vaughan TA, Ryan JM, Czaplewski NJ (2000) Mammalogy, 4th edn. Brooks Cole, Toronto

Vila B, Vourc'h G, Gillon D, Martin J, Guibal F (2002) Is escaping deer browse just a matter of time in *Picea sitchensis*? A chemical and dendroecological approach. Trees 16:488–496

Villalba JJ, Provenza FD (1999) Effects of food structure and nutritional quality and animal nutritional state on intake behaviour and food preferences of sheep. Appl Anim Behav Sci 63:145–163

Villalba JJ, Provenza FD (2000) Roles of novelty, generalization, and postingestive feedback in the recognition of foods by lambs. J Anim Sci 78:3060–3069

Villalba JJ, Provenza FD (2005) Foraging in chemically diverse environments: energy, protein, and alternative foods influence ingestion of plant secondary metabolites by lambs. J Chem Ecol 31:123–138

Vivås HJ, Sæther BE (1987) Interactions between a generalist herbivore, the moose *Alces alces*, and its food resources: an experimental study of winter foraging behaviour in relation to browse availability. J Anim Ecol 56:509–520

Vivås HJ, Sæther BE, Andersen R (1991) Optimal twig-size selection of a generalist herbivore, the moose *Alces alces*: implications for plant–herbivore interactions. J Anim Ecol 60:395–408

Vourc'h G, Vila B, Gillon D, Escarré J, Guibal F, Fritz H, Clausen TP, Martin J (2002) Disentangling the causes of damage variation by deer browsing on young *Thuja plicata*. Oikos 98:271–283

Wade MH, Peyraud JL, Lemaire G, Cameron EA (1989) The dynamics of daily area and depth of grazing of cows in a five day paddock system. In: Proceedings of the 16th International Grasslands Congress, French Grassland Society, Nice, pp 1111–1112

Walker BH (1987) A general model of savanna structure and function. In: Walker BH (ed) Determinants of tropical savannas. ICSU Press, Miami, pp 1–12

WallisDeVries MF, Laca EA, Demment MW (1999) The importance of scale of patchiness for selectivity in grazing herbivores. Oecologia 121:353–63

Welch BL, McArthur ED (1981) Variation of monoterpenoid content among subspecies and accessions of *Artemisia tridentata* grown in a uniform garden. J Range Manage 34:380–384

Welch D, Staines BW, Scott D, French DD, Catt DC (1991) Leader browsing by red and roe deer on young Sitka spruce trees in western Scotland. I. Damage rates and the influence of habitat factors. Forestry 64:61–82

White SM, Flinders JT, Welch BT (1982) Preference of pygmy rabbits (*Brachylagus idahoensis*) for various populations of big sagebrush (*Artemisia tridentata*). J Range Manage 35:724–726

Wiens JA (1976) Population responses to patchy environments. Annu Rev Ecol Syst 7:81–120.

Wilson JR (1984) Environmental and nutritional factors affecting herbage quality. In: Hacker JB (ed) Nutritional limits to animal production from pastures. Commonwealth Agricultural Bureau, Slough, UK, pp 111–131

Wilt FM, Geddes JD, Tamma RV, Miller GC, Everett RL (1992) Interspecific variation of phenolic concentrations in persistent leaves among six taxa from subgenus Tridentatae of Artemesia (Asteraceae). Biochem Syst Ecol 20:41–52

Woolnough AP, du Toit JT (2001) Vertical zonation of browse quality in tree canopies exposed to a size-structured guild of African browsing ungulates. Oecologia 129:585–590

Wright W, Illius AW (1995) A comparative study of the fracture properties of five grasses. Funct Ecol 9:269–278

Zhang X, States JS (1991) Selective herbivory of ponderosa pine by Abert squirrels: are-examination of the role of terpenes. Biochem Syst Ecol 19:111–115

Chapter 6
The Comparative Population Dynamics of Browsing and Grazing Ungulates

Norman Owen-Smith

6.1 Introduction

Among African ruminants a clear dietary distinction exists between grazers, feeding predominantly on grasses and sedges (graminoids) year-round, and browsers consuming mostly the leaves and shoots of woody plants as well as herbaceous dicots (or forbs) and fruits (Owen-Smith 1992, 1997). Even so-called mixed feeders can be divided between a subset obtaining the greater proportion of their diet from grasses year-round (mainly grazers, e.g., Thomson's gazelle (*Gazella thomsoni*), and mainly browsers dependent largely on woody and herbaceous dicots, especially during the dry season (e.g., Grant's gazelle *Gazella granti*). Just two species seem firmly intermediate, switching between consuming mostly grass in the wet season, and a predominance of browse during the dry season: impala (*Aepyceros melampus*) and nyala (*Tragelaphus angasi*). The grazer-browser distinction is replicated among the African rhinos, with the white rhino (*Ceratotherium simum*) being exclusively a grazer and the black rhino (*Diceros bicornis*) almost entirely a browser (Owen-Smith 1988). All extant zebras (*Equus* spp.) are exclusively grazers.

The grazer-browser separation is less clear among temperate zone ungulates. Although primarily grazers, bison (*Bison bison* and *B. bonasus*), feral cattle (*Bos primigenius*), and even feral horses consume a higher proportion of dicots than is typical of grazers in African ecosystems (Owen-Smith 1997; Putman 1986). Sheep and goats (Caprinae) consume a variable mix of grass and low browse (shrubs and herbs). Most deer (Cervidae) are mainly or entirely browsers, but red deer (*Cervus elaphus*) and North American elk (or wapiti; *C. e. canadensis*) as well as fallow deer (*Dama dama*) may obtain the greater portion of their summer diet from grass (Gebert and Verheyden-Tixier 2001). The lack of strict grazers occurs also in tropical Asia and South America, as well as among Australian macropods. Hence the contrast I will draw is between species that are exclusively or mainly grazers, and those that are largely or entirely browsers. Ungulates occupying tropical forests, deserts or tundra, mountainous regions, and tropical Asia plus South America, will largely be ignored, because of the absence of grazers for comparison in these environments. Accordingly, this review is focussed mostly on a comparison between African ungulates inhabiting savannas, grasslands and shrublands, and their temperate zone counterparts from North American or European woodlands, grasslands, and steppe.

The topics of relevance to this chapter include the spatio-temporal dynamics of the food resource, population abundance levels and their temporal variability, spatial and demographic structure of populations, density dependent or independent processes affecting population dynamics, and regional synchrony in population fluctuations. Special attention will be paid to the reputed propensity of ungulate populations to show eruptive dynamics, i.e., rapid growth to densities severely impacting vegetation resources, followed by population crashes (Riney 1964; Caughley 1970, 1976a; McShea et al. 1997; Forsyth and Caley 2006). This phenomenon is often used as justification for culling or other management intervention to curtail population growth or abundance.

The specific questions to be addressed are these:

1. What distinctions in food resource dynamics in quantity and quality underlie potential differences in population features between grazers and browsers?
2. How do the abundance levels attained by grazers and browsers differ, independently of influences from body mass?
3. Does the spatial, social, and demographic structure of populations differ substantially between these feeding categories?
4. Are the feedback mechanisms regulating population dynamics different in any consistent way between these feeding types?
5. Which feeding category is most susceptible to eruptive or oscillatory dynamics?
6. Do these feeding types differ in their susceptibility to broad- or fine-scale weather influences on population dynamics?

6.2 Spatial and Temporal Dynamics of Grass and Browse

In African savanna ecosystems, grass growth is determined largely by rainfall, occurring mostly within a restricted wet season. During the dry season, grass leaves and stems become converted progressively into dead tissues or 'necromass'. Figure 6.1 compares the seasonal phytomass dynamics of the grass layer for a subtropical South African savanna, with a single six-month rainy season, with that for equatorial East African grasslands having one major and one minor rainy period. All of the sites were subject to grazing, albeit at somewhat different levels. The grass layer showed a general pattern of green biomass accumulation during the main rainy season, followed by its progressive conversion to dead leaves and stems. Under light grazing, the total standing crop of potential food for grazers declines little between the growing season and the end of the dry season, as shown by the South African site (Fig. 6.1A). However, the amount of green leaf remaining in this material by the late dry season is quite small, and diluted among the standing dead biomass. In very dry years, the green fraction may fade to zero, except in bottomlands where the soil retains some moisture (Prins 1988). Dry season fires alter this pattern by removing most of the above-ground material, thereby promoting a flush of leafy regrowth if there is sufficient soil moisture (van de Vijver et al. 1999). Without burning or heavy grazing, the

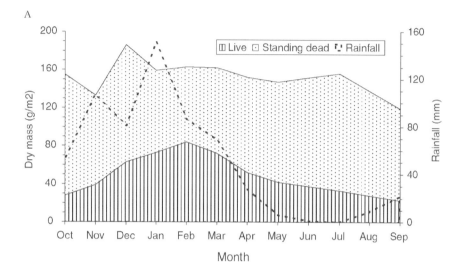

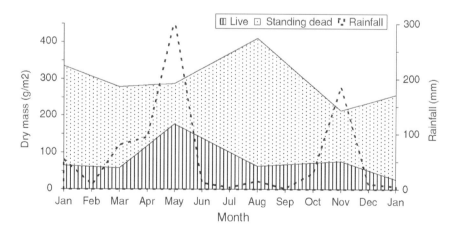

Fig. 6.1A–D Monthly biomass dynamics of green (live) and brown (dead) grass forage in representative regions. **A** Lightly grazed grass layer in Burkea savanna in Nylsvley Nature Reserve, South Africa; annual rainfall during study period 600 mm (from Grunow et al. 1980, averaging across years). **B** Grazed grassland in Nairobi National Park, Kenya; annual rainfall during study period 800 mm (from Boutton et al. 1988a, averaging across sites).

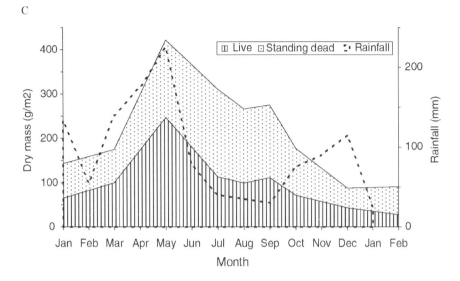

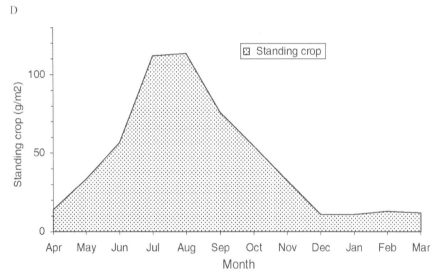

Fig 6.1 (continued) **C** Grassland in Masai Mara Reserve, Kenya, heavily grazed especially by migratory ungulates between July and December; annual rainfall during study period 1200mm (from Boutton et al 1988a, averaging across sites). **D** Grassland on Isle of Hirta, Scotland, heavily grazed by sheep; mean annual rainfall 1200mm (from Milner and Gwynne 1974, averaging across sites)

accumulated dead plant parts tend to be carried over into the next growing season, as is evident in Fig. 6.1A and B. In temperate regions, moisture from winter snow or rain generally enables grass growth to commence as soon as temperatures rise sufficiently in spring, leading to a peak in standing phytomass by mid-late summer (Fig. 6.1D). Total phytomass production seems to be less than that of the African savanna sites, perhaps because of slower growth, but a much larger fraction may remain green, especially under wet maritime conditions. Nevertheless, extensive fires in late summer were formerly a feature of temperate grasslands, as well as many woodland types, and remain so where not controlled by humans.

Annual variability in rainfall leads to quite wide fluctuations between years in herbaceous production in African savanna and grassland ecosystems (Le Houerou et al. 1988; O'Connor et al. 2001). For the semi-arid grassland depicted in Fig. 6.2, the coefficient of variation (CV) in peak phytomass was nearly double that in rainfall, and grassland judged to be in poor condition, from its cover and species composition, produced less forage for the same rainfall (O'Connor et al. 2001). Forage production seems somewhat less variable annually in mesic grasslands of the Northern hemisphere than in these tropical or subtropical grasslands (Knapp and Smith 2001).

The nutritional quality of grass forage in African savanna and grassland ecosystems, as indexed by crude protein (nitrogen content × 6.25), drops substantially after the end of the rainy season even in remaining green leaves (Fig. 6.3). There are periods when the crude protein content of the grass falls below 5 or 6%, the assumed minimum maintenance requirement for a large ruminant (Robbins 1993). East African sites exhibited a higher grass quality during the short rains than during the main rainy season (Fig. 6.3A), probably as a result of the nitrogen flux supporting initial regrowth (Prins 1988). Crude protein levels in temperate (predominantly C3) grasses are generally higher than in tropical (mostly C4) grasses, although the sheep-fertilised meadows on Hirta (Fig. 6.3B) are perhaps an extreme example

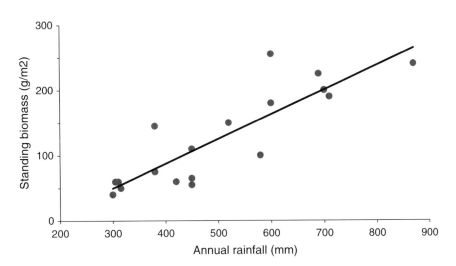

Fig. 6.2 Between-year variability in grass standing biomass at the end of the growing season for a *Themeda* grassland in good condition in the Free State in relation to annual rainfall; mean annual rainfall 560 mm (from O'Connor et al. 2001)

(but see Ydenberg and Prins 1981). Individual grass species can differ quite widely in their nutritional value, with better quality grasses remaining above the 5% crude protein level for most of the year and poor quality species falling below this level except for a few months during the wet season (Fig 6.3C; see also Prins and

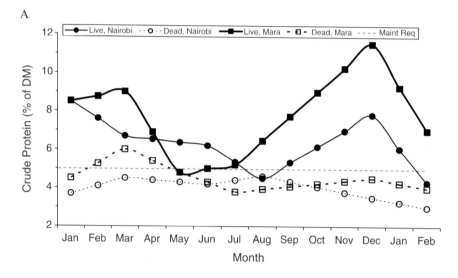

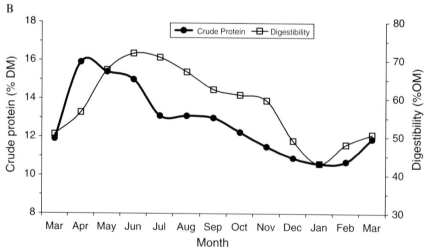

Fig. 6.3A–E Seasonal variability in the nutritional quality of grass forage. **A** Changes in crude protein contents in live (green) and dead (brown) grass at two sites in Kenya (from Boutton et al. 1988b). **B** Seasonal changes in crude protein content and digestibility of grass ('pinch samples') on Isle of Hirta, Scotland (from Milner and Gwynne 1974)

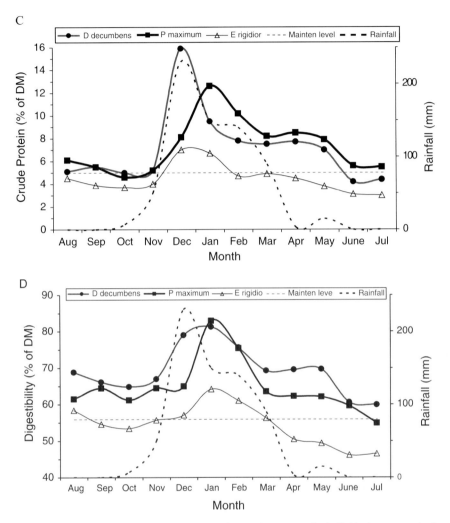

Fig. 6.3 (continued) **C** Seasonal changes in crude protein contents for individual grass species in Botswana, comparing good (*Panicum maximum*), intermediate (*Digitaria decumbens*), and poor (*Eragrostis rigidior*) species (Mosienyane 1979). **D** Seasonal changes in digestibility comparing individual grass species in Botswana (Mosienyane 1979).

Beekman 1989). Changes in digestibility largely reflect the changes in crude protein (Figs. 6.3B, C, and D), because the build-up of indigestible fibre affects both.

Through selective foraging for grass species and green leaves over dry leaves and stems, grazers may maintain the nutritional value of the forage they consume above the minimum need throughout the year, even in predominantly poor quality grassland (Fig. 6.3E). The nutritional value of the prevalent grass species is affected

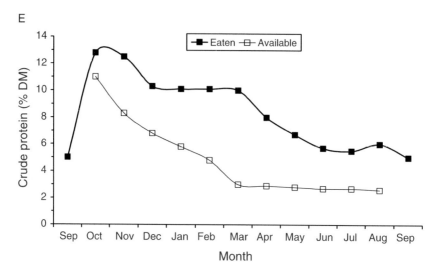

Fig. 6.3 (continued) **E** Comparison between crude protein contents in the grass consumed by cattle (from fistula samples) and that in the available grassland in the Nylsvley Nature Reserve (from Zimmerman 1980 cited by Scholes and Walker 1993)

by soil fertility, dependent largely on clay minerals retaining cations against the forces of leaching. Nevertheless, even in regions with predominantly poor soils there are local sites where better quality grasses prevail, such as under tree canopies and in drainage sumps where nutrients accumulate (Ludwig et al. 2004).

In African savannas the majority of woody plants are deciduous, shedding their leaves during the dry season. The evergreen or semi-evergreen species that retain leaves through the dry season tend to have sclerophyllous foliage, defended chemically as well as high in fibre. Accordingly, an acute bottleneck in the leafy browse remaining on woody plants can develop by the end of the dry season (Fig. 6.4A and B). Forbs may contribute some herbaceous browse, but mostly wither away during the dry season. Leaves shed from higher layers of trees can provide additional forage, especially for small to medium-sized browsers, but soon become scattered and degraded (Owen-Smith and Cooper 1985). Browse availability can vary widely across the landscape, as well as between ecosystems, depending on the tree canopy cover. In Serengeti, the peak browse biomass in open woodland of the midslope region was about one tenth of that for upland regeneration thicket (Fig. 6.4B). Drainage line thickets produced almost three times as much browse as the upland woodland, and retained more forage through the dry season due to a greater evergreen component (Pellew 1983).

Because most shoot and leaf growth on woody plants takes place prior to or early in the wet season, browse production depends more on the rainfall of the previous season than on the current season's rain, and appears more constant over

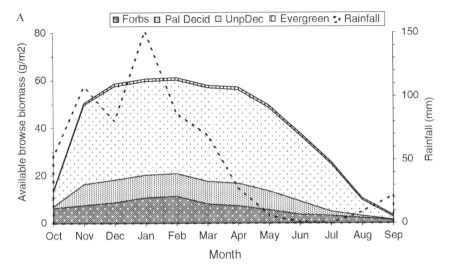

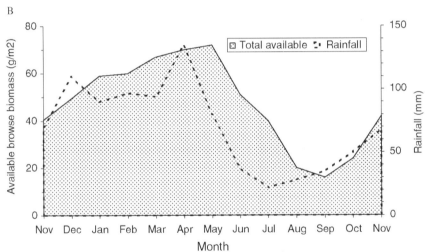

Fig. 6.4A, B Seasonal biomass dynamics of available browse. **A** Burkea savanna in Nylsvley Nature Reserve, South Africa, amount available below 2.5 m to kudu, distinguishing palatable deciduous, unpalatable deciduous and evergreen tree species plus forbs (from Owen-Smith and Cooper 1989). **B** Regenerating acacia thicket on ridge crest in Serengeti National Park, Tanzania, available below 5.75 m to giraffe (from Pellew 1983)

the years than grass production (Rutherford 1984). Leaf retention through the dry season is influenced by rain falling in the latter part of the wet season. In drought years, leaf abscission occurs early and the leaf flush at the start of the next season is delayed, prolonging the period of food stress for browsers. Herbaceous dicots respond more directly to rainfall, and some species initiate growth ahead of the rains (personal observations). Fires can promote a flush of forbs, although these later become shaded out by grass regrowth.

In temperate regions, deciduous trees and shrubs are of variable nutritional value (Robbins and Moen 1975), while evergreen species are generally less nutritious due to high fibre or secondary chemical contents (Renecker and Hudson 1988; Klein 1990). During winter, browsers depend largely on dormant buds of deciduous species, terpene-rich needle-like leaves of conifers, or evergreen shrubs such as heather (Belovsky 1981; Saether and Anderson 1990; Tixier and Duncan 1996; Lothan et al. 1999). Highest browse availability occurs at the margins of woodland patches, and in woodlands regenerating after fire. The production of summer browse within reach of moose (*Alces alces*) in the form of leaves of deciduous shrubs amounted to 10–30 gDM/m^2 in southwestern Quebec, similar to values reported for other forests at comparable latitudes in North America and Europe (Crete and Jordan 1982). Browse remaining available during winter in the form of twigs of deciduous species, plus needles and twigs of balsam fir, amounted to about 20% of the summer estimates.

Mediterranean-type vegetation shows a different seasonal pattern, with most woody plant growth occurring during the wet winter. Evergreen trees and shrubs constituted most of the browse consumed by red deer during the dry summer (Pappageourgiou 1978; Bugalho and Milne 2003).

Crude protein levels in tree leaves tend to be higher and more consistent than those in grasses, especially for the thorny *Acacia* species that occur widely in African savannas (Fig. 6.5). Unpalatable deciduous species prevalent on infertile

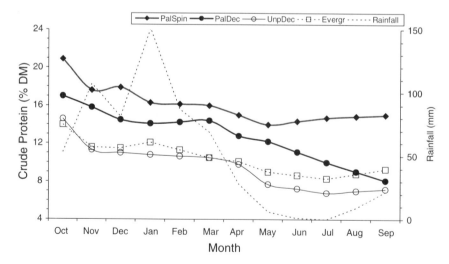

Fig. 6.5 Seasonal changes in the nutritional quality of tree foliage in Nylsvley Nature Reserve, distinguishing (1) palatable spinescent species, mainly Mimosaceae; (2) palatable deciduous species, including Combretaceae and Tiliaceae; (3) unpalatable deciduous species, Caesalpiniaceae and other families; and (4) evergreen species, various families (from Owen-Smith 1994)

soils offer protein levels not much different from those in evergreen foliage. Although crude protein contents in leaves rarely drop below 5% of dry mass, the digestibility of tree leaves tends to be somewhat less than that of grasses. The fibre component is more lignified, while tannins and other secondary chemicals may additionally reduce digestibility (Bryant et al. 1991; Owen-Smith 1993a). Herbaceous plants with weak supporting stems, especially annuals and creepers, offer some of the highest nutritional yields, but may also contain toxic chemicals (Levin 1976). The range in crude protein levels in the leaves of northern trees in late summer seems basically similar to that recorded for African savannas, e.g., 12–16% for mountain maple (*Acer spicatum*) and 10% for balsam fir (*Abies balsamea*) in Quebec (Crete and Jordan 1982). Protein levels in twigs browsed by moose in winter are much lower, around 5–6.5%.

6.3 Population Density Levels

The triangular allometric relationship documented for birds by Brown and Maurer (1987), with density declining with changing body mass both below and above some pivotal size, is shown also by ungulates (Fig. 6.6). Above a mean body mass of around

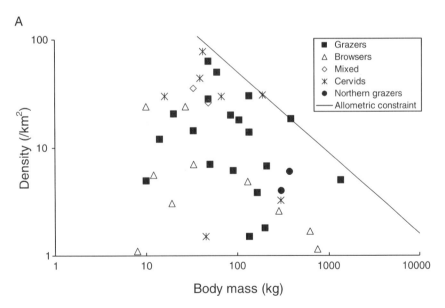

Fig. 6.6A–D Allometric scaling of abundance levels for various ungulate species, assessed via log–log plots relative to body-mass (M) distinguishing feeding categories. *Points* represent averages for each species across localities where data available. **A** Local numerical densities

35 kg (standardised as three-quarters of adult female mass at population level), the upper bound to the local or "ecological" population density declines with increasing body mass, following the allometric trend in energy metabolism as suggested by

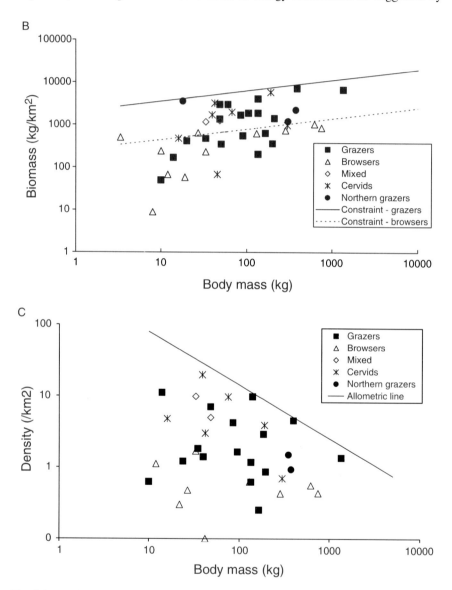

Fig. 6.6 (continued) D within the extent of a home range. *Line* represents allometric scaling assuming each species uses a constant proportion of the food produced: $D = aM^{-0.75}$. **B** Local biomass (B) densities; *lines* represent the equivalent allometric scaling, i.e., $B = aM^{0.25}$, for grazers and browsers separately. **C** Regional numerical densities across the extent of the park or other survey area, including unoccupied regions; *line* as in A.

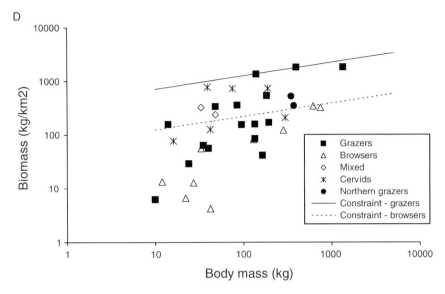

Fig. 6.6 (continued) **D** Regional biomass densities, *lines* as in B

Damuth (1981). Below this mass, numerical abundance also decreases as body mass is reduced, even when assessed at a local scale. Dikdik (*Madoqua kirki*) are a notable exception: the local density within pair territories covering less than a hectare amounts to over 100 animals per km^2. Furthermore, many species do not approach the abundance levels attained by other species of similar size, as shown by the numerous points falling well below the outer bounds to the triangular distribution. Highest numerical densities are manifested by medium-sized grazers occupying floodplain or otherwise locally productive habitats in Africa, as well as by similar-sized deer in North America. Wildebeest (*Connochaetes taurinus*) achieve this density within the Serengeti ecosystem, but not consistently elsewhere. The highest ecological density of around 200 animals per km^2 is for feral Soay sheep (*Ovis aries*) on Hirta (in the Scottish St. Kilda archipelago).

In the lower body size range, grazers and browsers appear not to differ in the density levels that they show at either local or regional scale (Fig. 6.6). However, the maximum abundance levels attained by larger browsers seem to be an order of magnitude lower than those exhibited by similar-sized grazers in African ecosystems. This is most clearly evident when local abundance is plotted as biomass rather than numerical density (Fig. 6.6B). Biomass density rises initially with increasing body mass before reaching the upper asymptote imposed by the scaling of energy requirements with body size. Impala and nyala, the only clearly intermediate feeders, seem intermediate also in their biomass density. Among temperate-zone species, mixed feeders like red deer and wapiti, as well as browsers like white-tailed deer (*Odocoileus virginianus*) and mule deer (*O. hemionus*), show local biomass densities matching those of African grazers such as wildebeest, buffalo (*Syncerus caffer*) and white rhino, taking into account metabolic scaling. However, the largest cervid, moose, falls closely along the browser line.

Crude densities assessed at a regional scale incorporate unoccupied regions, especially for populations with restricted distributions. Accordingly, regional density or biomass levels are generally lower than local ecological densities. Grazer-browser distinctions remain little altered, but the density of some of the smallest species is reduced more at the regional scale than is the case for larger species (Figs. 6.6B and D). Regional biomass densities generally increase with annual rainfall for most African ungulate species, both grazers and browsers (East 1984).

6.4 Demographic Patterns

Large grazers may aggregate in vast herds, greatly outnumbering those formed by similar-sized browsers. African buffalo herds can exceed 2,000 (Sinclair 1977; Prins 1996), while the largest eland (*Taurotragus oryx*) herd was just over 400 (Hillman 1987). Browsing ungulates under 35 kg in mean body mass are mostly solitary, apart from female–offspring associations, while grazing oribi (*Oerebia oerebi*) and mountain reedbuck (*Redunca fulvorufula*) form small herds of 5–12, and Thomson's gazelle aggregates of 50 or more. Only among certain small browsers do females maintain exclusive territories, sometimes shared with a male partner (e.g., dikdik, klipspringer *Oreotragus oreotragus*). Territories defended by males for mating purposes are a common feature of grazing ungulates, but not among browsers, probably because females of the larger browsers do not attain sufficient local densities to support such a strategy (Jarman 1974, Owen-Smith 1977).

Contrasting patterns have been recorded in the extent of the home range covered seasonally by African grazers and browsers. Grazers including buffalo (Funston et al. 1994, Ryan et al. 2006), white rhino (Owen-Smith 1975) and plains zebra (*Equus burchelli*; Klingel 1969) move over a wider area in the dry season than in the wet season, even excluding excursions to water. In contrast, browsers like eland (Hillman 1988), kudu (*Tragelaphus strepsiceros*; personal observations) and black rhino (Goddard 1967) contract their home ranges during the dry season. This could reflect spatial differences in seasonal predictability of food resources. During the dry season, grazers may opportunistically seek out areas where localised rainshowers have produced areas of green regrowth (Talbot and Talbot 1963). In contrast, trees do not respond to dry season rainfall, while the localities where evergreen foliage persists through the dry season are fixed in the landscape. Among temperate zone ungulates, moose (Cedarlund and Okarma 1988; Mysterud et al. 2001) and European red deer (Clutton-Brock et al. 1992; Mysterud et al. 2001) generally followed the browser pattern, with a contraction in home range during winter. However, roe deer (*Capreolus capreolus*; Kjellander et al. 2004; Mysterud et al. 2001), mule deer (Mysterud et al. 2001) and North American wapiti (Anderson et al. 2005), as well as bison (McHugh 1958), showed an expansion in home range during the winter months, while white-tailed deer showed no consistent pattern (Mysterud et al. 2001).

Large-scale migrations are a feature of certain grazer populations, notably wildebeest, zebra, topi (*Damaliscus lunatus*), and Thomson's gazelle (*Gazella thomsoni*) in the Serengeti (Maddock 1979), and elsewhere (Jewell 1972; Williamson et al. 1988) white-eared kob (*Kobus kob leucotis*) in Sudan (Fryxell and Sinclair 1988), and in former times springbok (*Antidorcas marsupialis*) in South Africa (Skinner 1993). Among temperate zone ungulates, extensive migrations are also shown by saiga antelope (*Saiga tatarica*; Bekonov et al. 1998) and Mongolian gazelles (*Procapra gutturosa*; Jiang et al. 2002), which feed partly on grasses as well as herbs and shrubs. Migratory wildebeest in the Serengeti vastly outnumber the resident wildebeest subpopulation (Fryxell et al. 1988). Where migratory routes have been blocked by fences or settlements, substantial reductions in population have resulted, e.g., for wildebeest in the southern Kalahari (Williamson and Mbano 1988), Etosha (Gasaway et al. 1996), and Kruger Park (Whyte and Joubert 1988). Mainly browsing eland roam widely, but without any regular migratory pattern (Hillman 1988). Most African browsers are fairly sedentary, without seasonally distinct home ranges, but so are many grazer populations. Temperate zone browsers and mixed feeders commonly shift their home ranges seasonally along altitudinal gradients in mountainous regions, to take advantage of differences in plant phenology (Craighead et al. 1972; Schoen and Kirchoff 1985; Albon and Langvatn 1992; Histol and Hjeljord 1994; Mysterud 1999).

Population composition in terms of age and sex classes is largely an outcome of the recruitment rate, governed by litter size and phase of population growth. All African bovids produce just a single offspring annually, irrespective of whether they are grazers or browsers (although springbok can reproduce twice during an annual cycle; Skinner and Louw 1996). Among cervids, species in the New World subfamily Odocoilinae commonly produce twins or triplets, while Old World deer in the Cervinae mostly give birth to single offspring. Sheep and goats are mostly polytocous, although the extent of twinning differs among populations. Differences in litter size, and hence in the juvenile proportion in the population, determine the maximum population growth rate that can be sustained. This is around 25% per year for species producing a single young, but up to 50% per year for roe deer, white-tailed deer and moose (Anderson and Linnell 2000; McCullough 1997). Most ungulates show a strongly female-biased sex ratio in the adult segment, indicating higher male than female mortality, but no grazer-browser distinction is evident (Owen-Smith 1993b; Berger and Gompper 1999; Owen-Smith and Mason 2005).

6.5 Regulation of Abundance

For small antelope, territories maintained by both sexes may limit abundance relative to localised food resources, to the extent that these territories cannot be compressed. Among larger ungulates, direct contests for food are rare, because of the wide distribution and generally low quality of vegetation resources (Prins 2000).

Intraspecific competition arises largely indirectly through the effects of exploitation for subsequent food availability during the dormant season. There may also be longer-term consequences for vegetation structure and composition affecting the productive potential of plants. Because interference with feeding may only be manifested after crowding has reached high levels, and indirect consequences via vegetation impacts are delayed, a convex or plateau-and-ramp pattern of reduction in the population growth rate with increasing density is generally expected (Fowler 1981, 1987; McCullough 1992, 1999). However, the trend in population growth rate may appear approximately linear around the zero growth level, depending on how the density feedback affects different population segments (Owen-Smith 2006).

Density influences on population growth can be obscured by errors in the abundance estimates, climatic fluctuations and variable age structure, as well as by non-linear or delayed responses (Bonenfant et al., unpublished manuscript). The density effect may be revealed only after controlling for the influence of variable rainfall on food availability (Owen-Smith 1990). Density dependence is most clearly evident in body mass (Solberg et al. 2004) and the consequences of growth for rates of reproduction, in particular for the age at first parturition (Bonenfant et al. unpublished manuscript). Mortality losses among juveniles and old adults also generally rise with increasing population density, while the survival of prime-aged females changes little until high abundance levels are attained (Gaillard et al. 2000). The high sensitivity of juvenile survival to food shortfalls as well as impacts of predation may obscure the density dependence in population growth (Owen-Smith et al. 2005). Most of the change in population growth rate towards the zero growth level may be due to the reduction in adult survival, but with a shift in the age structure of the population towards older animals with lowered survival chances also contributing (Festa-Bianchet et al. 2003; Owen-Smith et al. 2005; Owen-Smith 2006).

No consistent grazer-browser distinction in the sensitivity of population growth to either density or rainfall was evident among ungulate species in the Kruger Park (Mills et al. 1995; Ogutu and Owen-Smith 2003, Owen-Smith and Mills 2006). Browsing kudus responded positively to rainfall, along with buffalo, waterbuck (*Kobus ellipsiprymnus*) and other less common grazers. Giraffe (*Giraffa camelopardalis*) seemed as little affected by rainfall variability as grazing wildebeest and zebra, while impala appeared intermediate.

Predation may either mask the influence of resource availability on population dynamics, or accentuate it, depending on the extent to which the mortality loss imposed is additive and how it depends on nutritional status (Chapter 10 in Owen-Smith 2002a; Owen-Smith and Mills, submitted). Predation pressure (Klein 1965; Prins and Iason 1989), and in some situations disease impacts (Prins and Weyerhauser 1987), may hold ungulate populations below the food ceiling, or cause lagged population responses to changing density levels (Owen-Smith and Mills, submitted). The risk of predation can amplify the effects of nutritional deficits on population growth by affecting foraging behaviour (Sinclair and Arcese 1995). Migratory populations may escape predator control, while resident populations can be held in a "predator pit" below the abundance level that could be sustained by resources (Fryxell et al.1988; Gasaway et al. 1992), especially if human hunting is imposed in addition to predation (Messier and Crete 1985). An alternative prey species may be reduced to low abundance by a predator population sustained by a more abundant species, e.g., for non-migratory caribou (*Rangifer tarandus*) and moose, with wolves (*Canis lupus*) as the predator (Seip 1992; McLoughlin et al. 2003), and for less common ungulate species in the Kruger Park, including both grazers and browsers (Owen-Smith and Mills 2006).

6.6 Population Dynamics

Large herbivores have the potential to "irrupt" to population levels where their consequent feeding impact on vegetation resources depresses food availability, precipitating a population crash (Caughley 1976a). A book addressing "the science of overabundance" focussed on white-tailed deer and mule deer in North America as the exemplars (McShea et al. 1997). Most of the examples of population irruptions and crashes come from island situations where dispersal was precluded, and usually involve animals introduced in the absence of predators (McCullough 1997; Kaji et al 2004; Forsyth and Caley 2006). The moose population on Isle Royale (Lake Superior) has shown repeated oscillations in abundance despite the presence of wolves (Peterson 1999), but the effectiveness of the wolves as predators is limited by other factors (Peterson et al. 1998). Although theoretical models suggest that extreme peak densities could result in persistent vegetation degradation, observations indicate no change in subsequent peak abundance levels attained by oscillating deer populations (McCullough 1997).

While the above deer are mainly browsers, feral Soay sheep which are largely grazers, have shown persistent oscillations in abundance over a greater-than-twofold range in the St Kilda islands (Clutton-Brock et al. 1991). In contrast, mixed-feeding red deer inhabiting a similar island environment on Rum (Inner Hebrides, Scotland) have remained relatively stable in abundance (Clutton-Brock et al. 1997; Clutton-Brock and Coulson 2002). Contributing to the susceptibility to irruptions of both the deer and the sheep is a high reproductive potential. White-tailed deer and mule deer, as well as moose, commonly produce twin offspring as well as undergoing first parturition at a young age. Soay sheep also first reproduce at an early age, but twin offspring are infrequent (Clutton-Brock et al. 1997). Nevertheless, the last two months of gestation precede the spring regrowth of grasses, so that in favourable years the high recruitment potential coupled with the absence of any density feedback during summer elevates the population well above the level supported by the food resources by late winter, precipitating a die-off. Supporting the high growth potential of the sheep is the high quality of the grasses growing in the lush meadows fertilised by sheep manure. Furthermore, heather may be less effective as a buffer resource for the sheep than for red deer (Chapter 13 in Owen-Smith 2002a).

Since browsers appear to be limited mainly by the amount of food remaining during the adverse season (Sect. 6.2 above), they should be more prone to fluctuations in abundance than grazers, limited more by seasonal changes in food quality (Chapter 13 in Owen-Smith 2002a). Nevertheless, feral cattle and ponies, which are mainly grazers, as well as mixed-feeding red deer, showed unusually high (15–30%) winter mortality in the Oostvadersplassen Reserve in Netherlands, following population growth towards high density levels (S. van Wieren, personal communication). These deaths occurred during exceptionally cold weather, with ice crusts restricting access to forage. Severe cold weather was also directly responsible for precipitating population crashes by Soay sheep (Coulson et al. 2001) and white-tailed deer (Patterson and Power 2002).

For African ungulates, the extreme weather conditions leading to major die-offs entail a lack of rainfall, restricting food availability. During the 1982/1983 El Nino-related drought, populations of six grazing ungulates crashed to 20% or less of their former abundance in the Klaserie Private Nature Reserve, adjoining the Kruger Park (Walker et al. 1987). In this case the two large browsers (kudu and giraffe) were less affected, dropping by 30–50% in abundance, while mixed-feeding impala showed a 75% decline. Contributing to the severe mortality was the close spacing of waterpoints on this private land, enabling animals to access almost all remaining grass, until virtually no forage remained. Trees started producing new foliage ahead of the rains, alleviating food stress for the browsers in this crucial period. Ungulate populations within the vast Kruger Park, where waterpoint spacing was somewhat wider, decreased much less during this same drought (Walker et al. 1987). Wildebeest and zebra were almost unaffected, with the greatest declines by 30–40% shown by browsing kudu and intermediate-feeding impala, as well as grazing waterbuck. Over the 13-year period spanned by park-wide surveys of Kruger Park, which included this and a subsequent extreme drought, the coefficient of variation (CV) in census totals was consistently larger for the three browsers than for grazers of similar body size (Fig. 6.7). Impala showed a relatively low CV, probably partly because sampling error was lowered as a consequence of their large population total, but notably they were also subject to predation by numerous predator species (Owen-Smith and Mills, submitted).

During an earlier drought in 1965, starvation-related mortality reduced populations of browsing giraffe and kudu by about a third in the Timbavati Private Nature Reserve, bordering Kruger Park, while no starvation occurred among grazing wildebeest, zebra, and impala (Hirst 1969). In the southern Kalahari, a drought during 1985 was associated with 35% mortality among mainly browsing eland, compared with 12–18% among grazing wildebeest and hartebeest (*Alcelaphus buselaphus*), and little mortality among springbok (Knight 1995).

Kudu are notoriously susceptible to die-offs when cold and wet weather occurs near the end of the dry season (Wilson 1970, Owen-Smith 2000). For this sub-tropical species, extreme cold is a maximum daily temperature under 15°C, coupled with wind and rain accentuating the loss in body heat at a stage when body reserves are low. Kudu need to forage for extended periods during this generally hot time of the year when food availability is at a minimum, and thus must tolerate high temperatures

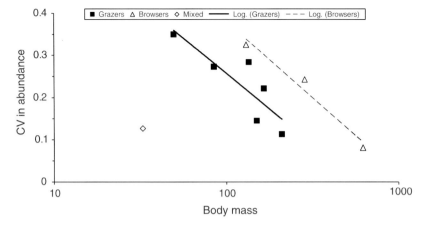

Fig. 6.7 Coefficient of variation in annual census totals for ten ungulate species in the Kruger Park over 1980–93 in relation to body mass, distinguishing feeding types. *Semi-log trend-lines* are indicated

at the expense of insulation from cold (Owen-Smith 1998). Die-offs following cold weather have been recorded for browsing giraffe (Walker et al. 1987) and intermediate-feeding nyala (Keep 1973). Among grazers, a population crash associated with cold weather has been documented for reedbuck (*Redunca arundimum*; Ferrar and Kerr 1971), but among African buffalo cold-related mortality was restricted to calves (Jolles personal communication).

Ungulate population die-offs have been recorded even within large, unfenced protected areas in East Africa. A drought in 2000 resulted in the death of at least 1,500 buffalo, representing 30% of the prior population, in the Ngorongoro Crater in Tanzania (Estes et al. 2006). However, declines by similar proportions in wildebeest and zebra appeared to be due more to movements of animals out of the crater rather than to mortality. The severe drought of 1993 led to a drop by 75% in the buffalo population within Maasai-Mara and adjoining group ranches in Kenya (Ottichilo et al. 2000), and brought about a 22% decline in the migratory wildebeest population in the Serengeti (Mduma et al. 1999).

Overall, these observations suggest that African browsers are susceptible to frequent but relatively minor die-offs, keeping populations below levels where severe crashes might be precipitated. In contrast, grazers may be subject to less frequent but potentially more substantial population declines during major droughts. Observations on domestic ungulates suggest that the vulnerability of grazers to large-scale mortality during droughts may be dependent on soil fertility, controlling the general palatability of grasses (Chapter 13 in Owen-Smith 2002a). On fertile soils where grasses are mostly palatable, almost all of the grass layer may be consumed in a drought year, so that cattle die through lack of food (Scoones 1993; Hatch and Stafford Smith 1997). On infertile soils, the less palatable grasses that are prevalent tend to remain uneaten, and thus serve as a buffer slowing starvation during drought periods. Bottomlands where grasses retain green leaves through the dry season may serve as key resource areas buffering herbivore populations against severe declines (Prins 1988; Scoones 1995; Illius and O'Connor 1999, 2000). Evergreen woody plants, also predominant especially along river margins, may serve similarly to buffer population fluctuations in browsers like kudu and giraffe (Owen-Smith and Cooper 1989). Notably, the population crash of moose on Isle Royale was associated with conditions under which their main winter food, balsam fir (*Abies balsamea*), had been severely depressed in abundance by previous browsing pressure (Risenhoover and Maass 1987). However, the contribution of different vegetation components can be somewhat more complex, with different plant types serving as reserve, bridging or buffer resources under different conditions (Owen-Smith and Cooper 1989; Chapter 11 in Owen-Smith 2002a; Knoop and Owen-Smith 2006).

An additional problem with domestic livestock, mostly grazers, in savannas is a shift in vegetation composition towards woody plants when high stocking levels and consequent heavy grazing suppress fires—the problem of bush encroachment (Walker et al. 1981). High abundance levels reached by wild grazers in the absence of predation in the Hluhluwe Reserve in South Africa seem to have been the primary cause of the transformation of large areas of open grassland into thicket, leading to greatly increased abundance of browsers and mixed feeders at the expense of grazers (Brooks and Macdonald 1983).

Megaherbivores like rhinos and elephants are largely invulnerable to predation as adults, and can have a corresponding 'mega impact' on vegetation. This raises concerns that their population dynamics could show cyclic oscillations (Caughley 1976b; Owen-Smith 1981, 1988). The outcome depends largely on lags in the recovery of plant populations to these impacts, and on how vegetation impacts are distributed in space and over time. Modelling suggests that sufficient functional heterogeneity in vegetation quality and forage retention during the adverse season, assisted by opportunities for dispersal, could suppress the risk of oscillations (Owen-Smith 2002a, 2002b, 2004).

6.7 Weather Patterns and Population Fluctuations

Large-scale weather patterns related to shifts in atmospheric pressure belts as a result of changes in oceanic circulation patterns affect environmental conditions, and hence potentially the population dynamics of herbivores (Saether 1997). The El Nino - Southern Oscillation (ENSO) linked to sea surface temperature in the eastern Pacific Ocean generates either high rainfall or extreme droughts in different regions when it intensifies at irregular intervals. Within Africa, there is additionally a quasi-decadal alternation between wet and dry phases in rainfall affecting the summer rainfall region of South Africa, with conditions expressed in tropical East Africa being opposite to those in southern

Africa (Tyson and Gatebe 2001). In north temperate latitudes, the North Atlantic Oscillation (NAO) and counterpart North Pacific Oscillation (NPO) generate an alternation in weather patterns between years of relatively warm conditions with high winter precipitation and hence deeper snow cover, and colder winters with less precipitation and hence less snow, although the regional consequences vary with altitude as well as latitude (Stenseth et al. 2002a, 2003).

These weather patterns can act on populations either by promoting mortality through the physiological stress imposed by temperature or wind extremes, or by affecting resource availability. High rainfall generally increases vegetation growth and hence food availability, while deep snow as a result of high precipitation during winter makes food less accessible. In African savannas, a lack of rainfall during the normally dry season can greatly reduce the amount of green leaf persisting, and hence food quality during this critical period (Mduma et al. 1999; Ogutu and Owen-Smith 2003). Temperature changes can also affect the timing of plant growth in spring in temperate latitudes, with potential effects on ungulate populations (Post and Stenseth 1999).

Such broad-scale weather patterns can induce regional synchrony in population trends, as documented for mixed-feeding red deer (Forschhammer et al. 1998) and browsing roe deer in Norway (Grotan et al. 2005), as well as browsing moose and white-tailed deer in North America (Post and Stenseth 1998). Deep snow also increases susceptibility to predation by wolves (Post and Stenseth 1998; Hebblewhite 2005). For Soay sheep in the St Kilda islands, where snow is not a factor, warm winters were adverse through being wet and windy, thereby increasing mortality (Milner et al 1999) and at the same time synchronising populations on nearby islands (Grenfell et al. 1998).

In African savanna ecosystems, high rainfall is associated with increased populations of most ungulates (Mills et al. 1995; Owen-Smith and Ogutu 2003; Ogutu and Owen-Smith 2003, 2005). Hence individual species within the diverse assemblage of grazers and browsers in the Kruger Park fluctuated largely in parallel in response to rainfall variability. However, wildebeest and zebra performed better during the dry phase of the rainfall oscillation, due to changing susceptibility to predation (Mills et al. 1995; Ogutu and Owen-Smith 2005; Owen-Smith and Mills, submitted). In years of low rainfall there is less grass cover for stalking lions, and herds can aggregate in more extensive short grassland, thereby diluting predation pressure (Smuts 1978). This contrasts with the positive response to rainfall of a zebra population in northern Kenya, where predation pressure is somewhat lower (Georgiadis et al. 2003).

Ellis and Swift (1988; see also Ellis et al. 1993) suggested that in regions where the coefficient of variation in rainfall is 30% or greater, herbivore populations do not reach any equilibrium with the vegetation, and fluctuate widely in response to droughts while rarely approaching carrying capacity. This standpoint was based largely on observations on domestic livestock managed by nomadic pastoralists in northern Kenya. Illius and O'Connor (2000) pointed out that key resource areas during the dry season may largely maintain herbivore populations and incur the brunt of grazing or browsing impacts, although populations may be largely decoupled from resource limitations during the wet season. Towards high rainfall conditions where the coefficient of variation in annual rainfall drops below 25%, herbivore dynamics might be more closely coupled to vegetation dynamics, but forage quality then becomes a potential limitation, especially for grazers.

6.8 Summary and Conclusions

Grazing and browsing ungulates depend largely on different components of the vegetation, which show contrasting dynamics in quantity and quality seasonally and between years. For grazers, the nutritional value of the dry grass forming the bulk of the potential food persisting through the winter or dry season is low. For browsers, the small amount of green foliage remaining towards the end of the dormant season shows relatively less decline in nutrient contents than grasses. These resource differences might be expected to produce distinctions in population dynamics between grazers and browsers. However, the situation is not quite so simple. Some green grass persists in moister regions of the landscape, while in fertile ecosystems a large fraction of the dry grass may be of adequate nutritional value. Much of the tree foliage remaining by the end of the dormant season is evergreen and defended by digestibility-reducing chemicals. Furthermore, fires during the dry season can greatly alter food availability for grazers.

Distinctions exist between grazers and browsers in maximum population density levels, but only among larger species, and some grazers attain abundance levels no greater than those of equivalent-sized browsers. Browsers and mixed feeders in temperate woodlands can reach biomass levels comparable to those of abundant grazers in African savannas. Grazers commonly expand their home ranges during the dry season or winter months, and some populations migrate seasonally in response to changing food and surface water availability. Browsers are generally somewhat more sedentary and may contract their ranges when food availability is restricted.

Density feedbacks largely arise indirectly via seasonal or annual depression of food resources, and can potentially be delayed. No fundamental distinctions between grazers and browsers are apparent here. Extreme or persistent oscillations in abundance seem to be a feature of northern ungulates with high reproductive potentials, or island situations where constrictions in spatial extent limit the ability of herbivores to exploit resource heterogeneity. Quite severe die-offs in years when adverse weather conditions restrict food availability, and perhaps also induce thermal stress, occur even among populations inhabiting large protected areas, but may be exacerbated when movements are restricted. Long term consequences for the capacity of the vegetation to support these populations have not been documented, except where the suppression of fire by heavy grazing has resulted in the transformation of savanna into thicket. African browsers appear to show more frequent but relatively minor die-offs, while grazers may show greater population crashes at longer intervals when severe droughts occur, especially in fertile ecosystems where food quality is not much of a limitation. African browsers seem more susceptible to mortality through hypothermia than syntopic grazers.

Large-scale climatic patterns can induce variability in populations either through physiological stress or by affecting food availability, and be responsible for regional synchrony in population fluctuations. Synchrony may be disrupted by local variability in weather effects and relative vulnerability to predation under different conditions. Hence the consequences of the distinctions in quality, quantity and temporal dynamics of grass and browse resources are complex, and do not affect abundance levels of the ungulate species feeding on these resource types or their population fluctuations and distribution in any simple way.

Acknowledgements I am indebted to Herbert Prins and Iain Gordon for helpful suggestions that greatly improved this chapter.

References

Albon SD, Langvatn R (1992) Plant phenology and the benefits of migration in a temperate ungulate. Oikos 65:502–613

Anderson DP, Forester JD, Turner MG, Frair JL, Merrill EH, Fortin D, Mao JS, Boyce MS (2005) Factors influencing female home range sizes in elk in North American landscapes. Landscape Ecol 20:257–271

Anderson R, Linnell JDC (2000) Irruptive potential in roe deer: density-dependent effects on body mass and fertility. J Wildl Manage 64:698–706

Bekonov AB, Grachevand IuA, Milner-Gulland EJ (1998) The ecology and management of the saiga antelope in Kazakhstan. Mammal Rev 28:1–52

Belovsky GE (1981) Food plant selection by a generalist herbivore: the moose. Ecology 62:1020–1030

Berger J, Gompper ME (1999) Sex ratios in extant ungulates: products of contemporary predation or past life histories? J Mammal 80:1084–1113

Bonenfant C, Gaillard J-M, Loe LE, Loison A, Blanchard P, Garel M, Pettorelli N, Owen-Smith N, du Toit J, Duncan P (unpublished manuscript, submitted) Empirical evidence of density dependence in ungulates.

Boutton TW, Tieszen LL, Imbamba SK (1988a) Biomass dynamics of grassland vegetation in Kenya. Afr J Ecol 26:89–101

Boutton TW, Tieszen LL, Imbamba SK (1988b) Seasonal changes in the nutrient content of East African grassland vegetation. Afr J Ecol 26:103–116

Brooks PM, Macdonald IAW (1983) In: Owen-Smith RN (ed) The Hluhluwe-Umfolozi Reserve: an ecological case history. Haum, Pretoria, pp.51–77

Brown JH, Maurer BA (1987) Evolution of species assemblages: effects of energetic constraints and species dynamics on the diversification of the North American avifauna. Am Nat 130:1–17

Bryant JP, Provenza FD, Pastor J, Reichardt PB, Clausen TP, du Toit JT (1991) Interactions between woody plants and browsing mammals mediated by secondary metabolites. Annu Rev Ecol Syst 22:431–446

Bugalho MN, Milne JA (2003) The composition of the diet of red deer in a Mediterranean environment: a case of summer nutritional constraint. Forest Ecol Manage 181:23–29

Caughley G (1970) Eruption of ungulate populations with emphasis on Himalayan tahr in New Zealand. Ecology 51:53–72

Caughley G (1976a) Plant-herbivore systems. In: May RM. (ed) Theoretical ecology. Blackwell, Oxford, pp 94–113
Caughley G (1976b) The elephant problem - an alternative hypothesis. E Afr Wildl J 14:265–283
Cedarlund GN, Okarma H (1988) Home range and habitat use of adult female moose. J Wildl Manage 52:336–343
Clutton-Brock TH, Price OF, Albon SD, Jewell PA (1991) Persistent instability and population regulation in Soay sheep. J Anim Ecol 60:593–608
Clutton-Brock TH, Guinness FE, Albon SD (1992) Red deer. behaviour and ecology of two sexes, 2nd edn. Edinburgh University Press, Edinburgh
Clutton-Brock TH, Illius AW, Wilson K, Grenfell BT, MacColl A, Albon SD (1997) Stability and instability in ungulate populations: an empirical analysis. Am Nat 149:196–219
Clutton-Brock TH, Coulson T (2002) Comparative ungulate dynamics: the devil is in the detail. Phil Trans R Soc Lond B 357:1285–1298
Coulson T, Catchpole EA, Albon SD, Morgan BJT, Pemberton JM, Clutton-Brock TH, Crawley MJ, Grenfell BT (2001) Age, sex, density, winter weather, and population crashes in Soay sheep. Science 292:1528–1531
Craighead JJ, Atwell G, O'Gara BW (1972) Elk migrations in and near Yellowstone National Park. Wildl Monogr No. 29
Crete M, Jordan PA (1982) Production and quality of forage available to moose in southwestern Quebec. Can J Forest Res 12:151–159
Damuth J (1981) Home range, home range overlap and energy use among animals. Biol J Linn Soc 15:185–19
East R (1984) Rainfall, soil nutrient status and biomass of large African savanna mammals. Afr J Ecol 22:245–270
Ellis JE, Swift DM (1988) Stability of African pastoral ecosystems: alternative paradigms and implications for development. J Range Manage 41:450–459
Ellis JE, Coughenour MB, Swift DM (1993) Climatic variability, ecosystem stability and the implications for range and livestock development. In: Behnke RH, Scoones I, Kerven C (eds) Range ecology at disequilibrium. Overseas Development Institute, London, pp 31–41
Estes RD, Atwood JL, Estes AB (2006) Downward trends in Ngorongoro Crater ungulate populations 1986–2005: conservation concerns and the need for ecological research. Biol Cons 131:106–120
Ferrar AA, Kerr MA (1971) A population crash of the reedbuck in Kyle National Park, Rhodesia. Arnoldia (Rhodesia) 5:1–9
Festa-Bianchet M, Gaillard J-M, Cote SD (2003) Variable age structure and apparent density dependence in survival of adult ungulates. J Anim Ecol 72:640–649
Forschhammer M, Stenseth NC, Post E, Langvatn R (1998) Population dynamics of Norwegian red deer: density dependence and climatic variation. Proc R Soc Lond B 265:341–350
Forsyth DM, Caley P (2006) Testing the irruptive paradigm of large-herbivore dynamics. Ecology 87:297–303
Fowler CW (1981) Density dependence as related to life history strategy. Ecology 62:602–610
Fowler CW (1987) A review of density dependence in populations of large mammals. In: Genoways HH (ed) Current Mammalogy Vol. 1. Plenum, New. York, pp 401–441
Fryxell JM, Greever J, Sinclair ARE (1988) Why are migratory ungulates so abundant? Am Nat 131:781–798
Fryxell JM, Sinclair ARE (1988) Seasonal migration by white-eared kob in relation to resources. Afr J Ecol 26:17–31
Funston PJ, Skinner JD, Dott HM (1994) Seasonal variation in movement patterns, home range and habitat selection of buffaloes in a semi-arid habitat. Afr J Ecol 32:100–114
Gaillard J-M, Festa-Bianchet M, Yoccoz NG, Loison A, Toigo C (2000) Temporal variation in fitness components and dynamics of large herbivores. Annu Rev Ecol Syst 31:367–393
Gasaway WC, Boertje RD, Grangaard DV, Kelleyhouse DG, Stephenson RO, Larsen DG (1992) Predation limiting moose at low densities in Alaska and Yukon and implications for conservation. Wildl Monogr No. 120

Gebert C, Verheyden-Tixier H (2001) Variations of diet composition of red deer in Europe. Mammal Rev 31:189–201

Georgiadis N, Hack M, Turpin K (2003) The influence of rainfall on zebra population dynamics: implications for management. J Appl Ecol 40:125–136

Goddard J (1967) Home range behaviour and recruitment rates of two black rhinoceros populations. E Afr Wildl J 5:133–150

Grenfell BT, Wilson K, Finkenstadt BF et al (1998) Noise and determinism in synchronised sheep dynamics. Nature 394:674–677

Grotan V, Saether B-E, Engen S, Solberg EJ, Linnelll JDC, Anderson R, Broseth H, Lund E (2005) Climate causes large-scale spatial synchrony in population fluctuations of a temperate herbivore. Ecology 86:1472–1482

Grunow JO, Groeneveld HT, du Toit HC (1980) Above ground dry matter dynamics of the grass layer in the Nylsvley tree savanna. J Ecol 68:877–889

Hatch GP, Stafford Smith DM (1997) The bioeconomic implications of various drought management strategies for a communal cattle herd in a semi-arid savanna of KwaZulu-Natal. Afr J Range Forage Sci 14:17–25

Hebblewhite M (2005) Predation by wolves interacts with North Pacific Oscillation (NPO) on a western North American elk population. J Anim Ecol 74:226–233

Hillman JC (1987) Group size and association patterns of the common eland. J Zool Lond 213:641–663

Hillman JC (1988) Home range and movement of the common eland in Kenya. Afr J Ecol 26:135–148

Hirst SM (1969) Populations in a Transvaal lowveld nature reserve. Zool Afr 4:199–230

Histol T, Hjeljord O (1994) Winter feeding strategies of migrating and non-migrating moose. Can J Zool 71:1421–1428

Illius AW, O'Connor TG (1999) On the relevance of nonequilibrium concepts to arid and semi-arid grazing systems. Ecol Applic 9:798–813

Illius AW, O'Connor TG (2000) Resource heterogeneity and ungulate population dynamics. Oikos 89:283–294

Jarman PJ (1974) The social organization of antelope in relation to their ecology. Behaviour 48: 15–267

Jewell PA (1972) Social organisation and movements of topi during the rut at Ishasha, Queen Elizabeth Park, Uganda. Zool Afr 7:233–255

Jiang Z, Takatsuki S, Li J, Wang W, Gao Z, Ma J (2002) Seasonal variation in foods and digestion of Mongolian gazelles in China. J Wildl Manage 66:40–58

Kaji K, Okada H, Yamanaka M, Matsuda H, Yabe T (2004) Irruption of a colonizing sika deer population. J Wildl Manage 68:889–899

Keep, ME (1973) Factors contributing to a population crash of nyala in Ndumu Game Reserve. Lammergeyer (Natal) 19:16–23

Kjellander P, Hewison AJM, Liberg O, Angibault J-M, Bideau E, Cargnelutti B (2004) Experimental evidence of density dependence of home-range size in roe deer: a comparison of two long-term studies. Oecologia 139:478–485

Klein DR (1965) Ecology of deer range in Alaska. Ecol Monogr 35:259–284

Klein DR (1990) Variation in quality of caribou and reindeer forage plants associated with season, plant part, and phenology. Rangifer, Special Issue No. 3, pp.123–129

Klingel H (1969) The social organisation and population ecology of the plains zebra. Zool Afr 4:249–263

Knapp AK, Smith MD (2001) Variation among biomes in temporal dynamics of aboveground primary production. Science 291:481–485

Knight MH (1995) Drought-related mortality of wildlife in southern Kalahari and the role of man. Afr J Ecol 33:377–394

Knoop M-C, Owen-Smith N (2006) Foraging ecology of roan antelope: key resources during crucial periods. Afr J Ecol 44:228–236

Le Houerou HN, Bingham RL, Skerbek W (1988) Relationship between the variability of primary production and the variability of annual precipitation in world arid lands. J Arid Envir 15:1–18

Levin DA (1976) Alkaloid-bearing plants: an ecogeographic perspective. Am Nat 110:261–284
Lothan J, Staines BW, Gorman ML (1999) Comparative feeding ecology of red and roe deer in a Scottish plantation forest. J Zool Lond 247:409–418
Ludwig F, De Kroon H, Berendse F, Prins HHT (2004) The influence of savanna trees on nutrient, water and light availability and the understorey vegetation. Plant Ecol 170:93–105
Maddock L (1979) The "migration" and the grazing succession. In: Sinclair ARE, Norton-Griffiths M (eds) Serengeti: dynamics of an ecosystem. Univ Chicago Press, Chicago, pp 104–129
McCullough DR (1992) Concepts of large herbivore population dynamics. In: McCullough DR, Barrett RH (eds) Wildlife 2001: populations. Elsevier, Amsterdam, pp 967–984
McCullough DR (1997) Irruptive behavior in ungulates. In: McShea WJ, Underwood HB, Rappole JH (eds) The science of overabundance. Smithsonian, Washington DC, pp 69–99
McCullough DR (1999) Density dependence and life-history strategies of ungulates. J Mammal 80:1130–1146
McHugh T (1958) Social behavior of American buffalo. Zoologica–New York 43:1–40
McLoughlin PD, Dzus E, Wynes B, Boutin S (2003) Declines in populations of woodland caribou. J Wildl Manage 67:755–761
McShea WJ, Underwood HB, Rappole JH (1997) The science of overabundance. Smithsonian, Washington DC
Mduma SAR, Sinclair ARE, Hilborn R 1999 Food regulates the Serengeti wildebeest: a 40-year record. J Anim Ecol 68:1101–1122
Messier F, Crete M (1985) Moose–wolf dynamics and the natural regulation of moose populations. Oecologia 65:503–512
Mills MGL, Biggs HC, Whyte IJ (1995) The relationship between rainfall, lion predation and population trends in African herbivores. Wildl Res 22:75–88
Milner C, Gwynne D (1974) The Soay sheep and their food supply. In: Jewell PA, Milner C, Morton Boyd J (eds) Island survivors. The ecology of the Soay sheep of St. Kilda. Athlone Press, London, pp 273–325
Milner JM, Elston DA, Albon SD (1999) Estimating the contributions of population density and climatic fluctuations to interannual variation in survival of Soay sheep. J Anim Ecol 68:1235–1247
Mosienyane B (1979) Nutrient evaluation of Botswana range grasses for beef production. Thesis, Cornell University, Ithaca, NY
Mysterud A (1999) Seasonal migration patterns and home range of roe deer along an altitudinal gradient in southern Norway. J Zool Lond 247:479–486
Mysterud A, Perez-Barbieria FJ, Gordon IJ (2001) The effect of season, sex and feeding style on home range area versus body mass scaling in temperate ungulates. Oecologia 127:30–39
O'Connor TG, Haines LM, Snyman HA (2001) Influence of precipitation and species composition on phytomass of a semi-arid grassland. J Ecol 89:850–860
Ogutu J, Owen-Smith N (2005) Oscillations of large herbivore populations: are they due to predation or rainfall? Afr J Ecol 43:332–339
Ottichilo WK, De Leeuw J, Skidmore AK, Prins HHT, Said MY (2000) Population trends of large non-migratory wild herbivores and livestock in the Masai Mara Ecosystem, Kenya, between 1977 and 1997. Afr J Ecol 38:202–216
Owen-Smith N (1975) The social ethology of the white rhinoceros. Z Tierpsychol 38:337-384
Owen-Smith N (1977) On territoriality in ungulates and an evolutionary model. Q Rev Biol 52:1–38
Owen-Smith N (1981) The white rhinoceros overpopulation problem, and a proposed solution. In: Jewell PA, Holt S, Hart D (eds) Problems in management of locally abundant wild mammals. Academic Press, New York, pp 129–150
Owen-Smith N (1988) Megaherbivores. The influence of very large body size on ecology. Cambridge University Press, Cambridge
Owen-Smith N (1990) Demography of a large herbivore, the greater kudu, in relation to rainfall. J Anim Ecol 59:893–913

Owen-Smith N (1992) Grazers and browsers: ecological and social contrasts among African ruminants. In: Spitz F, Janeau G, Aulagnier S (eds) Ungulates / Ongules 91. SFEPM-IRGM, Toulouse, pp 175–181

Owen-Smith N (1993a) Woody plants, browsers and tannins in southern African savannas. S Afr J Sci 89:505–510

Owen-Smith N (1993b) Comparative mortality rates of male and female kudus: the costs of sexual size dimorphism. J Anim Ecol 62:428–440

Owen-Smith N (1994) Foraging responses of kudus to seasonal changes in food resources: elasticity in constraints. Ecology 75:1050–1062

Owen-Smith N (1997) Distinctive features of the nutritional ecology of browsing versus grazing ruminants. Z Saugetierkd 62 Suppl.II:176–191

Owen-Smith N (1998) How high ambient temperature affects the daily activity and foraging time of a subtropical ungulate, the greater kudu. J Zool Lond 246:183–192

Owen-Smith N (2000) Modeling the population dynamics of a subtropical ungulate in a variable environment: rain, cold and predators. Natur Res Mod 13:57–87

Owen-Smith N (2002a) Adaptive herbivore ecology. From resources to populations in variable environments. Cambridge University Press, Cambridge

Owen-Smith N (2002b) Credible models for herbivore - vegetation systems: towards an ecology of equations. S Afr J Sci 98:445–449

Owen-Smith N (2004) Functional heterogeneity within landscapes and herbivore population dynamics. Landscape Ecol 19:761–771

Owen-Smith N (2006) Demographic determination of the shape of density dependence for three African ungulate populations. Ecol Monogr 76:93–109

Owen-Smith N, Cooper SM (1985) Comparative consumption of vegetation components by kudus, impalas and goats in relation to their commercial potential as browsers in savanna regions. S Afr J Sci 81:72–76

Owen-Smith N, Cooper SM (1989) Nutritional ecology of a browsing ruminant, the kudu, through the seasonal cycle. J Zool Lond 219:29–43

Owen-Smith N, Mason DR (2005) Comparative changes in adult versus juvenile survival affecting population trends of African ungulates. J Anim Ecol 74:762–773

Owen-Smith N, Mills MGL (2006) Manifold interactive influences on the population dynamics of a multi-species ungulate assemblage. Ecol Monogr 76:73–92

Owen-Smith N, Mills MGL (submitted) Shifting prey selection generates contrasting herbivore dynamics with a large-mammal predator-prey web. Ecol Monogr

Owen-Smith N, Ogutu JO (2003) Rainfall influences on ungulate population dynamics in the Kruger National Park. In: du Toit JT, Rogers KH, Biggs HC (eds) The Kruger experience: ecology and management of savanna heterogeneity. Island Press, Washington, DC, pp 310–331

Owen-Smith N, Mason DR, Ogutu JO (2005) Correlates of survival rates for ten African ungulate populations: density, rainfall and predation. J Anim Ecol 74:774–788

Pappageorgiou NK (1978) Food preference, feed intake and protein requirements of red deer in central Greece. J Wildl Manage 42:940–943

Patterson BR, Power VA (2002) Contributions of forage competition, harvest, and climate fluctuations to changes in population growth of northern white-tailed deer. Oecologia 130:62–71

Pellew RA (1983) The giraffe and its food resource in the Serengeti. I. Composition, biomass and production of available browse. Afr J Ecol 21:241–267

Peterson RO (1999) Wolf–moose interactions on Isle Royale: the end of natural regulation? Ecol Applic 9:10–16

Peterson, RO, Thomas NJ, Thurber JM, Vucetich JM, Waite TA (1998) Population limitation and the wolves of Isle Royale. J Mammal 79:828–841

Post E, Stenseth NC (1998) Large-scale climatic fluctuation and population dynamics of moose and white-tailed deer. J Anim Ecol 67:537–543

Post E, Stenseth NC (1999) Climatic variability, plant phenology, and northern ungulates. Ecology 80:1322–1339

Prins HHT (1988) Plant phenology patterns in Lake Manyara National Park, Tanzania. J Biogeogr 15:465–480
Prins HHT (1996) Ecology and behaviour of the African buffalo. Chapman & Hall, London
Prins HHT (2000) Competition between wildlife and livestock in Africa. In: Prins HHT, Grooterhuis JG, Dolan TT (eds) Wildlife conservation by sustainable use. Kluwer, Dordrecht, pp.51–80
Prins HHT, Beekman JH (1989) A balanced diet as a goal for grazing: the food of the Manyara buffalo. Afr J Ecol 27:241–259
Prins HHT, Iason GR (1989) Dangerous lions and nonchalant buffalo. Behaviour 108:262–297
Prins HHT, Weyerhaus FJ (1987) Epidemics in populations of wild ruminants: anthrax and impala, rinderpest and buffalo in Lake Manyara National park, Tanzania. Oikos 49:28–38
Putman RJ (1986) Grazing in temperate ecosystems: large herbivores and their ecology in the New Forest. Croon Helm, London
Renecker LA, Hudson RJ (1988) Seasonal quality of forages used by moose in the aspen-dominated boreal forest, Alberta. Holarctic Ecol 11:111–118
Riney T (1964) The impact of introductions of large herbivores on the tropical environment. IUCN public, new series no. 4, pp 261–273
Risenhoover KL, Maass SA (1987) The influence of moose on the composition of Isle Royale forests. Can J For Res 17:357–364
Robbins CT (1993) Wildlife feeding and nutrition. Academic Press, New York
Robbins CT, Moen AN (1975) Composition and digestibility of several deciduous browses in the north-east. J Wildl Manage 39:337–341
Rutherford MC (1984) Relative allocation and seasonal phasing of growth of woody plant components in a South African savanna. Progr Biometeorol 3:200–221
Ryan SJ, Knechel CU, Getz WM (submitted) Seasonal and interannual variation in home range and habitat selection of African buffalo: a long-term study in the Klaserie Private Nature Reserve, South Africa. J Wildl Manage 70:764–776.
Saether B-E (1997) Environmental stochasticity and population dynamics of large herbivore: a search for mechanisms. Trends Ecol Evol 12:143–149
Saether B-E, Andersen R (1990) Resource limitation in a generalist herbivore, the moose: ecological constraints on behavioural decisions. Can J Zool 68:993–999
Schoen JW, Kirchoff MD (1985) Seasonal distribution and home-range patterns of Sitka black-tailed deer on Admiralty Island, Southeast Alaska. J Wildl Manage 49:96–103
Scholes RJ, Walker BH (1993) An African savanna. Synthesis of the Nylsvley study. Cambridge University Press, Cambridge
Scoones I (1993) Why are there so many animals? Cattle population dynamics in the communal areas of Zimbabwe. In: Behnke RH, Scoones I, Kerven C (eds) Range ecology at disequilibrium. Overseas Development Institute, London, pp. 62–76
Scoones I (1995) Exploiting heterogeneity: habitat use by cattle in dryland Zimbabwe. J Arid Envir 29:221–237
Seip DR (1992) Factors limiting woodland caribou populations and their interrelationships with wolves and moose in southeastern British Columbia. Can J Zool 70:1494–1503
Sinclair ARE (1977) The African buffalo. A study of resource limitation in populations. University of Chicago Press, Chicago
Sinclair ARE, Arcese P (1995) Population consequences of predation-sensitive foraging: the Serengeti wildebeest. Ecology 76:882–891
Skinner JD (1993) Springbok treks. Trans Roy Soc S Afr 48:291–305
Skinner JD, Louw GN (1996) The springbok. Transvaal Museum Monograph No. 10
Smuts GL (1978) Interrelations between predators, prey, and their environment. BioScience 28:316–320
Solberg EJ, Loison A, Gaillard J-M, Heim M (2004) Lasting effects of conditions at birth on moose body mass. Ecography 27:677–687
Stenseth NC, Mysterud A, Ottersen G, Hurrell JW, Chan KS, Lima M (2002) Ecology and climatology: ecological effects of climatic fluctuations. Science 297:1292–1298

Stenseth NC, Ottersen G, Hurrell JW, Mysterud A, Lima M, Chan KS, Yoccoz NG, Adlandsvik B (2003) Studying climate effects on ecology through the use of climate indices: the North Atlantic Oscillation, El Nino Southern Oscillation and beyond. Proc R Soc Lond B 270:2087–2096

Talbot LM, Talbot MH (1963) The wildebeest in western Masailand, East Africa. Wildl Monogr No. 12

Tixier H, Duncan P (1996) Are European roe deer browsers? A review of variations in the composition of their diets. Rev Ecol (Terre Vie) 51:3–17

Tyson PD, Gatebe CK (2001) The atmosphere, aerosols, trace gases and biogeochemical change in southern Africa: a regional integration. S Afr J Sci 97:106–118

van de Vijver CADM, Poot P, Prins HHT (1999) Causes of increased nutrient concentrations in post-fire regrowth in an East African savanna. Plant Soil 214:173–185

Walker BH, Ludwig D, Holling CS, Peterman RS (1981) Stability of semi-arid savanna grazing systems. J Ecol 69:473–498

Walker BH, Emslie RH, Owen-Smith N, Scholes RJ (1987) To cull or not to cull: lessons from a southern African drought. J Appl Ecol 24:381–402

Whyte IJ, Joubert SCJ (1988) Blue wildebeest population trends in the Kruger National Park and the effects of fencing. S Afr J Wildl Res 18:78–87

Williamson D, Mbano B (1988) Wildebeest mortality during 1983 at Lake Xau, Botswana. Afr J Ecol 26:341–344

Williamson D, Williamson J, Ngwamotsoko KT (1988) Wildebeest migration in the Kalahari. Afr J Ecol 26:269–280

Wilson VJ (1970) Data from the culling of kudu in the Kyle National Park, Rhodesia. Arnoldia (Rhodesia) 4:1–26

Ydenberg RC, Prins HHT (1981) Spring grazing and the manipulation of food quality by barnacle geese. J Appl Ecol 18:443–453

Chapter 7
Species Diversity of Browsing and Grazing Ungulates: Consequences for the Structure and Abundance of Secondary Production

Herbert H.T. Prins and Hervé Fritz

7.1 Introduction

There are two fundamentally different ways to look on the way consumers use their resources. The first is that different species partition the resource in a particular manner, so that some species get a particular part and other species another part. In this view, the sum of the shares is equal to the total. The second way of looking at resource partitioning is shaped by thoughts about niche differentiation: species differ in their ability to extract particular resources from a continuum, and because they are different there is an additive effect if different species utilise an area together. Specialization leads to increased partitioning of resources which leads to a higher total offtake. We are interested in the question, then, whether increased species richness of vertebrates that make use of the vegetation—that is, ungulate grazers and browsers—leads to an increased offtake of the plant biomass in an area, and subsequently whether, if true, this translates into higher herbivore biomass and/or productivity.

This question is of interest for several reasons. The first is from a theoretical point of view. Much work has been done since the niche concept was developed in the 1920s by Grinell, and later examined closely again in the 1950s and 1960s by Hutchinson and McArthur. The quest for finding order among all the competing species may have come from a corporatist world view and a harking back to the guild structure of earlier societal organisation. Yet, the niche concept has been an enormous stimulus to ecology, and has long been shaping our thinking. In the 1980s and 1990s, however, the niche concept met increasing resistance among ecologists, culminating in Hubbell's neutral theory (Hubbell 2001). This may have been a reflection, again, of society's drift towards neo-liberalism and its associated notions about a totally free market. The neutral theory implies functional redundancy among species, and hence absence of impacts of changes in diversity on functional processes at the community or ecosystem scale. At the other extreme, niche theory postulates that all species differ to some extent in the resources they use. This implies functional complementarity among species, and hence, for instance, increased productivity and other ecosystem processes with diversity (Tilman et al. 1997; Loreau 1998). So, theory is moving hither and thither; a confrontation with facts can be useful to ascertain whether the direction in which theoreticians are driving us is the right one or not, because only encounter with reality will help us decide which theory we'd best embrace. However, as often, apparently opposed theory may in fact apply to different ecological situations; a mutually exclusive situation does not necessarily exist (Holyoak and Loreau 2006; Leibold and McPeek 2006).

From a purely thermodynamic and energy-capture-rate viewpoint, one also may reason that a part of the Earth's surface has a particular capacity for harnessing the energy of photons into chemical energy through the action of chlorophyll. That process depends on the conversion efficiency of plants (actually only about 3%) and the amount of leafy material. Neither grazing nor browsing will affect the conversion efficiency per se, but grazing or browsing could increase the amount of living phytomass (McNaughton et al. 1988; du Toit et al. 1990), and it could be envisaged that a particular combination of different herbivore species could result in a higher primary production, and because of that a higher secondary production. Even if primary production remained constant, different combination of herbivores may use it more thoroughly, leaving less of it to decomposers, also leading to a higher biomass and secondary production.

The second reason to be interested in the question of the efficiency of harvesting has to do with management. Many people in many societies are interested in whether harvests can be optimised. Much of the Earth's primary production is inedible for humans, and we use grazers and browsers to transfer this primary production into resources that are of direct use to us: meat, milk, hides, bones, hooves, and organs all have a direct or indirect use in human society. Again, we thus ask ourselves whether a judicious combination of different herbivore species could result in a higher secondary production. Much agriculture is done through the monoculture of a crop, even though De Wit (1960) and others have shown that from a production point mixed crops are more productive. The reason that farmers choose for monoculture has to do with management, not with productivity. Our management question has direct effect on the relevance of biodiversity: is a high diversity of grazers and browsers 'good'? We know it is pleasing to see many different forms of herbivores in an area—indeed, this is the basis for much eco-tourism—but does high biodiversity have other benefits? In this chapter we will first ask the question, 'What causes species richness, especially of ungulates?', and then we will try to answer the question 'Does increasing the species richness of the herbivore community lead to a higher secondary production and more efficient use of the vegetation?' We will pay special attention to whether combining browsers with grazers leads to a higher offtake than either grazers alone or browsers alone. In our chapter we will not concentrate on '*ad hominum*' (*ad animalum*?) types of explanation in which every species is so uniquely adapted to its niche that general patterns cannot be found. Indeed, we keep in mind that 'fiber digestion is not significantly different between browsers and grazers, although fiber digestion is positively related to herbivore size' (Robbins et al. 1995), and that 'after controlling for the effects of body mass, there is little difference in digestive strategy [among] (African) ruminants with different morphological adaptations of the gut' (Gordon and Illius 1994).

7.2 Suggested Causes of Species Richness

Finding the cause of species richness is like the quest for the Holy Grail. The question is difficult to formulate precisely, because it entails evolution and phylogenetics (see Janis, Chapter 2), special adaptations (see Clauss, Chapter 3), past and present competition (see Duncan and Poppi Chapter 4; Searle and Shipley, Chapter 5) and different population dynamics (see Owen-Smith, Chapter 6), yet the literature discloses different possibilities, and here we single out the following (Box 7.1):

Oindo et al. (2001) and Oindo (2002) looked in great detail through the use of satellite remote sensing and weather data at the spatial distribution of species richness in Kenya, Tanzania, and Uganda. They looked at different groups of organisms, including ungulates, and came to the conclusion that by looking at the normalised vegetation index (NDVI, a good proxy for phytomass production), variability in primary production correlates well with species richness. Also Janis et al. (2000) search for the cause of ungulate species richness in primary production, but they think it is linked to the average levels of plant productivity: 'Both maximum species richness of all ungulates and the proportion of browsers declined steadily in the ungulate communities through the middle Miocene, to levels comparable to those of the present by the late Miocene. We suggest that the early Miocene [17 Ma] browser-rich communities may reflect higher levels of primary productivity in Miocene vegetation, compared with equivalent present-day vegetation types. The observed decline in species richness may represent a gradual decline in primary productivity, which would be consistent with one current hypothesis of a mid-Miocene decrease in atmospheric CO_2 concentrations from higher mid-Cenozoic values'. Note that this is a correlative conclusion, and that a mechanism for the link between species richness and primary productivity is not suggested. Olff, Ritchie and Prins (2002) modelled a possible cause for species richness on the basis of existing theory and available data from across Africa. They then made predictions concerning the found relation to other continents, and tested it for North American data. They came to the conclusion that 'More plant-available moisture reduces the nutrient content of plants but increases productivity,

---

Box 7.1. Biodiversity, that is, species richness, of ungulate assemblages is enhanced by:

Low intra-annual variation in NDVI = primary production (*Oindo 2002; Oindo et al. 2001*)),

High inter-annual average NDVI (*Oindo 2000; Oindo et al. 2001*)

High primary production (*Janis et al. 2000*)

High soil nutrients in combination with intermediate precipitation (*Olff, Ritchie & Prins 2001*)

High spatial heterogeneity of ecosystems (*du Toit & Cumming 1999*)

whereas more plant-available nutrients increase both of these factors. Because larger herbivore species tolerate lower plant nutrient content but require greater plant abundance, the highest potential herbivore diversity should occur in locations with intermediate moisture and high [soil] nutrients. ... Thus gradients of precipitation, temperature and soil fertility might explain the global distribution of large herbivore diversity' (Olff et al. 2002). A fourth explanation for ungulate species richness looks at a different putative mechanism, namely spatial heterogeneity: 'This exceptional fauna diversity and herbivore biomass density is *directly* linked [italics added] to the high spatial heterogeneity of African savanna ecosystems. The dependence of herbivore dietary tolerance on body size translates into important size-related differences between savanna ungulate species in terms of habitat specificity, geographical range, and the share of community resources exploited' (du Toit and Cumming 1999). How this direct link works is not explained, but that is here beside the point. The point is that species richness is thought to be caused by something—it is not a random phenomenon. The exact cause is not clear yet and at this moment the approach of Prins and Olff (1998) and Olff et al. (2002), which was strongly inspired by Hutchinson (1957), is the most causal-analytical.

In all studies but that of Janis et al. (2000), the patterns and the likely processes do not distinguish between browsers and grazers, and mostly concentrate on body size. There may be commonalities between browsers and grazers that only body size explains, e.g., volatile fatty acid (VFA) in digestive physiology (Gordon and Illius 1994) but there are differences in body-size distribution between the two dietary types. Grazers tend to be larger on average than browsers, but body size spans equally for both dietary groups, with the largest ruminant in fact being a browser (namely, the giraffe). Body size is unlikely to encompass all traits associated with browsers and grazers, as many other traits vary between trophic guilds, although body size certainly also co-varies partly with most of them (reviews Gordon 2003; Fritz and Loison 2006). Obviously, the different reaction to the distribution of their primary resources will also affect the relative abundance and diversity of browsers and grazers, as suggested by Janis et al. (2000), and possibly ultimately secondary production. As a matter of fact, browser and grazer biomasses respond similarly to annual rainfall (same slopes) in African savanna ecosystems, but browsers remain lower in biomass for a given annual rainfall (Fig. 7.1). This is even more true when mixed-feeders such as elephants are removed from the comparison (mixed-feeders are split into the grazer and browser component according to their graze/browse share in the diet). The fact that browse resources are more dispersed and less abundant in rangelands may explain this pattern.

Interestingly, the numbers of species in the grazer and browser guilds in savannas show the same quadratic shape with annual rainfall but with a lower species richness in browsers (Fig. 7.2). This suggest that the environmental determinants of species richness in grazers and browsers may be similar, but the pool of species may be smaller in savanna browsers due to the lower browse production in savannas (compared to forest, for instance) and possibly also paleo-historical changes in landscape that induced more losses in browsers (e.g., Janis et al. 2000 for North America).

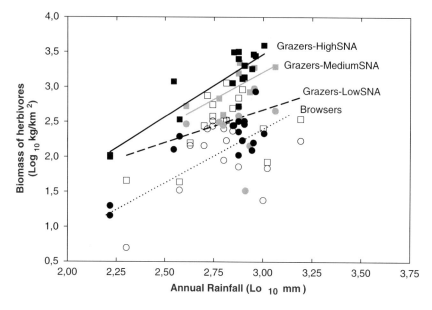

Fig. 7.1 Relationship between annual rainfall, soil nutrient status and the abundance of the two major feeding guilds, grazers (*squares*) and browsers(*circles*). Soil nutrient availability (SNA) was only significant for grazers

7.3 The Effect of Species Richness on Ecosystem Functioning: An Overview

So, species richness of assemblages of browsers, grazers, mixed-feeders, or combinations thereof is different at different places and at different times (see also Janis, Chapter 2). What does this mean to ecosystem functioning? In Box 7.2 we present some hypotheses that have been floating around in the literature already for a long time but which are still of great interest:

In this chapter we are only interested in the first effect; the increased species richness of herbivores on nutrient retention, resilience, resistance have to our knowledge never been tested, and here we do not deal with the buffering effect but with the increased productivity. Of course, we are interested to know if increased species richness leads to increased secondary productivity, not on if it increases primary productivity.

With plants it has been shown about 50 years ago that increased species richness leads to increased production (De Wit 1960), but with the new interest in questions concerning ecosystem functioning the same insights are now re-appraised and reformulated. 'There is evidence that biodiversity loss can lead to reductions in biomass production, … Under the unperturbed conditions, the species-poor systems achieved lower biomass production than the species-rich systems' (Pfisterer and Schmidt 2002). Other studies report the same, e.g., 'Recent studies on grasslands demonstrate that species losses and subsequent changes in diversity can … alter

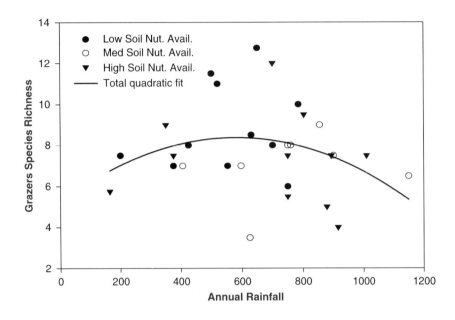
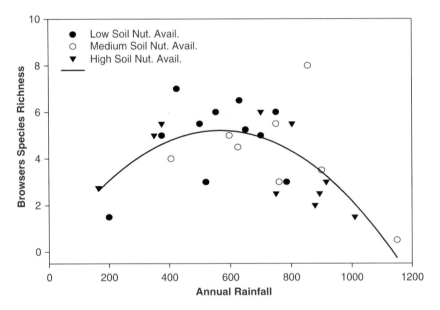

Fig. 7.2 The relationship between species richness (number of species) as a function of annual rainfall and soil nutrient status, for grazers (*upper graph*) and browsers (*lower graph*). Only the quadratic relationship was significant, not the soil nutrient status. Rainfall only explains 10% of the observed variance in grazers whereas it explains 41% in browsers

> **Box 7.2.** Increased biodiversity, measured as increased species richness, causes:
> Increased productivity (Tilman *et al*. 2001, Engelhard & Ritchie 2002)
> Increased nutrient retention (Hiremath & Ewel 2001, Cardinale *et al*. 2002)
> Increased resilience (<u>not</u> supported: Engelhardt & Kadlec 2001: no effect)
> Increased resistance (<u>not</u> supported: Pfisterer & Schmidt 2002: even opposite)
> Reduced temporal variability in ecosystem processes; buffering effect, including an 'insurance effect' (probable: Yachi & Loreau 1999, Loreau 2000, Loreau *et al*. 2001).

ecosystem functioning (e.g., productivity' Engelhardt and Kadlec 2001), and 'Plant diversity and niche complementarity had progressively stronger effects on ecosystem functioning ... with 16-species plots attaining 2.7 greater biomass than monoculture. Diversity effects were neither transient nor explained solely by a few productive or unviable species. ... Even the best chosen monocultures cannot achieve greater productivity or carbon stores than higher-diversity sites' (Tilman et al. 2001). These findings have been well summarised as 'Positive short-term effects of species diversity on ecosystem processes, such as primary productivity and nutrient retention, have been explained by two major types of mechanisms: (1) functional niche complementarity (the complementarity effect; Engelhard and Ritchie 2002 call this 'the niche differentiation effect'), and (2) selection of extreme trait values (the selection effect; Engelhard and Ritchie 2002 call this the 'sampling effect'). In both cases, biodiversity provides a range of phenotypic trait variation. In the complementarity effect, trait variation then forms the basis for a permanent association of species that enhance collective performance. In the selection effect, trait variation comes into play only as an initial condition, and a selection process then promotes dominance by species with extreme trait values' (Loreau 2000; see also Bond and Chase 2002). All these studies dealt with plants or plankton, not with animals or even vertebrates. Partly that is because vertebrate studies are more difficult to conduct, but partly it is because plant ecologists have discovered that good experimental studies that are well designed may lead to answers much faster than observational studies.

A higher ecosystem nutrient use efficiency (the ratio of net primary productivity to soil nutrient supply), fostered by higher plant species diversity, is an integrative measure of ecosystem functioning (Hiremath and Ewel 2001). We maintain that the ecosystem nutrient use efficiency is defined too narrowly in modern ecosystem studies that are dominated by plant ecologists. We maintain that *secondary productivity*, that is, the ratio of secondary primary productivity to nutrient supply from plant resources is also an integrative measure of ecosystem functioning (and by the same token tertiary productivity by predators). So, does higher consumer diversity lead to higher secondary productivity? As higher diversity at a given trophic level may affect the stocks conditioning the fluxes between ecosystem compartments, as well as productivity, we also investigated the role of species diversity in the biomass of herbivore assemblages.

7.4 Herbivore Diversity and the Use of Primary Production

One of the fundamental assumptions behind the prediction relating diversity to biomass or to productivity is that an increase in species richness induces a more efficient, or more complete, use of the primary production. We thus first investigated whether sympatric species of ungulates used resources differently, before exploring the possible effect of species diversity on herbivore biomass and productivity.

7.4.1 Diet Overlap and Feeding Niches

Many studies have been conducted, on domestic herbivores, wild herbivores, and on combinations of wild and domestic ones, on the issue of diet overlap. These studies generally point out that there is some degree of niche segregation among different types of herbivore. Here we give a short overview of different type of results that have been found.

In semi-arid temperate grassland diet overlap between red deer *Cervus elaphus* and cattle varied greatly depending on availability of palatable fractions of herbs, shrubs and grasses; red deer were better shrub users (28–50% in diet) than cattle (6–12%) (Pordomingo and Rucci 2000). In a Louisiana pine range white-tailed deer *Odocoileus virginianus* and cattle diet overlap was 11–31%. Deer mostly used browse and herbs, cattle graminoids (Thill and Martin 1986). In California dietary overlap between black-tailed deer *Odocoileus hemionus* and elk *Cervus elaphus* was lowest in wet winter months (dietary N highest, standing crop lowest), and overlap highest in dry summer months (dietary N lowest, standing crop highest) (Gogan and Barrett 1995). In Colorado the diet overlap between mule deer *Odocoileus hemionus* and elk was 3% in winter and 48% in summer; between elk and cattle it was 30–50% in summer, while at the same time of the year it was 12–38% between mule deer and cattle (Hansen and Reid 1975). Also in Colorado diet overlap between mule deer and cattle was 2–11%; between mule deer and horse it was 2–11%, which was indicative more of a complementary than of a competitive relationship (Hubbard and Hansen 1976). Again in Colorado diet overlap between mule deer and cattle was 1–22%; the authors observed that 'When cattle are forced from a grass-dominated diet to browse forage on overgrazed ranges, diet overlap and forage competition between deer and cattle increase'. (Lucich and Hansen 1981). Diet overlap was also studied among domestic cattle, sheep, bison (*Bison bison*), and pronghorn (*Antilocapra americana*) in Colorado, leading to the conclusion that diet overlap appeared to depend on recent evolutionary history and on body size, though values were strongly influenced by forage quantity and quality (Schwartz and Ellis 1981). In West Virginia the diet overlap between cattle and sheep was 76%, while between cattle and goats 75% (Cox-Ganser 1990).

European studies found the same sort of dietary overlaps between grazers or browsers. In Northern Fennoscandia diet overlap between moose (*Alces alces*) and roe deer (*Capreolus capreolus*) was 21–34%, between moose and red deer 32%, between red deer and sheep 59–64%, between sheep and reindeer (*Rangifer tarandus*) 55% and finally between sheep and goat it was 77%. Neither difference in feeding type nor body mass successfully predicted diet overlap (Mysterud 2000). In Scotland: 'Deer showed no change in the proportion of grass in their diet in the presence or absence of sheep, but … the diet of sheep contained a significantly higher proportion of grasses when they were grazing with red deer (52% versus 38%)' (Cuartas et al. 2000). Goats grazed *Myrica*, *Juncus* and *Molinea* more than sheep, while the sheep preferred *Caluna* which was not significantly grazed by goats, but the overlap was considerable (Fisher et al. 1994). Also in the Netherlands overlap between red deer and other ungulates was large (in summer 70% and in winter 62% for cattle and red deer, and 58% in summer and 77% in winter for ponies and red deer (Van Wieren 1996). In the French Vosges an analysis of stomach contents showed an overlap in diet between red deer and roe deer ranging 28–55% in winter months and 26–51% in summer months (Storms et al. 2006). In the Camargues in France horses and cattle largely overlapped in their niches (58–77%), both for habitat and food (Ménard et al. 2002).

Latin American studies also find differing degrees of dietary overlap. In Mexico diet overlap between white-tailed deer and cattle was 51% (Gallina 1993). In the dry areas of Brazil, during the wet season the diets of goats and sheep was quite different but by the end of the dry-wet transition period intake of grasses and woody plants was similar. There was a high similarity of diets (Araujo Filho et al. 1996). Interestingly, there was not a strong dietary overlap between sheep and the indigenous llama (*Lama glama*) in Bolivia (Genin et al. 1994); neither was there too much overlap between diets of indigenous Venezuelan deer (*Odocoileus virginianus*) with cattle or capybara (*Hydrochoerus hydrochaeris*) because 93% of their diet originated from the wooded fringe area. The capybara did not compete with cattle in the extensive intermediate area, and the taller and drier herbage was preferred by horses and cattle but not by capybara. However, in the natural habitat of the capybara, the lowest region, there was substantial diet overlap (Escobar and Gonzalez 1976).

Studies like these have also been conducted in Africa. Dekker (1997) gives diet overlaps for Messina in South Africa, and du Toit et al. (1995) found that the dietary overlap between sheep and goats in the Karoo differed; it was 95–96% during the growing season, and 79–86% during the dormant season. In Senegal cattle selected a very different diet from goats, and sheep were intermediate; the differences in diet among animal species declined in the dry season (Nolan et al. 1996). Other studies highlighted the diet overlap or dietary difference between domestic species and indigenous ones. The diet and feeding height of kudu (*Tragelaphus strepsiceros*) and goats and of black rhinoceros (*Diceros bicornis*) and goats overlapped to a large extent. Overlap in diet between giraffe (*Giraffa camelopardalis*) and goats was extensive but overlap in feeding height was small; goats and eland (*Taurotragus oryx*), despite feeding at similar heights, generally consumed different species (Breebaart et al. 2002). Makhabu (2005) on the Chobe riverfront, Botswana, found around 20% overlap among elephant (*Loxodonta africana*) and three other browsers (giraffe, kudu, impala; *Aepyceros melampus*) both in dry and wet season, but that the overlap in plant use ranged from 56% to 76% among the three other species in the wet season and from 57% to 82% in the dry season. However, although plant parts use also overlapped in a similar way (49–72%), giraffe, kudu, and impala overlapped less in feeding heights, especially in the dry season (4%–32%). In Tanzania cattle overlapped with zebra (*Equus burchellii*) in the early wet season and with wildebeest (*Connochaetes taurinus*) in the early dry season; in the wet season, cattle showed overlap in resource use with both zebra and wildebeest (Voeten and Prins 1999). Other studies focussed on natural assemblages without domestic species. In Kenya diet overlap between ungulates and very small herbivores is low (French 1985), but between large herbivores the overlaps are large. For instance, in Lake Nakuru National Park, Mwasi (2002) found the following overlaps: between impala and African buffalo (*Syncerus caffer*), late wet season 58%, short dry season 81% and early wet season 75%; between impala and common zebra, late wet season 83%, short dry season 55%, and early wet season 82%. In Uganda there were significant seasonal differences in the diet of most of the herbivores, including buffalo, Uganda kob (*Kobus kob*), topi (*Damaliscus lunatus*), warthog (*Phacochoerus aethiopicus*), waterbuck (*Kobus ellipsiprimnus*), and hippopotamus (*Hippopotamus amphibus*); and there was greater separation in the longer dry season (Field 1972). In the Democratic Republic of Congo Hart (1986) found that the diets of all species (six varieties of duiker and chevrotain) in the upland forest converged when high quality fruits and seeds were abundant, diets diverged when high quality food was scarce. During scarcity some species showed habitat segregation, other segregated along lines of fruit specialisation. Diet overlap occurred in the mixed forest when both food abundance and diversity were low. In Mozambique, Prins et al. (2006) found considerable overlap between duiker antelopes and suni (*Neotragus moschatus*) in the wet season (63–83%) but less in the dry season (21–38%). Overall, only 10% of dietary items were species-exclusive in any given season. Only a few studies into diet overlap among different herbivores were conducted in Asia, but they show the same picture. In Ladakh, for example, Mishra (2001) reported the following overlaps from the high-altitude grasslands there: between blue sheep (*Pseudois nayaur*) and domestic yak, summer 61%, winter 52%; between blue sheep and donkey, summer 43%, winter 96%; between goat and yak, summer 84%, winter 72%; and finally between goat and donkey, summer 66%, winter 92%. Also, ibex (*Capra sibirica*) has a very similar diet and habitat to goats and sheep in these systems, suggesting competition to explain its absence from pastoral zones (Bagchi et al. 2004). Conversely, in Nepal, the overlap in plant use was very low among blue sheep, argali sheep (*Ovis ammon hodgsoni*) and domestic goat (1–8%), although broad diet composition in terms of grass, forbs, and browse were less different (overlap in categories from 20% to 76%); forb and browse species use discriminated the ungulate species (Shrestha et al. 2005). In the more semi-arid areas in India, the diet similarities among nilgai (*Boselaphus tragocamelus*), chital (*Axis axis*), and chinkara (*Gazella bennetti*) were very high in the dry season, but chital used different habitat, more similar to sambar (*Cervus unicolor*), being fairly well segregated from the three other species (Bagchi et al. 2003).

A general picture does not appear from these studies. Most of them have been descriptive. The general conclusion is that if different herbivore species, whether they are browsers or grazers, utilise a given area then there is generally considerable overlap in diet but there is some segregation, too. Many conclusions concerning competition or the lack thereof have been drawn from these studies in diet overlap or dietary segregation. The picture emerging from these studies is, however, disconcertingly unclear. As a general rule it appears to emerge that if there are more herbivore species, then a wider array of plant species are being consumed by the assemblage in total.

7.4.2 Postulated Advantages of Mixed-Species Feeding

It is interesting to note that scientists who have studied niche overlap have often stressed niche segregation, and from their studies have made inferences about ecosystem functioning and secondary productivity without providing the necessary productivity data to underscore their contentions. Milton (2000) formulated it as a very clear hypothesis as 'diversification of livestock (through grazer-browser combinations) tends to stabilise or *enhance* utilisable secondary production' (italics added). Sometimes this hypothesis is implicit in the studies we review, for example, 'Diets differed only by 4-5% during the growing season. *This margin was considered too small to recommend combining small stock breeds in an effort to ensure greater utilisation efficiency through multiple use of the vegetation*' (du Toit et al. 1995; italics added). Other studies are much more explicit. For instance, Breebaart et al. 2002) propose 'a mixed farming system which includes goats, eland and giraffe as a useful management tool for using savanna vegetation more efficiently', and Owen-Smith (1985) concluded that 'the kudu is a prime candidate for the inclusion alongside cattle in mixed species ranching enterprises in most regions of savanna vegetation', and Genin et al. (1993) wrote the absence of 'a strong dietary overlap between [sheep and llama], suggested that mixed grazing could allow a better utilization of the vegetation'. A study conducted on cattle ranches in Zimbabwe came to the conclusion that 'there is a definite need for a browser to utilise woody plants and to balance the present monospecies ranching system. ... it is concluded [on basis of the evaluation of the attributes of eland and other browsers] that eland are very well adapted to complement cattle in the ranching industry' (Lightfoot and Posselt 1977), and in a South African review it was concluded that 'The full production potential of the thornveld areas can be achieved with cattle farming as the primary enterprise and with goats playing a secondary role' (Aucamp 1976).

Jewell (1980) was unambiguous when he stated that 'natural communities of game animals exhibit a high standing crop biomass because their ecological separation, particular in their utilisation of food resources renders many species complementary. A high density of one species may facilitate energy flow and the success of another herbivorous species'. Also Nolan et al. (1999) in a review of 87 references explicitly state 'Complementary grazing behaviour patterns among different animal types improve individual animal performance and output per unit of area ...'. The conclusion thus seems to be rather exact and unquestionable that higher consumer diversity leads to higher secondary productivity. However, even though many animal ecologists have concluded, sometimes speculated, on the issue of secondary productivity, very few if any controlled studies have been done in which the herbivore assemblage was manipulated while productivity was measured. From our review, it is clear that we have to turn to the agricultural literature to explore this relationship further; and it is also evident that conclusions about a positive diversity–biomass relationship in herbivores require more rigorous analyses across sites of varying diversity, controlling for environmental parameters.

7.5 Mammalian Herbivore Species Richness Links to Secondary Productivity and Biomass

Why would an increased diversity of consumers lead to an increased productivity of those consumers? There are at least two main mechanisms that could be involved: one is that the efficiency of biomass transfer from the one trophic level to the next is influenced by the number of species in the consumer as well as in the producer guild; second is that the apparent increase in total use of the primary production, as documented above, increases the consumer biomass irrespective of transfer efficiency, or increases the availability of plant resources through facilitation (Cardinale et al. 2002). Agriculture experiments provide simple cases to explore the functional link between herbivore diversity and secondary productivity, mainly to test the possible changes in biomass transfer efficiency, as the diversity of plants and consumers is often too low to explore adequately the effect of niche complementarities on secondary biomass, and possibly productivity. This latter issue is best investigated in wild ungulate assemblages, or eventually in pastoral herds, more diverse and faced with a heterogeneous primary production.

7.5.1 Domestic Herbivore Diversity and Secondary Productivity

We evaluated many agricultural experiments. Table 7.1 shows that very often the combination of two grazers leads to an increase of total secondary productivity of the system, but a combination of a browser and a grazer does not (Table 7.2): actually that combination quite often led to a *decrease*. Mixed grazing can thus increase secondary production, but this does not occur always. The exact interaction between different herbivores and vegetation dynamics appears to be of great importance: an increase is found when sheep and cattle are grazed together, but it rarely occurs when sheep and goats or goats and cattle are grazed together. If secondary productivity does not merely depend on the combination of different classes of species ('browsers' versus 'grazers') but if species-specific idiosyncratic differences

Table 7.1 Effect of mixed grazing by two grazing species (sheep and cattle) on secondary productivity (kg/ha per year or per grazing season): very frequently, the combination of two grazing species leads to an increase total productivity

Sheep		Sheep + cattle		Cattle	Reference
		++	>	+	Dickson et al. 1981
+	=	+			Nolan and Connolly 1989
		++	>	+	Nolan and Connolly 1989
		++	>	+	de Boer and Hanekamp 1992
		++	>	+	Logan et al. 1991
++	=	++	>	+	Olson et al. 1999
		++	>	+	Martinez et al. 2002
+	=	+	=	+	Hamilton 1976
+	<	++			Abaye et al. 1994

Table 7.2 Effect of mixed grazing by grazers (sheep or cattle) and browsers (goats) on secondary productivity (kg/ha per year or per grazing season): only rarely the combination of a grazer and a browser lead to an increased total productivity; normally it does not

Goats		Goats + cattle		Cattle	Reference
		++	>	+	Martinez et al. 2002
		+	<	++	Donaldson 1979
		+	<	++	Leite et al. 1995

Sheep		Sheep + goat		Goats	
+	=	+	=	+	Wilson and Mulham 1980 (normal)
++	>	+	<	++	Wilson and Mulham 1980 (drought)
		+	<	++	Donaldson 1979

between different species are important, it is important to look at the exact effects on the resource when species graze together.

Ecological theory suggests that small grazers outcompete larger ones and that larger grazers facilitate smaller ones (Illius and Gordon 1987, Prins and Olff 1998, Huisman and Olff 1998). We evaluated this by assessing which species was gaining (in terms of productivity) when grazed in combination in comparison to when it was husbanded in a single-species setting, and which species lost (Table 7.3). There is no clear picture emerging: sometimes a small species benefits from a large one but sometimes a large one to the detriment of the smaller; sometimes a grazer benefits and sometimes a browser. The explanation does not lie in (1) functional niche complementarity or in (2) selection of extreme trait values (Tilman 1999; Loreau 2000; Loreau et al. 2001; Engelhardt and Ritchie 2002). Secondary production increase is mainly found in systems where grass and clover grow together but rarely in other areas!

It appears as if the biodiversity effect in these cases works indirectly through mediating plant competition. If grazer A modifies the competitive interaction between two resources, in this case grasses and clover, and if the other grazing species B makes better use of that second resource than grazer A, and if the productivity

Table 7.3 On basis of *individual* performance of animals (measured as weight gain per day or weight reached at the end of the season), it was assessed which species gained from mixed grazing, and which species lost. The column remarks briefly describe the system in which the experiment was conducted

System	Who benefits?	Remarks	Reference
Sheep & Goats	Sheep	Goats increase clover	del Pozo et al. 1998
Sheep & Goats	Sheep	Goats increase clover	Hardy and Tainton 1995
Cattle & Sheep	Sheep	Cattle increase clover	McCall et al. 1986
Cattle & Sheep	Sheep	Lolium-Trifolium	Abaye et al. 1994
Cattle & Goats	Cattle	Goats increase clover	Osoro et al. 2000
Cattle & Goats	Cattle	Mopani veld (If after goat)	Donaldson 1979
Cattle & Goats	Goats	Mopani veld (If after cattle)	Donaldson 1979

System	No Benefit?	Remarks	Reference
Cattle & Sheep	Cattle	Normal years, Australia	Hamilton 1976
Cattle & Sheep	Cattle	Sweden	Brelin 1979
Cattle & Sheep	Cattle	Mixed bush grass, RSA	Mitchell 1985
Cattle & Sheep	Cattle	Sourveld, RSA	Hardy and Tainton 1995
Cattle & Sheep	Cattle	Irrigated land, Martinique	Mahieu et al. 1997
Cattle & Sheep	Cattle	Rangeland, Utah	Olson et al. 1999
Cattle & Sheep	Sheep	Rangeland, Utah	Olson et al. 1999
Goats & Sheep	Sheep	Dry lands, Brazil	Araujo Filho et al. 1982
Goats & Sheep	Sheep	Dry lands, Brazil	Leite et al. 1995
Goats & Sheep	Goats	Dry lands, Brazil	Leite et al. 1995

System	Who Suffers?	Remarks	Reference
Cattle & Sheep	Cattle	Cattle forced to poor food	Clark 1980
Cattle & Sheep	Cattle		Martinez et al. 2002
Goat & Cattle	Cattle	Dry lands, Brazil	Araujo Filho et al. 1982
Goat & Sheep	Sheep	Australia, drought years	Wilson and Mulham 1980

of the benefiting species B is higher than the loss of species A, then the total productivity of A + B can increase. The key lies in understanding the competitive interaction between the resources (the plant species) that comprise the primary production. Consumers can shift the competitive balance between the species comprising the first trophic level, which may affect secondary production, but it is not a rule that increased diversity of consumers leads to increased productivity of the consumer assemblage. Agricultural experiments show that different trophic levels are not governed by the same general rules. There are two caveats though, one is that agricultural experiments are set in contexts of low diversity/heterogeneity of the primary production, hence the complementarities are likely to be reduced. The second is that these experiments use herbivores that have been selected over hundreds of generations to be extremely efficient in converting plant productivity into secondary production—and perhaps they are equally efficient! In that case diversity does not increase productivity. The often observed mixed herding strategies are then merely a case of risk spreading. An apparent opposition between agriculture context and theoretical prediction about diversity and ecosystem functioning is not new, as it also occurred in plant studies, for which conditions or motivations were not necessarily those relevant for testing the theoretical models (Vandermeer et al. 2002). The next step is therefore to investigate the patterns exhibited by wild herbivore assemblages of different species richness.

7.5.2 Diversity–Biomass Relationship in Wild Assemblages

The literature review on diet overlaps from wild herbivore studies makes it likely that a more complete use of the primary production takes place if there are more herbivore species, which could thus translate into an increased secondary production. This is actually supported by the experiments conducted on the Dos Arroyos Ranch in the Sonoran Desert, which shows that increased diversity seems indeed to lead to increased secondary productivity (Mellink 1995), and by the results from Fritz and Duncan (1994), which suggest an effect of species number on the biomass of wild African ungulate communities. We developed the comparative approach to specifically test for diversity-production relationship, but as productivity is difficult to access in wild herbivore assemblages, we mostly investigated patterns relating species diversity to biomass. To be able to have enough variations in species richness, only African savanna ungulate assemblages provided an adequate case study.

In the current theoretical framework, the predictions are that when the consumers of a given trophic level are generalist, the increase in diversity may not induce an increase consumer biomass or productivity, as species would not be complementary (Long and Morin 2005; Jiang and Morin 2005; Gamfeldt et al. 2005). Conversely, if the consumers are mainly made up of specialists, then theory predicts that diversity should be positively linked with production as species would then be more complementary in their resource use. In the context of mammalian herbivore assemblages, this certainly calls for distinguishing the relative roles of body size (generalists tend to be bigger) and diet types: grazers (more often generalist), browsers (more specialist), and mixed-feeders (the ultimate generalists?). There then seems to be a place for the concept of 'feeding types' (sensu Hofmann 1973 and later work).

We should thus expect that the diversity–production relationship should be observed in browsers, in which the number of selective specialist species is higher. Accordingly, the speciation rate (particularly for *Tragelaphinae*; Vrba 1987) and also the complementarity of niche is potentially greater as the vertical dimension of the niche can be discriminating (e.g., du Toit 1990; Makhabu 2005). In large mammalian herbivores, however, there are species using resources that will rarely if ever be used by other mammalian species, but also that will transform the environment because of their body size (Owen-Smith 1988). The most striking example is the elephant, which is able to consume primary production in the form of branches and bark, uneaten by others, and which also modifies the environment. The elephant in fact may have enough impact to cause increased niche diversity. In this example, body size is a trait associated with two specific properties within the community, namely, a wide dietary niche because body size allows for the use of poorer quality food, and an impact on the environment that may promote diversity and abundance to some extent, yet possibly reduce them at very high densities (e.g., Fritz et al. 2002). Diversity could be positively linked with production in a community with elephants, although here the pattern may in fact only be due to one species, and not to species richness per se. Conversely, as the elephant is the ultimate generalist, ungulate assemblages may only exhibit a diversity–biomass relationship once elephants are accounted for.

In a detailed analysis performed on 30 protected areas we show here that the overall metabolic biomass of herbivores is affected by the number of species in the system (5% of the observed variance), an effect secondary to rainfall and soil nutrient availability (overall model $R^2=0.90$) confirming the initial results from Fritz and Duncan (1994) on a data set including pastoral areas. Interestingly, the analysis for pastoral sites exclusively (1–6 species) did not show any significant relationship, which is consistent with most results from agriculture experiments (see above). When investigating at the feeding guild level, we found that the metabolic biomass of grazers was not related to the number of species, as expected from theory, whether considering the whole community, the community without

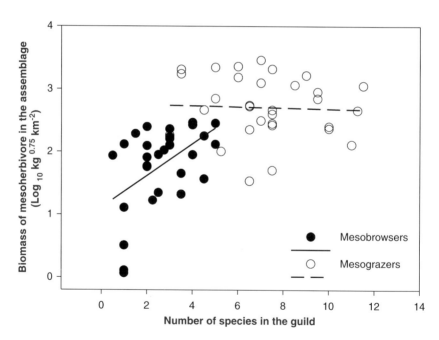

Fig. 7.3 Relationship between species diversity (number of species) and the biomass of mesoherbivores (<1000kg), grazers and browsers, in the assemblages. The relationship is only significant for mesobrowers. The equation is $y = 0.25x + 1.11$, $r^2 = 0.25$

elephants, or that without megaherbivore. For the browsers, species richness was only significant on the metabolic biomass of the mesobrowser guild (Fig. 7.3) and on the browser guild without elephants; the overall browser biomass was only influenced by rainfall, as expected from the fact that the biomass of browsers is largely dominated by elephants (again the ultimate generalist) and that elephant biomass is exclusively explained by rainfall (Fritz et al. 2002).

Our regional analysis thus show that there may be an effect of species diversity on herbivore biomass, at least in wild herbivores, and that this relationship supports theoretical predictions associated with the roles of specialists and generalists in food web and ecosystem functioning. The results from pastoral areas and agriculture experiments are in fact also in line with the specialist/generalist predictions, since domestic species have been selected (at least in extensive farming and pastoral systems) to use primary production efficiently and are often fairly generalist. Therefore, it is not surprising that species richness does not promote higher herbivore biomass in these systems.

7.6 Conclusions

In this chapter we concentrated on the possible role of ungulate species diversity in promoting higher secondary production or biomass. In savannas where we could perform a comparison, browsers' and grazers' diversity seem to respond in similar ways to environmental variables, although grazer species are more numerous, paralleling the fact that grazers generally contribute a greater share of the biomass in the assemblages. In general, reviewing the literature across most continents, we found that increasing herbivore species increased the use of primary production, at least through the use of a wider range of plants and plant parts. However, the suggestion that a greater number of herbivore species increases primary production is weak, and so is the possibility that herbivore diversity increases efficiency in biomass transfer. The fact that some combination of herbivores may modify the competitive abilities of plants and then promote higher secondary productivity should be investigated more as a process interrelating herbivore diversity and production.

We show that a positive diversity–biomass relationship exists, although in mammalian herbivores it does not seem as significant as in primary producers. The diversity–productivity relationship seems circumstantial in domestic assemblages, but could not be tested or examined on wild assemblages, for which we only examined standing biomass. The most important point in this comparison regarding the theoretical framework about diversity, food web, and ecosystem functioning is that secondary consumers' levels dominated by specialists will show a relationship, whereas not if generalists dominate. In showing that specialist browsers seem to be complementary in their use of their primary resources, we not only confirm field observations about possible niche complementarity, but also suggest that systems richer in browsers may use primary production more thoroughly, at least in savanna-type ecosystems. This echoes the debate on mixed-species production systems, especially in wild or mixed wild/domestic assemblages (e.g., Cumming 1993), but underlines the point that composition of species has to be carefully thought about.

The lack of relationship between species diversity and metabolic biomass in grazers brings into question the function of such a great diversity in African savannas, and certainly calls for investigating the 'insurance hypothesis' that species diversity buffer the ecosystem processes against environmental fluctuations because different species respond differently, and hence lead to functional compensations (Yachi and Loreau 1999). A comparative analyses of three African systems recently showed that species diversity, indeed, seemed to promote the stability of secondary production and biomass over long time periods (Prins et al., unpublished results). The same insurance strategy may well be the reason for multispecies pastoral herds, as generalist domestic species do not show striking niche complementarity.

References

Abaye AO, Allen VG, Fontenot JP (1994). Influencing of grazing cattle and sheep together and separately on animal performance and forage quality. J Anim Sci 72:1013–1022

Araujo Filho JA de, Gadelha JA, Viana OJ (1982) Complementary grazing by cattle, sheep and goat on the "Caatinga" of northeast Brazil. Proc Third International Conference on Goat Production and Disease, Dairy Goat Journal Pub Co, Scottsdale, AZ pp 532

Araujo Filho, AJ de, Gadelha JA, Leite ER, Souza PZ, Crispim SMA, Rego MC (1996) Botanical and chemical composition of the diet of sheep and goats grazing together in the Inhamuns region, Ceara. Rev Soc Brasil Zootech 25:383–395

Aucamp AJ (1976) The role of the browser in the bushveld of the Eastern Cape. Proc of the Grassland Soc Southern Afr 11:135–138

Bagchi S, Goyal SP, Sankar K (2003) Niche relationships of an ungulate assemblage in a dry tropical forest. J Mammal 84:981–988

Bagchi S, Mishra C, Bhatnagar YC (2004) Conflicts between traditional pastoralism and conservation of Himalayan ibex (*Capra sibirica*) in the Trans-Himalayan mountains. Anim Conserv 7:121–128

Bond EM, Chase JM (2002) Biodiversity and ecosystem functioning at local and regional spatial scales. Ecol Lett 5:467–470

Breebaart L, Brikraj R, O'Connor TG (2002) Dietary overlap between Boer goats and indigenous browsers in a South African savanna. Afr J Range Forage Sci 19:13–20

Brelin B (1979) Mixed grazing with sheep and cattle compared with single grazing. Swed J Agr Res 9:113–120

Cardinale BJ, Palmer MA, Collins SL (2002) Species diversity enhances ecosystem functioning through interspecific facilitation. Nature 415:426–429

Clark L (1980) Sheep and cattle on saltbush. Rural Res 107:13–15

Cuartas P, Gordon IJ, Hester AJ, Perez Barberia FJ, Hulbert IAR (2000) The effect of heather fragmentation and mixed grazing on the diet of sheep *Ovis aries* and red deer *Cervus elaphus*. Acta Theriol 45:309–320

Cumming DHM (1993) Multispecies systems: progress, prospects and challenges in sustaining range animal production and biodiversity in East and southern Africa. Proceedings of the World Conference on Animal Production, Edmonton, Canada, vol 1, pp 145–159

Cox-Ganser JM (1990) Comparative grazing behavior of cattle, goats and sheep. Thesis, West Virginia University

de Boer J, Hanekamp WJA (1992) Combined grazing system of yearling heifers and sheep. Rapport Proefstation voor de Rundveehouderij, Schapenhouderij en Paardenhouderij 135

Dekker B (1997) Calculating stocking rates for game ranches: substitution ratios for use in the Mopani Veld. Afr Jour Range Forage Sci 14:62–67

del Pozo M, Osoro K, Celaya R (1998) The effects of complementary grazing by goats on sward composition and on sheep performance managed during lactation in perennial ryegrass and white clover pastures. Small Ruminant Res 29:173–184

De Wit CT (1960) On competition. Landbouwpublicaties, Wageningen, NL

Dickson IA, Frame J, Arnold DP (1981) Mixed grazing of sheep versus cattle only in an intensive grassland system. Anim Prod 33:265–272

Donaldson CH (1979) Goats and/or cattle on mopani veld. Proc Grassland Soc Southern Afr 14:119–123

du Toit JT (1990) Feeding-height stratification among African browsing ruminants. Afr J Ecol 28:55–61

du Toit JT, Bryant JP, Frisby K (1990) Regrowth and palatability of acacia shoots following pruning by African savanna browsers. Ecology 71:149–154

du Toit JT, Cumming DHM (1999) Functional significance of ungulate diversity in African savannas and the ecological implications of the spread of pastoralism. Biodivers Conserv 8:1643–1661

du Toit PCV, Blom CD, Immelman WF (1995) Diet selection by sheep and goats in the arid Karoo. Afr J Range Forage Sci 12:16–26

Escobar A, Gonzalez JE (1976) Study on the competitive consumption of large herbivores of the flooded area of the Llanos with special reference to the capybara (*Hydrochoerus hydrochaeris*). Agron Trop 26:215–277

Engelhardt KAM, Kadlec JA (2001) Species traits, species richness and the resilience of wetlands after disturbance. J Aquat Plant Manage 39:36–39

Engelhardt KAM, Ritchie ME (2002) The effects of aquatic plant species richness on wetland ecosystem processes. Ecology 83:2911–2924

Field CR (1972) The food habits of wild ungulates in Uganda by analyses of stomach contents. E Afr Wildl J 10:17–42

Fisher GEJ, Scanlan S, Waterhouse A (1994) The ecology of sheep and goat grazing in semi-natural hill pastures of Scotland. In: 't Mannetje L, Frame J (eds) Grassland and society, Wageningen Pers, Wageningen, NL, pp 286–289

French NR (1985) Herbivore overlap and competition in Kenya rangeland. Afr J Ecol 23:259–286

Fritz H, Duncan P (1994) On the carrying capacity for large ungulates of African savanna ecosystems. P Roy Soc Lond B Bio 256:77–82

Fritz H, Duncan P, Gordon J, Illius AW (2002) Megaherbivores influence trophic guilds structure in African ungulate communities. Oecologia 131:620–625

Fritz H, Loison A (2006) Large herbivores across biomes. In: Danell K, Bergström R, Duncan P, Pastor J (eds) Large herbivore ecology, ecosystem dynamics and conservation. Cambridge University Press, Cambridge, pp 19–49

Gallina, S (1993) White-tailed deer and cattle diets at La Michellia, Durango, Mexico. J Range Manage 46:487–492

Gamfeldt L, Hillebrand H, Jonsson PR (2005) Species richness changes across two trophic levels simultaneously affect prey and consumer biomass. Ecol Lett 8:696–703

Genin D, Villca Z, Abasto P (1994) Diet selection and utilization by llama and sheep in high-altitude arid rangeland of Bolivia. J Range Manage 47:245–248

Gogan PJP, Barrett RH (1995) Elk and deer diets in a coastal prairie-scrub mosaic, California. J Range Manage 48:327–335

Gordon IJ (2003) Browsing and grazing ruminants: are they the same beasts? Forest Ecol Manag 181:13–21

Gordon IJ, Illius AW (1994) The functional significance of the browser–grazer dichotomy in African ruminants. Oecologia 98:167–175

Hamilton D (1976) Performance of sheep and cattle grazing together in different ratios. Aust J Exp Agr Anim Husb 16:5–12

Hansen RM, Reid LD (1975) Diet overlap of deer, elk, and cattle in southern Colorado. J Range Manage 28:43–47

Hardy MB, Tainton NM (1995) The effects of mixed species grazing on the performance of cattle and sheep in Highland Sourveld. Afr J Range Forage Sci 12:97–103

Hart, JA (1986) Comparative dietary ecology of a community of frugivorous forest ungulates in Zaire. Thesis, Michigan State University

Hiremath AJ, Ewel JJ (2001) Ecosystem nutrient use efficiency, productivity, and nutrient accrual in model tropical communities. Ecosystems 4:669–682

Holyoak M, Loreau M (2006) Reconciling empirical ecology with neutral community models. Ecology 87:1370–1377

Hubbell SP (2001) The unified neutral theory of biodiversity and biogeography. Princeton University Press, Princeton, NJ

Hofmann, RR (1973) The ruminant stomach: stomach structure and feeding habits of East African game ruminants. East African Literature Bureau, Nairobi

Hubbard RE, Hansen RM (1976) Diets of wild horses, cattle, and mule deer in the Piceance basin, Colorado. J Range Manage 29:389–392

Huisman J, Olff H (1998) Competition and facilitation in multi-species plant–herbivore systems of productive environments. Ecol Lett 1:25–29

Hutchinson GE (1957) Homage to Santa Rosalia or why are there so many kind of animals? Am Nat 93:145–159

Illius AW, Gordon IJ (1987) The allometry of food intake in grazing ruminants. J Anim Ecol 56:989–999

Janis CM, Damuth J, Theodor JM (2000) Miocene ungulates and terrestrial productivity: where have all the browsers gone? Proc Nat Acad USA 97:7899–7904

Jewell PA (1980) Ecology and management of game animals and domestic livestock in African savannas. In: Human ecology in savanna environments. Academic Press, London, pp 353–381

Jiang L, Morin PJ (2005) Predator diet breadth influences the relative importance of bottom-up and top-down control of prey biomass and diversity. Am Nat 165:350–363

Leibold MA, McPeek MA (2006) Coexistence of the niche and neutral perspectives in community ecology. Ecology 87:1399–1410

Leite ER, de Araujo Filho JA, Pinto FC (1995) Combined grazing with goats and sheep in lowered caatinga: performance of pasture and of animals. Pesqui Agropecu Brasil 30:1129–1134

Lightfoot CJ, Posselt J (1977) Eland (*Taurotragus oryx*) as a ranching animal complementary to cattle in Rhodesia. 2. Habitat and diet selection. Rhod Agr J 74:53–61

Logan JL, Jennings PG, McLaren LE (1991) Mixed grazing of cattle and sheep. Proc VI World Red Poll Congress, Jamaica Agricultural Development Foundation, Kingston, pp 99–101

Long ZT, Morin PJ (2005) Effects of organism size and community composition on ecosystem functioning. Ecol Lett 8:1271–1282

Loreau M (1998) Biodiversity and ecosystem functioning: a mechanistic model. Proc Nat Acad Sci USA 95:5632–5636

Loreau, M (2000) Biodiversity and ecosystem functioning: recent theoretical advances. Oikos 91:3–17

Loreau M, Naeem S, Inchausi P et al (2001) Ecology, biodiversity and ecosystem functioning: current knowledge and future challenges. Science 294:804–808

Lucich GC, Hansen RM (1981) Autumn mule deer foods on heavily grazed cattle ranges in northwestern Colorado. J Range Manage 34:72–73

Mahieu M, Aumont G, Michaux Y et al (1997) Mixed grazing sheep/cattle on irrigated pastures in Martinique (FWI). Prod Anim 10:55–65

Makhabu SW (2005) Resource partitioning within a browsing guild in a key habitat, the Chobe riverfront, Botswana. J Trop Ecol 21:641–649
Martinez A, Osoro K, Lemaire G (2002) Yearling calves live weight gains and productivity under single or mixed spring grazing with goat or sheep. In: Durand JL, Emile JC, Huyghe C (eds) Multi-function grasslands: quality forages, animal products and landscapes. Proc 19th European Grassland Federation Meeting, Versailles, pp 1050–1051
McNaughton SJ, Ruess RW, Seagle SW (1988) Large mammals and process dynamics in African ecosystems. Biosci 38:794–800
Mellink E (1995) Use of Sonoran rangelands: lessons from the Pleistocene. In: Steadman DW, Mead JI (eds) Late quaternary environments and deep history: a tribute to Paul S. Martin. Mammoth Site of Hot Springs, Hot Springs, SD, pp 50–60
Ménard C, Duncan P, Fleurance G, Georges JY, Lila M (2002) Comparative foraging and nutrition of horses and cattle in European wetlands. J Appl Ecol 39:120–133
Milton SJ (2000) Theme: interactions between diversity and animal production in natural rangelands. Afr J Range Forage Sci 17:1–3, 7–9
Mishra C (2001) High altitude survival: conflicts between pastoralism and wildlife in the Trans-Himalaya. Thesis, Wageningen University
Mitchell TD (1985) Goats in land and pasture. In: Copland JW (ed) Goat production and research in the tropics. Australian Centre for International Agricultural Research, Canberra, pp 115–166
Mwasi SM (2002) Compressed nature: co-existing grazers in a small reserve in Kenya. Thesis, Wageningen University
Mysterud A (2000) Diet overlap among ruminants in Fennoscandia. Oecologia 124:130–137
Nolan T, Connolly J (1989) Mixed versus mono-grazing by steers and sheep. Anim Prod 48:519–533
Nolan T Pulina G, Sikosana JLN, Connolly J (1999) Mixed animal type grazing research under temperate and semi-arid conditions. Outlook Agr 28:117–128
Oindo B (2002) Patterns of herbivore species richness in Kenya and current ecoclimatic stability. Biodivers Conserv 11:1205–1221
Oindo B, Skidmore AK, Prins HHT (2001) Body size and abundance relationship: an index for diversity for herbivores. Biodivers Conserv 10:1923–1931
Olff H, Ritchie ME, Prins HHT (2002) Global environmental controls of diversity in large herbivores. Nature 415:901–904
Olson KC, Wiedmeier RD, Browne JE, Hurst RL (1999) Livestock response to multispecies and deferred-rotation grazing on forested rangeland. J Range Manage 52:462–470
Osoro K, Martinez A, Celaya R, Vassalo JM (2000) The effects of mixed grazing with goats on performance of yearling calves in perennial ryegrass with clover pastures. In: Rook AJ, Penning PD (eds) Grazing management: the principles and practice of grazing, for profit and environmental gain, within temperate grassland systems. BBSRC Institute of Grassland and Environmental Research, Aberystwyth, UK, pp 115–116
Owen-Smith N (1985) The ecological potential of the kudu for commercial production in savanna regions. J Grassland Soc Southern Afr 2:7–10
Owen-Smith N (1988) Megaherbivores. The influence of very large body size on ecology. Cambridge University Press, Cambridge
Pfisterer AB, Schmidt B (2002) Diversity-dependent production can decrease the stability of ecosystem functioning. Nature 416:84–86
Pordomingo AJ, Rucci T (2000) Red deer and cattle diet composition in La Pampa, Argentina. J Range Manage 53:649–654
Prins HHT, de Boer WF, van Oeveren H, Correira A, Mafuca J, Olff H (2006) Co-existence and niche segregation of three small bovids in southern Mozambique. Afr J Ecol 44:186–198
Prins HHT, Olff H (1998) Species richness of African grazer assemblages: towards a functional explanation. In: Newbery DM, Prins HHT, Brown ND (eds) Dynamics of tropical communities. BES Symp vol 37, Blackwell, Oxford, p 449–490

Robbins CT, Spalinger DE, Van Hoven W (1995) Adaptations of ruminants to browse and grass diets: are anatomical-based browser–grazer interpretations valid? Oecologia 103:208–213

Schwartz CC, Ellis JE (1981) Feeding ecology and niche separation in some native and domestic ungulates on the shortgrass prairie. J Appl Ecol 18:343–353

Shrestha R, Wegge P, Koirala RA (2005) Summer diets of wild and domestic ungulates in Nepal Himalaya. J Zool 266:111–119

Storms D, Said S, Fritz H, Hamann J-L, Saint-Andrieux C, Klein F (2006) Influence of hurricane Lothar on red and roe deer winter diets in the northern Vosges, France. Forest Ecol Manag 237:164–169

Thill RE, Martin A (1986) Deer and cattle diet overlap on Louisiana pine–bluestem range. J Wildl Manage 50:707–713

Tilman D (1999) The ecological consequences of changes in biodiversity: a search for general principles. Ecology 80:1455–174

Tilman D, Lehman CL, Thomson KT (1997) Plant diversity and ecosystem productivity: theoretical considerations. Proc Nat Acad Sci USA 94:1857–1861

Tilman D, Reich PB, Knops J, Wedin D, Mielke T, Lehman C (2001) Diversity and productivity in a long-term grassland experiment. Science 294:843–845

Vandermeer J, Lawrence D, Symstad A, Hobbie S (2002) Effect of biodiversity on ecosystem functioning in managed ecosystems. In: Loreau M, Naeem S, Inchausti P (eds) Biodiversity and ecosystem functioning. Oxford University Press, Oxford, pp 221–233

Van Wieren SE (1996) Do large herbivores select a diet that maximizes short term digestible energy intake? Forest Ecol Manag 88:149–156

Voeten MM, Prins HHT (1999) Resource partitioning between sympatric wild and domestic herbivores in the Tarangire region of Tanzania. Oecologia 120:287–294

Vrba, ES (1987) Ecology in relation to speciation rates: some case histories of Miocene–recent mammal clades. Evol Ecol 1:283–300

Wilson AD, Mulham WE (1980) Vegetation changes and animal productivity under sheep and goat grazing on an arid belah (*Casuarina cristata*)–rosewood (*Heterodendrum oleifolium*) woodland in western New South Wales. Aust Rangeland J 2:183–188

Yachi S, Loreau M (1999) Biodiversity and ecosystem productivity in a fluctuating environment: the insurance hypothesis. Proc Nat Acad Sci USA 96:1463–1468

Chapter 8
Impacts of Grazing and Browsing by Large Herbivores on Soils and Soil Biological Properties

Kathryn A. Harrison and Richard D. Bardgett

8.1 Introduction

Herbivores can have a wide range of effects on terrestrial ecosystems. Some of these effects are direct, such as the removal and consumption of herbage—which can vary some 100-fold across terrestrial ecosystems from less than 1% to greater than 60% (McNaughton et al. 1989)—treading on soil and vegetation, and the return of excreta (Floate 1981). Herbivores also have important indirect effects on ecosystems, altering rates of nutrient cycling and changing nutrient availability to plants (Bardgett and Wardle 2003; Bardgett 2005). These indirect effects of herbivores on ecosystems are mediated by feedbacks that occur between plants and below-ground decomposer communities, and especially soil microbes, which play a central role regulating nutrient availability to plants.

This chapter examines some of the many ways that above-ground herbivores might indirectly influence ecosystem properties through their influence on below-ground organisms and processes. We also examine how these changes in soil biological properties may in turn feed back to affect above-ground primary production in both positive and negative ways. The chapter will highlight the different mechanisms by which herbivores affect decomposer organisms and their activities over various temporal and spatial scales, ranging from short-term responses at the individual plant level, to long-term responses at the level of the plant community. The effects of different types of herbivory, namely grazing and browsing, on soil biological properties will also be discussed. The mechanisms highlighted will be illustrated with specific examples from a range of different ecosystems, thereby providing a global perspective on herbivore effects on soils.

8.2 Herbivore Effects on Nutrient Dynamics

The effects of herbivory on plant community structure and function have been widely studied (e.g., McNaughton 1984; Haukioja et al. 1990; Dyer et al. 1993; Collins et al. 1998; Lehtilä et al. 2000). However, the indirect effects of herbivory on below-ground properties have often been overlooked. In recent years, it has become apparent that there exists an important link between above-ground and below-ground processes, and that these links are affected by herbivores. A review by Bardgett et al. (1998) was one of the first to address this issue, giving an account of how foliar herbivory influences soil organisms and decomposer food webs. Historically, much work in this area has been done in grassland ecosystems (Floate 1970a, b; Bardgett et al. 1997; McNaughton et al. 1997; Tracy and Frank 1998), although work has also been done on salt marshes (Van Wijnen and Van der Wal 1999), oak savanna (Ritchie et al. 1998), boreal forests (Pastor et al. 1993; Stark et al. 2000), arctic tundra (Van der Wal et al. 2004) and New Zealand forests (Wardle et al. 2001).

Many of the indirect effects of herbivory have been identified, however, the processes involved are complex, and effects vary on a case-by-case basis (Bardgett and Wardle 2003; Bardgett 2005). The majority of effects, such as increased rhizodeposition and the production of secondary metabolites do, however, culminate in a soil process response, mediated by soil microbes and microfauna; this might ultimately alter rates of nutrient cycling. In many terrestrial ecosystems, nitrogen (N) is the main nutrient limiting plant growth (Vitousek and Howarth 1991). Therefore, the availability and recycling rates of nitrogen are important factors controlling primary productivity (Van Wijnen et al. 1999).

Previous studies have identified two hypotheses for herbivore effects on soil microbial properties and nutrient cycling: the accelerating effect and the decelerating effect (Ritchie et al. 1998; Bardgett et al. 1998; Fig. 8.1). The accelerating effect predicts that herbivory will stimulate soil microbial activity, thereby promoting rates of nutrient cycling. This effect is most commonly seen when herbivory takes the form of grazing on grassland species. In contrast, the decelerating effect predicts that herbivores will have a negative effect on soil biological properties and hence ecosystem productivity. Conversely, this effect is most commonly exhibited when herbivory takes the form of browsing on tree species. In this chapter we review some of the many ways in which above-ground herbivory can positively or negatively feed back on plant nutrition and productivity via changes in below-ground processes, such as nutrient cycling rates and decomposition.

8.3 Positive Feedback Effects of Above-Ground Herbivory

Positive effects of herbivory on soil biota and nutrient cycling occur when dominant plant species respond to grazing by exhibiting compensatory growth (Augustine and McNaughton 1998). This mechanism is most common in grasslands of high soil fertility (Fig. 8.2), where herbivory in the form of grazing positively affects the decomposer subsystem through preventing colonisation of later successional plants which produce poorer litter quality, as well as through returning carbon and nutrients to the soil in labile forms such as dung and urine, and as enhanced

A. Acceleration Effect

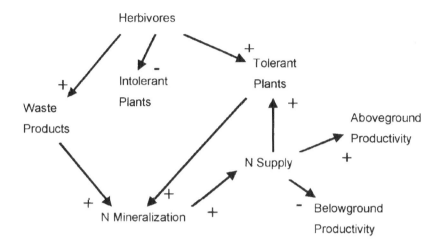

B. Deceleration Effect

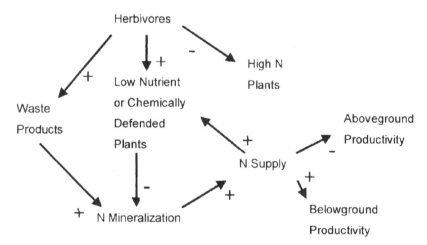

Fig. 8.1 Feedback loops illustrating (A) the acceleration effect and (B) the deceleration effect of herbivores on feedbacks between plant species and nutrient cycling. *Arrows* indicate net indirect effect of herbivores on the abundance of plants or the rate of the process (from Ritchie et al. 1998)

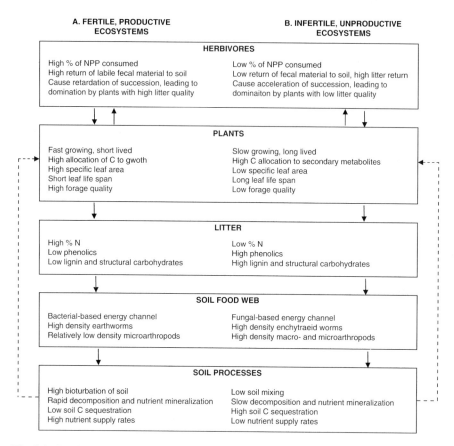

Fig. 8.2 Herbivore effects on ecosystems can vary depending on whether they impact on (A) fertile systems supporting high herbivory, typically in the form of grazing, or (B) infertile habitats supporting low herbivory, typically in the form of browsing. Herbivore-driven changes in plant species composition can indirectly feed back in a positive or negative way to influence the quality and quantity of resources (i.e., plant litter) entering the decomposed subsystem. The linkages between below-ground and above-ground systems feedback (*dotted line*) to the plant community positively (A) in fertile conditions and negatively (B) in infertile ecosystems (adapted from Wardle et al. 2004)

rhizodeposition (Holland and Detling 1990; Holland et al. 1996; Bardgett et al. 1997; McNaughton et al. 1997; Frank and Groffman 1998; Hamilton and Frank 2001). The mechanisms by which grazing by herbivores has been shown to positively feedback to increase nutrient cycling rates will now be described, with examples given of how these processes function in different ecosystems.

8.3.1 Urine and Dung Deposition

By depositing urine and dung, nitrogen is recycled in forms that are more available to plants and soil microbes (Frank and Evans 1997). Plant material that is removed by animals is digested and nutrients are released far quicker into soil from the resulting faeces than if released from litter, leading to increased nutrient availability and plant nutrient uptake. In many studies, the deposition of urine and dung has been shown to directly affect nutrient cycling and availability. For example, Stark et al. (2000) in a study of the effect of reindeer grazing in Scots pine forests in Finnish Lapland, stated that urine and faeces enhanced nutrient cycling, strengthening the positive effect that other identified factors, such as removal of lichen cover, had on nutrient cycling.

As increased soil microbial activity leads to accelerated nitrogen cycling in urine and faecal patches, a positive feedback mechanism may occur, whereby increased soil N availability leads to a greater tissue N concentration of plants. This in turn increases the probability that these areas will be re-grazed and thus receive extra excretory inputs, further enhancing recycling rates (Frank and Groffman 1998). McNaughton et al. (1997) found that urination from grazers, such as Thompson's (*Gazella thompsoni*) and Grant's (*Gazella granti*) gazelles in Serengeti National Park, Tanzania, enriched soils with N from urea, leading to a burst of organic matter mineralization that produced greater available mineral N in the soil than would occur solely from urea addition. This, together with other factors, such as increased leaf N concentration, led to a two-fold increase in rates of N mineralization in soils supporting dense resident animal populations compared with those where animals were uncommon.

The deposition of animal waste by herbivores has been shown to increase microbial biomass and stimulate microbial activity, which in turn increases nutrient cycling rates in grassland. (Bardgett et al. 1997, 2001; Tracy and Frank 1998). In British upland grasslands, urine and dung from grazing cattle and sheep have been shown to increase soil microbial activity and N cycling, and hence plant production (Floate 1970a, b). Similarly, Bardgett et al. (2001) found that microbial biomass was maximal at moderate levels of sheep grazing on semi-natural grassland, which was attributed to increased inputs of labile C to soil from root exudation and animal wastes, and also changes in litter quality due to vegetation change. Grazing of *Agrostis-Festuca* and *Nardus* dominated hill grasslands in Britain was also found to stimulate microbial biomass and activity, and the abundance of soil microfauna, which together regulate nutrient cycling (Bardgett et al. 1997). In dry grassland and shrub-grassland in Yellowstone National Park, Frank and McNaughton (1992) detected a positive association between dung deposition and above-ground primary production, and suggested that grazing and productivity are coupled to herbivore-facilitated nutrient cycling in this system. Tracy and Frank (1998) also found that grazing by elk (*Cervus elaphus*), bison (*Bison bison*), and pronghorn antelope (*Antilocarpa americana*) in Yellowstone caused increased microbial biomass and rates of N mineralization in soil. These authors hypothesized that microbial populations in grazed grassland were sustained mainly by inputs of labile C from dung deposition and increased root turnover or root exudation beneath grazed plants. In a tundra ecosystem in Spitsbergen, Van der Wal et al. (2004) found that soil microbial biomass C and N were stimulated by increasing the density of reindeer faeces (*Rangifer tarandus platyrhynchus*) applied to the ground and that this led to an increase in the standing crop of grasses; these findings were taken to indicate that large herbivores, such as reindeer, have the ability to manipulate their own food supply in a positive way.

Although urine and dung deposition undoubtedly have positive effects on below-ground properties and rates of nutrient cycling, it has been argued that it cannot explain the widely documented positive effects of herbivores at large spatial scales (Hamilton and Frank 2001). This is because, typically, waste patches remain localised within an ecosystem, influencing a relatively small surface area (Augustine and Frank 2001). Also, in some cases, the positive effects of animal wastes are insufficient to negate other, adverse effects of herbivory on nutrient mineralization and are therefore not apparent at the ecosystem scale (Pastor et al. 1993; Bardgett and Wardle 2003; Harrison and Bardgett 2003, 2004).

8.3.2 Alterations in Plant C and N Allocation

The allocation of C in plants is often closely linked to the allocation of N; therefore, these nutrients must be considered in tandem. The effects of herbivory on plant nutrient flow are evident as either short-term changes in plant C and N allocation and root exudation, or as long-term changes in root biomass. The following summarizes specific examples of how these responses may positively affect soil microbial communities and hence rates of nutrient cycling in soil.

8.3.2.1 Short-Term Responses

In the short-term, the quantity of resources supplied to the soil can be altered through effects of herbivory on plant C and N allocation and on root turnover and exudation patterns (Bardgett 2005). Above-ground herbivory of grasses has been shown to increase assimilate allocation to below-ground components of plants and decrease allocation to the shoots. For example, laboratory experiments on grasses have found that clipping can lead to an accumulation of assimilates in below-ground organs and increased root respiration and exudation (Bokhari and Singh 1974; Dyer and Bokhari 1976). Furthermore, increased root exudation, as a result of foliar herbivory, has been shown to greatly influence soil microbial community structure and stimulate soil microflora and microfauna and C use efficiency by microbes in the rhizosphere (Bardgett et al. 1998; Mawdsley and Bardgett 1997; Guitian and Bardgett 2001). This increased microbial productivity in the rhizosphere following herbivory may also favour higher level consumers such as enchytraeids and microbe-feeding nematodes in the soil microfood-web (Wardle 2002). These below-ground effects of herbivory can feed back positively on plant growth and plant N content. For example, Hamilton and Frank (2001) found that defoliation of *Poa pratensis* promoted root exudation of carbon, which was quickly assimilated by soil microbes increasing their biomass, in turn increasing soil N mineralization and plant N uptake, thereby increasing the growth and N status of the re-growing plant. Similarly, simulated herbivory in the form of browsing of tree seedlings has been shown to lead to enhanced N mineralization and inorganic N availability in rhizosphere soil, presumably owing to a stimulation of biological activity resulting from increased rhizodeposition (Ayres et al. 2004).

It has been established, therefore, that herbivores can alter plant C and N allocation leading to a positive feedback cycle that increases soil microbial activity and rates of nutrient cycling. Guitian and Bardgett (2000), in microcosm study, examined the response of three dominant grass species to different intensities of defoliation and found that defoliation of grasses that were tolerant to grazing (e.g., *Festuca rubra* and *Cynosurus cristatus*) led to an increased allocation of resources to shoots, whereas defoliation of grasses with a low tolerance to grazing (e.g., *Anthoxanthum odoratum*) led to an increase in the relative allocation of resources below ground. Mikola et al. (2001a, b) also found that defoliation of white clover (*Trifolium repens*), perennial ryegrass (*Lolium perenne*), and ribwort plantain (*Plantago lanceolata*) led to an increased allocation of resources to above-ground growth, which they suggested was typical of species adapted to intense, but infrequent defoliation. This implies that grazing causes alterations in the allocation of C and N within the plant, and that the direction the nutrients are allocated can depend on species.

8.3.2.2 Long-term Responses

The long-term effects of herbivory on plant C and N allocation are primarily apparent through alterations in root biomass and morphology. There is mixed evidence about how herbivory affects root productivity, with positive (Milchunas and Laurenroth 1993), negative (Guitian and Bardgett 2000; Mikola et al. 2001a; Ruess et al. 1998) and neutral (McNaughton et al. 1998) effects of herbivory on root biomass being detected. In the long-term, above-ground herbivory has been shown to reduce root biomass and change root morphology and architecture (Bardgett et al. 1998). For example, in a study on prairie dog colonies in Wind Cave National Park, South Dakota, grazing was found to decrease N allocation to the roots and decrease root biomass. This had the effect of decreasing N immobilization, due to reduced microbial growth, and increasing net N mineralization and plant available N in colonised prairie dog colonies (Holland and Detling 1990).

Enhancement of root biomass by grazing is also possible and has been shown to occur with increased rates of soil nutrient cycling. For example, a study carried out by Chaneton et al. (1996) on temperate sub-humid grassland in Argentina, found that grazing leads to the allocation of N to below-ground organs, increasing root biomass and the amount of underground nutrient circulation. Although herbivore-induced increases in soil nutrient cycling had previously been identified in a Serengeti grassland (McNaughton et al. 1997), other studies in this ecosystem showed that root biomass was not affected by grazing as there was no significant difference in mean root biomass on an annual basis between fenced and un-fenced plots (McNaughton et al. 1998). Herbivores, therefore, have been shown to have positive, negative and no effect on root biomass whilst still accelerating nutrient cycling rates. This indicates that a long-term ecosystem, response to herbivory may depend on the plant species, be they woody tree species, forbs or grasses, and on the ecosystem in question, and the relative impacts of other abiotic factors, such as topography and climate, on below-ground processes such as decomposition and rates of soil nutrient cycling.

8.3.3 Selective Foraging on Less Nutritious Species

Commonly, herbivores have been shown to selectively forage on *nutrient rich* plant species (Pastor et al. 1993; Ritchie et al. 1998; Van Wijnen et al.1999), which can ultimately lead to decreased rates of nutrient cycling (see Sect. 8.4.1). However, selective foraging on *less nutritious* species by herbivores has been shown to increase soil nutrient cycling rates; by selectively feeding on foliage from plants of lower nutritional content, herbivores by-pass the slower litter decomposition pathway and deposit more readily decomposable faecal material on soil, which is more rapidly made available for plant uptake. Stark et al. (2000) presented data concerning such an incidence. These authors investigated the effect of reindeer grazing in the lichen-dominated Scots Pine forests of Finnish Lapland. Lichen produces litter that decomposes slowly and hence, reindeer grazing reduced the amount of litter that was produced. By ingesting and metabolising lichen, the reindeer allowed nutrients to be released and made available for plant growth at a faster rate than if left on the ground to be decomposed by soil microbes.

8.3.4 Litter Deposition

As herbivores remove vegetation by grazing and browsing, litter production falls (Ruess and Seagle 1994). Biomass is, therefore, re-directed from the slow decomposition pathways in soil food webs to the herbivores, which recycle nutrients more efficiently and return plant-available nitrogen to the soil (in the form of urine and dung) at a faster rate (Coughenour 1991). By bypassing the decomposition process in this way, nutrient cycling rates are increased (Hamilton et al. 1998). Reducing litter build-up, therefore, can increase the productivity of an ecosystem (Milchunas and Laurenroth 1993). McNaughton et al. (1988) reviewed the effects of grazing by herbivores within the Serengeti and found that grasslands where the major ungulate herds concentrated during the wet season had nutrient turnover times of less than a year. However, fenced plots accumulated substantial quantities of litter, and turnover times exceeded five years. Therefore, the removal of grazing herbivores can slow nutrient cycling rates and reduce the productivity of an ecosystem.

8.3.5 Increased Soil Temperature

By removing vegetation, herbivores may increase soil microbial activity and hence nutrient cycling rates by allowing more light to reach the soil surface leading to an increase in soil temperature (Pastor et al. 1993). Areas that remain ungrazed have been found to have lower soil temperatures, due to the thick layer of litter present. This inhibits microbial activity and reduces net N mineralization (Frank and Groffman 1998). For example, exclusion of grazing by barnacle geese and reindeer over a period of seven years at Ny-Ålesund, Spitsbergen, caused an increase in the thickness of the moss layer, and a reduction in soil temperature of $0.9°$ C (Van der Wal et al. 2001). Conversely, areas affected by herbivory have less litter build up (see Sect. 8.3.5) allowing soil temperatures to rise, which stimulates microbial activity and increases net N mineralization. Similarly, studies in the high Arctic, Spitsbergen, have shown that reindeer grazing promotes soil N availability and plant productivity (Van der Wal et al. 2004). In this study, experimental faecal addition to moss dominated tundra was shown to increase grass growth and microbial biomass in soil, but also to lead to a reduction in the depth of the moss layer influencing soil temperature. These results took some three years to develop and the cause for this was thought to be a direct fertilizing effect of faeces in these strongly nutrient-limited situations, and to a suppressive effect of faeces on the depth of the moss layer, which strongly regulates soil temperature owing to its ability to hold moisture (Brooker and Van der Wal 2003). Studies on the effects of herbivory carried out on upland steppe in Yellowstone National Park (Coughenour 1991), in seasonally dry high country in New Zealand (McIntosh et al. 1997), and in oak savanna in Minnesota (Ritchie et al. 1998), have also found that grazing elevates soil temperature thereby accelerating decomposition and mineralization of organic matter by soil biota.

8.4 Negative Feedback Effects of Above-Ground Herbivory

Negative effects of herbivory on soil biota and rates of nutrient cycling most commonly occur in unproductive ecosystems, where browsing on leaves and woody material is the dominant form of herbivory (Fig. 8.2). Here, low consumption rates and selective foraging on nutrient-rich plants can lead to the dominance of defended plants that produce recalcitrant litter (Ritchie et al. 1998). Since most nutrients will be returned to the soil as recalcitrant plant litter, the net effect of herbivory in these low productivity ecosystems is to reduce soil biotic activity, nutrient mineralization, and supply rates of nutrients from soil, despite inputs of dung and urine (Pastor et al. 1993). Negative feedbacks may also occur where grazing and inappropriate stocking densities affect soil physical properties leading to reduced microbial activity and increased nutrient loss through erosion. These mechanisms will now be discussed in detail and illustrated with specific examples of ecosystems where these processes have been identified.

8.4.1 Selective Foraging on Nutrient-Rich Tissue

Plant species vary in the quality of foliage that they produce. Species (e.g., some woody plants and trees) that produce high amounts of defence compounds, which are less nutritious, also produce litter of poorer quality which is less readily decomposed by soil microorganisms. Since herbivore digestion and decomposition are both regulated by microorganisms with similar enzyme complements, digestion of plant material with high concentrations of secondary compounds will be slower than digestion of plant material of a better quality with a greater nutrient content (Pastor et al. 1993; Chesson 1997). It is reasonable to suggest, therefore, that herbivores might selectively browse on plant species with nutrient-rich tissue, and seek to avoid those species with foliage of high secondary compound concentration, and hence of a lower nutritional quality (Bardgett et al. 1998). In doing so, browsers may indirectly alter plant community structure encouraging the dominance of less nutritious plants with lower quality litter, which is slow to decompose. This shift would have negative impacts on below-ground processes governed by soil microbes, such as decomposition and soil nutrient cycling rates, ultimately reducing ecosystem productivity (Pastor et al. 1993; Kielland and Bryant 1998; Ritchie et al. 1998).

The importance of selective feeding by herbivores for soil processes was illustrated by a study of moose browsing in the boreal forests of the Isle Royale National Park in Lake Superior (Pastor et al. 1993). These authors showed that selective foraging by moose on hardwoods, with nutrient rich tissue, led to the dominance of less nutritious species such as spruce, which produce litter of lower quality and hence decomposability. This low quality litter was slow to decompose and built up on the soil surface, leading to a reduction in soil nitrogen mineralization (presumably due to low N availability from litter and also reduced soil temperature, which can lead to a less favourable microclimate for soil microbes and mesofauna) and hence a decline in the productivity of the ecosystem (Pastor et al. 1993). Many other studies on browsing and grazing have also found that selective foraging increases the dominance of less nutritious species and ultimately slows down rates of nutrient cycling. For example, Van Wijnen et al. (1999) found that grazing by geese, hares and rabbits on areas of salt marsh encouraged plants such as *Limonium vulgare* to dominate. *L. vulgare* is a less nutritious species with leaves of high tannin content which decompose slowly, hence decreasing rates of mineralization (Van Wijnen et al. 1999). Herbivory, in the form of browsing, on oak savanna in Minnesota by deer, rabbits, and a variety of insects, was also found to decrease nutrient cycling rates due to selective foraging by herbivores on nutrient-rich tissue (Ritchie et al. 1998). By selectively foraging on woody plants and legumes, prairie grasses such as *Andropogon geradi* and *Sorghastrum nutans* became dominant. The litter from these grasses was found to decompose slowly, thereby decelerating nutrient cycling rates in this ecosystem. Conversely, the exclusion of herbivores in these areas led to an increase in cover and biomass of legumes and woody plants. The nutrient-rich tissue of these plants decomposed relatively fast, therefore increasing rates of nutrient cycling (Ritchie et al. 1998).

Kielland and Bryant (1998) found that selective browsing by moose in Alaskan taiga led to a replacement of more nutritious deciduous species with less nutritious evergreens, causing slower nutrient turnover. In situations such as this, a positive feedback system may subsequently arise, whereby reduced rates of nutrient turnover and nutrient availability (due to the presence of less nutritious species) favour growth of nutrient-poor species, which reduce their loss of nutrients to the environment, allowing increased dominance in nutrient-poor habitats. This feedback strengthens the effects of the herbivores on the ecosystem and could lead to additional reductions in above-ground production and rates of nutrient cycling (Ritchie et al. 1998). In New Zealand's forests, however, although browsing by introduced mammals (feral goats and deer) significantly altered plant community composition, reducing nutritious broad-leaved species and promoting other less nutritious types, subsequent effects on measures of soil nutrient status were highly idiosyncratic, with an equal number of positive and negative effects being detected across thirty locations (Wardle et al. 2001).

8.4.2 *Production of Secondary Metabolites*

As outlined above, one factor determining the palatability and decomposability of plant material is the concentration of secondary metabolites within the leaves. Browsers can induce the production of secondary metabolites in foliage, which negatively impacts on soil biota due to reduced litter quality (Bardgett 2005). For example, severe defoliation of trees, such as that caused by periodic invertebrate attack, often results in reduced concentrations of N and increased concentrations of certain secondary metabolites (e.g., phenolics) in subsequently produced foliage (Rhoades 1985). Findlay et al. (1996) reported that damage to leaves from actions including herbivory results in nitrogen-complexation with bound phenolic material. These authors showed that cellular damage caused by spider mites to seedlings of *Populus deltoides* increased the concentration of polyphenols in foliage, resulting in a 50% reduction in the decomposition rate of subsequently produced litter. Compounds, such as polyphenols, are relatively resistant to decomposition as they inhibit the actions of soil microbes (Hättenschwiler and Vitousek 2000), and it has been suggested that they may also act to tie up nitrogen (Findlay et al. 1996). By inhibiting the activity of soil microbes, the presence of high concentrations of plant secondary metabolites can reduce nitrogen return via N mineralization, therefore decreasing rates of soil nutrient cycling.

8.4.3 Impact on Soil Physical Properties

Grazing, and inappropriate stocking densities, can have adverse effects on ecosystems, which include poaching of soils, compaction and erosion. These processes can, in turn, have negative consequences for both land and water quality as well as the exchange of greenhouse gases. For example, grazing was found to reduce soil organic matter C and N content and microbial biomass in a study carried out in Utah, USA, which looked at the destabilising effects of grazers on soil surfaces (Neff et al. 2005). These authors found strong evidence that grazing triggers wind erosion, largely due to the disruption of biological soil crusts and long-term changes in vegetation cover/composition, and results in significant nutrient loss in this semi-arid setting.

Herbivore effects on soil physical properties have also been shown to have negative feedback on soil biota, potentially decreasing nutrient cycling rates. For example, increased sheep stocking density on an Australian pasture (10, 20, and 30 sheep ha^{-1}) severely reduced numbers of Collembola in the surface soil (King and Hutchinson 1976; King et al. 1976). Similarly, reductions in collembollan numbers was associated with increased stocking density of a lowland perennial ryegrass (*Lolium perenne*) grassland (Walsingham 1976). These responses were attributed to changes in soil pore space, which was greatly reduced with increased sheep grazing. Similarly, browsing by introduced goats and deer in New Zealand's natural forests was found to have consistently adverse effects on mesofaunal and macrofaunal groups; this was attributed to the physical effects of the browsers, namely trampling and scuffing (Wardle et al. 2001). Herbivore induced alterations in soil microclimate have also been shown to have negative effects on the decomposer subsystem and hence nutrient cycling rates. Stark et al. (2000) found that removal of the protective cover of lichens by reindeer grazing led to the exposure of soil biota to a less favourable microclimate, which was predicted to have a significant negative affect on soil microbial processes in dry oligotrophic Scots pine forest in Fennoscandinavia. This was contrary to the observation that microbial activity was greater in grazed than ungrazed areas (see Sect. 8.3.5 also for effects of grazing on soil temperature). These authors suggested, however, that the relative dominance of these positive and negative effects of grazing may vary depending on season. Reindeer trampling, which compacts the soil structure, has also been shown to crush plant roots thereby reducing carbon input to the soil (Stark et al. 2003).

8.5 Conclusions

Herbivores impact on ecosystems via a variety of mechanisms, the dominance of which determines whether a positive or negative response is seen. These mechanisms operate at different spatial and temporal scales, for example, at the individual plant level (via the production of secondary metabolites), through to the ecosystem scale (via shifts in plant community structure), and at the short-term, for example hours to weeks (via root exudation) up to the long-term, for example months to years (via changes in root biomass).

Predicting whether herbivores have a positive or negative affect on processes such as decomposition and soil nutrient cycling in a particular ecosystem is therefore difficult, as the mechanisms involved invariably interact with each other, and also with abiotic factors such as climate and topography (Verchot et al. 2002). However, it is becoming clearer that it may be possible to predict the impacts herbivores have on soil nutrient cycling by looking at the type of ecosystem they are part of and the nature of their herbivory. For example, positive effects of herbivory on soil biota and nutrient cycling are more likely to occur when dominant plant species respond to grazing by exhibiting compensatory growth (Augustine and McNaughton 1998). This appears to be most common in productive grasslands, were herbivory in the form of grazing positively affects the decomposer subsystem through preventing colonisation of later successional plants which produce poorer litter quality, as well as through returning carbon and nutrients to the soil in labile forms as dung and urine, and as enhanced rhizodeposition (McNaughton et al. 1989; Holland and Detling 1990; Holland et al. 1996; Bardgett et al. 1997; Frank and Groffman 1998; Hamilton and Frank 2001; Bardgett and Wardle 2003). In contrast, negative effects of herbivory on soil properties and ecosystem productivity may be more likely to occur in unproductive ecosystems, where browsing, rather than grazing, is most common. In these systems, low consumption rates by herbivores and selective foraging on nutrient rich plants can lead to the dominance of defended plants that produce recalcitrant litter (Ritchie et al. 1998). The net effect of herbivory in these low productivity ecosystems is often to reduce soil biotic activity, nutrient mineralization, and supply rates of nutrients from soil, despite inputs of dung and urine (Pastor et al. 1993).

In this chapter we have illustrated how above- and below-ground linkages, such as those mediated by herbivores, can have important consequences on a number of ecosystem properties. It is clear, therefore, that future research in this area needs to address such interactions and feedbacks. Herbivores have repeatedly been shown to have positive, negative, and neutral effects on ecosystem processes, such as nutrient mineralization and decomposition, and it is this context dependency which now holds the greatest challenge for research in the future. It is only when we consider the interactions of herbivores and their environment at different temporal and spatial scales and the interactions of biotic and abiotic factors in the environment that we will be able to more accurately predict herbivore effects on ecosystems.

References

Augustine DJ, Frank DA (2001) Effects of migratory ungulates on spatial heterogeneity of soil nitrogen properties in a grassland ecosystem. Ecology 82:3149–3162

Augustine DJ, McNaughton SJ (1998) Ungulate effects on the functional species composition of plant communities: Herbivore selectivity and plant tolerance. J Wildl Manage 52:1165–1183

Ayres E, Heath J, Possell M, Black HIJ, Kerstiens G, Bardgett RD (2004) Tree physiological responses to above-ground herbivory directly modify below-ground processes of soil carbon and nitrogen cycling. Ecol Lett 7:469–479

Bardgett RD (2005) The biology of soil: a community and ecosystem approach. Oxford University Press. Oxford, UK, pp 242

Bardgett RD, Leemans DK, Cook R, Hobbs P (1997) Seasonality of the soil biota of grazed and ungrazed hill grasslands. Soil Biol Biochem 29:1285–1294

Bardgett RD, Wardle DA (2003) Herbivore mediated linkages between above-ground and below-ground communities. Ecology 84:2258–2268

Bardgett RD, Wardle DA, Yeates GW (1998) Linking above-ground and below-ground interactions: How plant responses to foliar herbivory influence soil organisms. Soil Biol Biochem 30:1867–1878

Bardgett RD, Jones AC, Jones DL, Kemmitt SJ, Cook R, Hobbs P (2001) Soil microbial community patterns related to the history and intensity of grazing in sub-montane ecosystems. Soil Biol Biochem 33:1653–1664

Bokhari UG, Singh JS (1974) Effects of temperature and clipping on growth, carbohydrate reserves and root exudation of western wheatgrass in hydroponic culture. Crop Sci 14:790–794

Brooker R, Van der Wal R (2003) Can soil temperature direct the composition of high Arctic plant communities? J Veg Sci 14:535–542

Chaneton EJ, Lemcoff JH, Lavado RS (1996) Nitrogen and phosphorus cycling in grazed and ungrazed plots in a temperate subhumid grassland in Argentina. JAppl Ecol 33:291–302

Chesson A (1997) Plant degradation by ruminants: parallels with litter decomposition in soils. In: Cadisch G, Giller KE (eds) Driven by nature: plant litter quality and decomposition. CAB International, Wallingford, UK, pp 47–66

Collins SL, Knapp AK, Briggs JM, Blair JM, Steinauer EM (1998) Modulation of diversity by grazing and mowing in native tallgrass prairie. Science 280:745–747

Coughenour MB (1991) Biomass and N responses to grazing of upland steppe on Yellowstone's northern winter range. J Appl Ecol 28:71–82

Dyer MI, Bokhari UG (1976) Plant–animal interactions: Studies of the effects of grasshopper grazing on blue grama grass. Ecology 57:762–772

Dyer MI, Turner CL, Seastedt TR (1993) Herbivory and its consequences. Ecol Appl 3:10–16

Findlay S, Carreiro M, Krischik V, Jones CG (1996) Effects of damage to living plants on leaf litter quality. Ecol Appl 6:269–275

Floate MJS (1970a) Mineralization of nitrogen and phosphorus from organic materials of plant and animal origin and its significance in the nutrient cycle in grazed upland hills and soils. J Brit Grassland Soc 25:295–302

Floate MJS (1970b) Decomposition of organic materials from hill soils and pastures II. Comparative studies on the mineralization of carbon, nitrogen and phosphorus from plant materials and sheep faeces. Soil Biol Biochem 2:173–185

Floate MJS (1981) Effects of grazing by large herbivores on N cycling in agricultural ecosystems. In: Clark FE, Rosswall T (eds) Terrestrial nitrogen cycles – processes, ecosystem strategies and management impacts. Ecol Bull Swedish Nature Science Research Council, Stockholm, pp 585–597

Frank DA, Evans RD (1997) Effects of native grazers on grassland N cycling in Yellowstone National Park. Ecology 78:2238–2248

Frank DA, Groffman PM (1998) Ungulate vs. landscape control of soil carbon and nitrogen processes in grasslands of Yellowstone National Park. Ecology 79:2229–2241

Frank DA, McNaughton SJ (1992) The ecology of plants, large mammalian herbivores, and drought in Yellowstone National Park. Ecology 73:2043–2058

Guitian R, Bardgett RD (2000) Plant and soil microbial response to defoliation in temperate semi-natural grassland. Plant Soil 220:271–277

Hamilton EW, Frank DA (2001) Can plants stimulate soil microbes and their own nutrient supply? Evidence from a grazing tolerant grass. Ecology 82:2397–2402

Hamilton EW, Giovannini EW, Moses, MS, Coleman JS, McNaughton SJ (1998) Biomass and mineral element responses of a Serengeti short-grass species to nitrogen supply and defoliation: Compensation requires a critical [N]. Oecologia 116:407–418

Harrison KA, Bardgett RD (2003) How browsing by red deer impacts on litter decomposition in a native regenerating woodland in the Highlands of Scotland. Biol Fert Soils 38:393–399

Harrison KA, Bardgett RD (2004) Browsing by red deer negatively impacts on soil nitrogen availability in regenerating native forest. Soil Biol Biochem 36:115–126

Hättenschwiler S, Vitousek PM (2000) The role of polyphenols in terrestrial ecosystem nutrient cycling. TREE 15:238–243

Haukioja E, Ruohomaki K, Senn J, Soumela J, Walls M (1990) Consequences of herbivory in the mountain birch (*Betula pubescens* spp *tortuosa*): Importance of the functional organisation of the tree. Oecologia 82:238–247

Holland JN, Cheng W, Crossley Jr DA (1996) Herbivore-induced changes in plant carbon allocation: Assessment of below-ground carbon fluxes using C-14. Oecologia 107:87–94

Holland EA, Detling JK (1990) Plant response to herbivory and below-ground nitrogen cycling. Ecology 71:1040–1049

Kielland K, Bryant JP (1998) Moose herbivory in Taiga: Effects on biochemistry and vegetation dynamics in primary succession. Oikos 82:377–383

King LK, Hutchinson KJ (1976) The effects of sheep stocking intensity on the abundance and distribution of mesofauna in pastures. J Appl Ecol 13:41–55

King LK, Hutchinson KJ Greenslade P (1976) The effects of sheep numbers on associations of Collembola in sown pastures. J Appl Ecol 13:731–739

Lehtilä K, Haukioja E, Kaitaniemei P, Laine RA (2000) Allocation of resources within mountain birch canopy after simulated winter browsing. Oikos 90:160–170

Mawdsley JL, Bardgett RD (1997) Continuous defoliation of perennial ryegrass (*Lolium perenne*) and white clover (*Trifolium repens*) and associated changes in the composition and activity of the microbial population of an upland grassland soil. Biol Fert Soils 24:52–58

McIntosh PD, Allen RB, Scott N (1997) Effects of exclosure and management on biomass and soil nutrient pools in seasonally dry high county, New Zealand. J Environ Manage 51:169–186

McNaughton SJ (1984) Grazing lawns: animals in herds, plant form and co-evolution. Am Nat 124:863–886

McNaughton SJ, Banyikwa FF, McNaughton MM (1997) Promotion of the cycling of diet-enhancing nutrients by African grazers. Science 278:1798–1800

McNaughton SJ, Banyikwa FF, McNaughton MM (1998) Root biomass and productivity in a grazing ecosystem: the Serengeti. Ecology 79:587–592

McNaughton SJ, Oesterheld M, Frank DA, Williams KJ (1989) Ecosystem-level patterns of primary productivity and herbivory in terrestrial habitats. Nature 341:142–144

McNaughton SJ, Ruess RW, Seagle SW (1988) Large mammals and process dynamics in African ecosystems. Herbivorous mammals affect primary productivity and regulate recycling balances. BioSci 38:794–800

Mikola J, Yeates GW, Barker GM, Wardle DA, Bonner KI (2001a) Effects of defoliation intensity on soil food-web properties in an experimental grassland community. Oikos 92:333–343

Mikola J Yeates GW, Wardle DA, Barker GM, Bonner KI (2001b) Response of soil food-web structure to defoliation of different plant species combinations in an experimental grassland community. Soil Biol Biochem 33:205–214

Milchunas DG, Laurenroth WK (1993) Quantitative effects of grazing on vegetation and soil over a global range of environments. Ecol Monogr 63:327–366

Neff JC, Reynolds, RL, Belnap, J, Lamothe, P (2005) Multi-decadal impacts of grazing on soil physical and biogeochemical properties in southeast Utah. Ecol Appl 15:87–95

Pastor J Dewey B, Naiman RJ, McInnes PF, Cohen Y (1993) Moose browsing and soil fertility in the boreal forests of Isle Royale National Park. Ecology 74:467–480

Rhoades DF (1985) Offensive–defensive interactions between herbivores and plants: their relevance in herbivore population dynamics and ecological theory. Am Nat 125:205–238

Ritchie ME, Tilman D, Knops JMH (1998) Herbivore effects on plant and nitrogen dynamics in oak savanna. Ecology 79:165–177

Ruess RW, Hendrick RL, Bryant JP (1998) Regulation of fine root dynamics by mammalian browsers in early successional Alaskan taiga forests. Ecology 79:2706–2720.

Ruess RW, Seagle SW (1994) Landscape patterns in soil microbial processes in the Serengeti National Park, Tanzania. Ecology 75:892–904

Stark S, Tuomi J, Strömmer R, Helle T (2003) Non-parallel changes in soil microbial carbon and nitrogen dynamics due to reindeer grazing in northern boreal forests. Ecography 26:51–59

Stark S, Wardle DA, Ohtonen R, Helle T, Yeates GW (2000) The effect of reindeer grazing on decomposition, mineralisation and soil biota in a dry oligotrophic Scots Pine forest. Oikos 90:301–310

Tracy BF, Frank DA (1998) Herbivore influence on soil microbial biomass and N mineralisation in a northern grassland ecosystem: Yellowstone National Park. Oecologia 114:556–562

Van der Wal R, Bardgett RD, Harrison KA, Stien A (2004) Vertebrate herbivores and ecosystem control: cascading effects of faeces on tundra ecosystems. Ecography 27:242–252

Van der Wal R, Brooker R, Cooper E, Langvatn R (2001) Differential effects of reindeer on high Arctic lichens. J Veg Sci 12:705–710

Van der Wal R, Pearce ISK, Brooker R, Scott D, Welch D, Woodin SJ (2003) Interplay between nitrogen deposition and grazing causes habitat degradation. Ecol Lett 6:141–146

Van Wijnen HJ, Van der Wal R (1999) The impact of herbivores on nitrogen mineralisation rate: Consequences for salt-marsh succession. Oecologia 118:225–231

Verchot LV, Groffman PM, Frank DA (2002) Landscape versus ungulate control of gross mineralisation and gross nitrification in semi-arid grassland of Yellowstone National Park. Soil Biol Biochem 34:1691–1699

Vitousek PM, Howarth RW (1991) Nitrogen limitation on land and in the sea – how can it occur. Biogeochem 13:87–115

Walsingham JM (1976) Effect of sheep grazing on the invertebrate population of agricultural grassland. Proc R Soc Dublin 11:297–304

Wardle DA, Barker GM, Yeates GW, Bonner KI, Ghani A (2001) Introduced browsing mammals in New Zealand natural forests: above-ground and below-ground consequences. Ecol Monogr 71:587–614

Wardle DA (2002) Communities and ecosystems: linking the aboveground and belowground components. Monogr Pop Biol 34. Princeton University Press, NJ

Wardle DA, Bardget RD, Klironomos JN, Setälä H, van der Putten WH, Wall DH (2004) Ecological linkages between aboveground and belowground biota. Science 304:1629–1633

Chapter 9
Plant Traits, Browsing and Grazing Herbivores, and Vegetation Dynamics

Christina Skarpe and Alison Hester

9.1 Introduction

Large browsing and grazing herbivores can have a profound influence on the physiognomy, composition, and function of vegetation, from the landscape scale to a single plant (Hobbs 1996; Augustine and McNaughton 1998). By selective foraging among populations of plant modules and genets and along resource gradients, herbivores exert differential pressures on plant populations at different spatial and temporal scales (Kielland and Bryant 1998; Allison 1990; Price 1991; Jia et al. 1995; Danell et al. 2003). In addition, herbivore behavioural costs (predation, shelter seeking, etc.) can strongly influence locational, and therefore foraging, choice (Schmitz 2003; Letourneau and Dyer 2004). Such differential herbivory pressures have consequences for vegetation composition, ecosystem properties and for plant evolution (Pastor and Naiman 1992; Jefferies et al. 1994; Danell et al. 2003). In this chapter we review some aspects of how large herbivores can modify vegetation, limiting the discussion to vascular plants and mammalian herbivores >ca. 5 kg in weight. In order to understand the interactions between large herbivores and vegetation we first show how various plant traits influence the foraging pattern by large herbivores, the amount of resources the plant loses in a foraging event, and the way the plant responds to the loss. This is important for the effects of herbivory on the competitive hierarchy among plants, which in turn govern vegetation dynamics and composition. We start with discussing how plant architecture, including that of trees/shrubs, forbs, and graminoids, influences patterns of foraging by large herbivores, the relative losses of resources by the plant, and the plant responses to those losses. We then discuss the resource economy of plants, how resources are acquired, and how they are allocated to growth and storage, and the implications of plant resource economy for plant performance following herbivory. This is followed by a discussion of plant strategies to resist herbivory, by either avoiding being eaten or minimising the negative effects of losses to herbivores. The significance of animal foraging behaviour is also touched upon, but this forms the main topic in several other chapters of the book. Finally, we discuss how the differences in plant properties and in herbivore utilisation among plant populations at different scales lead to shifts in competitive relations among plants and subsequently to shifts in vegetation properties, such as composition of populations, species, and functional types. Different conceptual models proposed to explain such shifts in vegetation are discussed.

9.2 Plant Architecture and Herbivory

9.2.1 Introduction

Vascular plant heights span five orders of magnitude, from minute, <1 cm, prostrate herbs to rainforest trees of more than 50 m. The associated range in biomass, related more to volume than to height, is even larger. The difference in plant height corresponds to a similar difference in the spatial distribution of photosynthetic biomass, flowers, and fruits of potential interest for foraging large herbivores (Harper 1977; Gill 1992, 2006; Bodmer and Ward 2006). Similarly, meristems are differently distributed in space, which is of importance for plant survival and resprouting following biomass removal (Raunkiær 1937; Haukioja and Koricheva 2000). The difference in size and architecture of plants is related to many other plant attributes, such as life history, longevity, and reproductive strategies. Size and architecture of plants also influence the proportion of the biomass that is edible and accessible for terrestrial large mammalian herbivores (Harper 1977; Crawley 1983; Haukioja and Koricheva 2000).

In order to understand plant strategies and responses in relation to herbivory, it is important to realise that plants are modular organisms, which are very different from mammals, for example. A genet, i.e., a plant originating from sexual reproduction, normally a seed, can in clonal species give rise to several ramets, looking more or less similar to the mother plant. Both genets and ramets are composed of smaller units, that we for simplicity also call modules, like shoots or tillers and leaves or needles (Harper 1977; Waller 1997). Under different conditions plant modules may behave either as members of the individual, exchanging resources with other modules, or as partly independent units, competing between themselves for resources. Such different behaviour has profound implications for the effect of herbivory on plants. With strong connectivity, resources can be drawn from many modules to ameliorate the loss of biomass in any part of the plant (Wilsey 2002). In contrast, with low connectivity resources are less mobile between modules, and the distribution of herbivory in the plant becomes much more important.

9.2.2 Trees and Shrubs

Common for all woody species, from minute arctic willows with only shoot tips and leaves exposed above the moss layer, to giant tropical rainforest trees, is the relatively large allocation of biomass to woody structures (Hytteborn 1975; Körner

1994). In most species, the primary function of woody stems and branches is to support the canopy and to transport water, mineral nutrients, and photosynthates between roots and leaves (Devlin 1966). The common explanation for the tall stems of trees is to lift the canopy higher than its neighbours in order to compete for light. Alternative explanations, for example for the occurrence of relatively tall trees in savanna or wooded steppe, where competition for light is negligible once canopies are above the grass layer, may be to reduce the risk of scorching by frequent grass fires, or to lift photosynthetic biomass out of reach of ground-living herbivores. Many species also store nutrients and carbohydrates in stems and branches (Honkanen et al. 1994; Vanderklein and Reich 1999; Millard et al. 2001). Woody biomass is relatively inedible for most mammalian herbivores and constitutes a protected resource of high importance for plant survival and resprouting after herbivory (van der Meijden et al. 1988).

Woody plants are composed of different kinds of shoots that can be defined in different ways (Kozlowski 1966). Here we simply separate between long shoots and short shoots. All new shoots grow from a meristem on an older shoot, often from the apical meristem in the tip of the old shoot. In addition long shoots bear several nodes with axillary meristems and leaves (Waller 1997). It is the long shoots that contribute to the growth and architecture of trees. Long shoots and their leaves are the modules most frequently browsed by large herbivores (Danell et al. 1994; Shipley et al. 1999), often with effects for the growth form of the tree (see below). Some tree genera also develop short shoots, which do not elongate, but produce one short internode and one leaf or a bouquet of leaves, or needles in conifers, each year (Atkinson et al. 1992). In species with plastic or indeterminate growth, such as most deciduous trees (Atkinson 1992; Raspe et al. 2000), short shoots may develop into long shoots following damage to other long shoots, e.g., by browsing (Danell et al. 1985). Thus, in such cases, short shoots might be regarded as resting buds with a means to be self-supporting with carbon. A third type of shoots are basal shoots or suckers, developing at the stem base primarily after severe damage to the plant, e.g., by herbivores or fire, and having the ability to develop into new ramets, ensuring persistence of the plant (Bond and Midgley 2001; Gill 2006). Some woody plants have clonal growth, implying that suckers may sprout from roots at some distance from the mother plant, so that one genet may give raise to many ramets. This is common in dwarf shrubs, e.g., *Vaccinium* spp. (Flower-Ellis 1971) but also occurs in many shrubs (Hester et al. 2006a) and a few trees, such as aspen (*Populus* spp.) (Peterson and Jones 1997). Ramets, as well as shoots and leaves, are modules of a genet. Modularity in woody plants is generally more complex than in herbs and often displays a lower degree of physiological integration (Beck et al. 1982; Sachs and Hassidim 1996; Peterson and Jones 1997; Haukioja and Koricheva 2000). Thus, responses to browsing may be localised to the modules closest to the damage (Danell et al. 2003), and the response by individual modules may differ profoundly from that of the whole tree or clone (Haukioja and Koricheva 2000).

Canopy architecture, i.e., the position of branches, shoots and leaves within a canopy, differs between plant species and affects how much light the leaves may capture (Waller 1997). Canopy architecture may also influence browsing by large herbivores, as widely dispersed shoots may give low foraging efficiency, whereas densely packed small shoots can be difficult for a browser to attack (Renaud et al. 2003). As shown later in this chapter, browsing herbivores may strongly change tree canopy architecture, for example by modifying numbers, distribution, and morphology of shoots, which, in turn, may affect future herbivory as well as plant photosynthesis rate, productivity, and competitive ability (Danell et al. 2003; Hester et al. 2006b). The strongest effect of large herbivores on tree vegetation is usually through selective browsing on young small individuals, determining the composition of the mature stand (Vourc'h et al. 2002; Danell et al. 2003).

9.2.3 Herbaceous Plants Other Than Graminoids

Herbaceous plants typically differ from most woody plants in being smaller, consisting of fewer modules and meristems and, it is believed, tend to be more short lived (Haukioja and Koricheva 2000), although little is known about the lifespan of most non-woody plants (Inghe and Tamm 1985). Independent of total life span, most herbs in seasonal climates have periodic shoot reduction to lower heights or to subterranean parts in the winter or the dry season (Raunkiær 1937). Hence, herbs tend to be less apparent (sensu Feeny 1976) for large herbivores both in space and time than are trees or shrubs. On the other hand many trees and shrubs have the advantage of growing above browse height of most terrestrial herbivores, unlike most herbaceous plants.

While woody plants allocate most biomass to support structures and roots and only a small proportion to photosynthetic tissue, herbs typically allocate about half their biomass to leaves and less than a quarter to stems, which are generally not strongly lignified (Körner 1994; Porter and Nagel 2000). If the amount of residual, inedible, biomass is of importance for plant resprouting after herbivory, herbs would generally have lower capacity for recovery than woody plants. In addition, herbs would have fewer meristems from where to resprout. On the other hand, the high allocation to photosynthetic tissue gives herbaceous plants a high potential growth rate (mass per mass unit) compared to woody plants (Haukioja and Koricheva 2000).

9.2.4 Graminoids

Grasses are by far the most important family of plants eaten by large herbivores, and the speciation of grazing large herbivores is closely linked with the development of grass-dominated vegetation during the Miocene (MacFadden 1997; Pérez-Barbería et al. 2004; Stromberg 2004). Apart from bamboo (*Bambusa* spp.), grasses, i.e., species of the family *Poaceae*, are herbaceous non-woody plants, generally from a few centimetres to a few metres high. In graminoids we include grasses and other grass-like monocotyledonous plants such as sedges and rushes. A fundamental difference between grasses and dicotyledonous plants, herbs as well as woody plants, is that dicotyledons grow from apical meristems that are vulnerable to herbivory, whereas growth in monocotyledons is from meristems (apical and intercalary) generally situated close to the ground. Thus, the latter are mainly below 'optimal' grazing height for most large herbivores (Illius and Gordon 1987) and protected between old leaf sheaths, which make them less accessible (Wolfson and Tainton 1999; Haukioja and Koricheva 2000).

The basic unit in a grass plant is the tiller, corresponding to the shoot in dicotyledonous plants (Wolfson and Tainton 1999). Vegetative grass tillers have a very short stem with an apical meristem close to the ground. As long as the meristem is not grazed, the tiller continues its growth after defoliation (Lemaire and Chapman 1996). What appears to be the stem is actually densely rolled young leaves, one within the other, from which the leaf blades develop, the oldest leaves at the base and the youngest at the top. The leaf blade grows from an intercalary meristem at the base, so that the tip is the oldest part of the leaf. Therefore, as long as the intercalary meristem is active, the leaf will continue its growth even if much of the blade is removed (Lemaire and Chapman 1996; Wolfson and Tainton 1999). Only when the apical meristem of the tiller turns into a reproductive meristem will the stem elongate into a culm carrying the inflorescence in its top. At that stage the meristem and inflorescences are vulnerable to herbivory, and therefore a high density of grazers eating inflorescences may almost entirely prevent sexual reproduction in grasses, giving an advantage to vegetatively reproducing species and genotypes, which subsequently often dominate in heavily grazed grasslands (Diaz et al. 1992; Briske 1996).

In most species of both perennial and annual grasses the initial tiller develops lateral basal meristems or axillary buds that can give rise to new tillers sprouting either immediately or following the death of the old tiller, e.g., after heavy grazing (Briske 1996). The character of these adventitious tillers determines the growth form of the plant. In tufted grasses the new tillers grow upwards and eventually form more or less dense tussocks. In clonal species some of the tillers form rhizomes or stolons, which develop roots and new tillers, i.e., new ramets, from axillary buds at the nodes, which in turn may develop new tillers, stolons, or rhizomes (Wolfson and Tainton 1999). In this way, one genet in many grasses and sedges consists of numerous ramets with many tillers and may cover a large area. There is some evidence to suggest that the stoloniferous growth form in grasses is an escape strategy (see 9.4.2) to grazing by large herbivores (Wolfson and Tainton 1999; Wilsey 2002). The reduction in erect tall grasses and the increase in decumbent and creeping morphotypes under grazing seem to depend both on the superiority of low growing genotypes and on phenotypic plasticity in many species (Georgiadis and McNaughton 1988; Jaramillo and Detling 1988; Painter et al. 1989; Briske 1996). A prostrate growth form allows the plant to maintain a large proportion of biomass and meristems below the grazing height for most large herbivores, facilitating resprouting after grazing (Diaz et al. 1992, Briske 1996, Wolfson and Tainton 1999). It also enables the plant to spread vegetatively in an environment where inflorescences may often be grazed, and to quickly colonise gaps and utilise the spatial heterogeneity in the sward imposed by grazing animals (Wilsey 2002). At the same time the grazing of (potentially) taller neighbours is a prerequisite for the plant to avoid being shaded out.

In many ways clonal grasses and other strongly clonal herbs resemble woody plants in possessing a large number and array of modules and meristems. The degree of connectivity and exchange of resources between grass modules have been subject to controversy, but the general conclusion is that they are more developed in graminoids than in many trees (Marshall and Sagar 1968; Langer 1972; Danckwerts and Gordon 1987; Peterson and Jones 1997). This provides flexibility in resource allocation and growth, and clonal grasses may be connected over large areas exchanging resources between ramets and tillers, enhancing the overall growth and resistance to grazing (Wilsey 2002). Stoloniferous graminoids are among the most grazing-tolerant plants, in some ecosystems sustaining consumption of up to 90% of the above-ground production (McNaughton 1985).

9.3 The Chemistry of Plants

9.3.1 *Photosynthesis*

The most fundamental chemical process in green plants is photosynthesis, by which chlorophyll is used as a catalyst to bind solar energy and carbon dioxide into energy-rich carbohydrates (Mooney 1997). A number of different pathways of photosynthesis has evolved, increasing the effectiveness of solar energy capture under different conditions. Two basic pathways of photosynthesis with relevance for herbivory are the C_3 and C_4 types, in which the compounds produced initially are 3-phosphoglyceric acid and four-carbon acids, respectively (Ellis et al. 1980). The C_3 pathway dominates in dicotyledonous plants and in non-tropical grasses, while C_4 photosynthesis is common in tropical grasses (Mooney 1997; Wolfson and Tainton 1999). C_4 photosynthesis requires high temperatures and in such climates its main advantage compared to C_3 photosynthesis is the low CO_2 compensation point, implying that photosynthesis is efficient even with partly closed stomata, and low (or non-existing) photorespiration (Mooney 1997; Wolfson and Tainton 1999). This gives C_4 plants higher water-use efficiency (units carbon fixed per unit water lost) than C_3 plants, making them more productive in hot dry climates than C_3 plants (Mooney 1997). However, C_4 plants tend to be more lignified and less digestible than C_3 plants, although Wilson and Hacker (1987) showed that the difference in digestibility between sympatric species differing in photosynthetic pathway was small. There is some evidence that C_4 grasses under some conditions may increase relative to C_3 grasses under grazing (Wang 2002). C_4 grasses support a large proportion of domestic and natural grazing ecosystems of the world, e.g., the African and American savannas.

Photosynthesis rate is strongly related to nitrogen (N) concentration in the leaf, and is generally highest in young leaves, declining with age as stomatal and mesophyll resistance increases (Ludlow and Wilson 1971). The photosynthetic rate of remaining mature leaves may, however, increase following herbivory (Hodgkinson 1974; Painter and Detling 1981; Wolfson and Tainton 1999). This may be caused by defoliation retarding the physiological aging of remaining leaves (Gifford and Marshall 1973), by increasing quantity and quality of light for the leaves as shading tissue is removed (Wallace 1990; Senock et al. 1991) or because competition between leaves for resources other than light is reduced (Wareing et al. 1968). Also the creation of strong sinks for photosynthates in resprouting meristems may contribute to this 'rejuvenation' (Wareing and Patrick 1975). However, the opposite response, with reduced photosynthetic rate of remaining leaves, has also been recorded (Wallace et al. 1985). As photosynthesis is the basic means of production in the plant, strategies related to herbivory often aim to protect or quickly restore photosynthetic capacity. This can be achieved, for example, by allocating resources to vegetative growth at the expense of reproduction and by developing leaves with high specific leaf area (Crawley 1983; Bergström et al. 2000; Skarpe and van de Wal 2002).

9.3.2 Energy and Nutrient Reserves

Carbohydrates produced in photosynthesis are stored in the leaf as vacuolar sugar or starch during the day and are translocated or used locally for growth and respiration during the night (Chapin et al. 1990). Young developing leaves and shoots depend initially on carbohydrates transferred from other parts of the plant. As leaf growth ceases, the leaf's requirement for carbohydrates declines, and it becomes self supporting and a net exporter of energy to young growing tissue in the same shoot or tiller, and eventually to other sinks in the plant (Chapin et al. 1990). Clearly, these movements of carbohydrates and nutrients within the plant have implications for the severity, timing and location of damage on the plant from herbivore impacts (Crawley 1983; Rosenthal and Kotanen 1994; Fitter 1997).

Nutrients are taken up from the soil solution by plant roots, often in association with mycorrhiza (Begon et al. 1996; Fitter 1997). Nitrogen is often taken up in the form of nitrate or ammonium. However, Ruess and McNaughton (1984) found that some African grasses from a heavily grazed area used urea N from the urine of grazing animals, which apparently stimulated above ground growth in these grasses more than did ammonium N. Since then it has been found that many plants, some of them from environments with intense herbivory, are taking up organic N (Falkengren-Grerup et al. 2000; Bardgett et al. 2003). Nutrients are transferred from roots to leaves, where they are used for growth and other life functions, such as formation of proteins (Nambiar 1987; Chapin et al. 1990). Many studies have shown that herbivores are attracted to plant material with high concentration of protein N (Danell et al. 1991a, b; Crawley 1997). Such selective removal of plant material can significantly reduce new growth, through reduced photosynthetic capacity as well as loss of nutrients and carbohydrate reserves (Crawley 1983; Rosenthal and Kotanen 1994; Crawley 1997; Millard et al. 2001).

Most perennial plants build up energy and nutrient stores over different time scales to buffer temporal asynchrony between resource uptake and demand (Chapin et al. 1990). There are differences in plant strategies, and species described as adapted to frequent disturbances (e.g., by herbivory) have been found to invest more in stored nutrient resources than species apparently not adapted to disturbance (Bloom et al. 1985; Kruger and Reich 1997), and may also re-absorb a larger proportion of nutrients from leaves before abscission (Pugnaire and Chapin 1993). Similarly, plants in resource-poor environments invest proportionately more in storage than do plants in resource-rich environments (Millard and Proe 1991; Millard 1996). In seasonal environments stores of energy in the form of starch or non-structural carbohydrates are built up late in the growing season, when growth ceases (Langstrom et al. 1990; Holopainen et al. 1995) and resources are re-absorbed from senescing leaves (Thomas and Stoddart 1980).

Following Chapin et al. (1990), we define storage as 'any resources that build up in the plant and can be mobilized later to support biosynthesis for growth or other functions'. This includes resources specifically allocated to reserves (sensu Chapin et al. 1990) in competition with growth and other needs of the plant. It also includes accumulation of 'surplus' resources, e.g., in situations when photosynthesis or nutrient absorption continues while growth is restricted (Chapin et al. 1990; Herms and Mattson 1992; Steinlein et al. 1993). Resources employed in growth may also be recycled and used for new growth elsewhere in the plant (Heilmeier et al. 1986). Opportunity costs for storage, of course, decline if the stored resources have other functions in the plant, for example, when used in defence compounds that can turn over and support structural biosynthesis when needed (Coley et al. 1985). Plants use recently acquired carbohydrates and nutrients before those that are stored, if both are available, suggesting that there is an opportunity cost to storage, involving both the growth of storage tissue (rhizomes, tubers, bulbs, etc.) and the filling of these tissues with storage products (Thomas and Stoddart 1980; Chapin et al. 1990; Steinlein et al. 1993) and/or the retrieval of the stored products. Woody plants use stems and branches for storage; many plants additionally use roots or other underground structures such as rhizomes, tubers and bulbs (Coleman et al. 1991; Millard et al. 2001; Fig. 9.1). Many evergreen species do not seem to store nutrients, but resources are left in situ in last generation's leaves/needles, and are translocated from there to new developing buds (Nambiar and Fife 1991; Eckstein et al. 1998; Millard et al. 2001), leaving them vulnerable to loss as a result of herbivory (Millard et al. 2001; Hester et al. 2004). Storage organs are often protected against large herbivores by chemical defence and/or by being situated underground. Large herbivores, however, often track the nutrient rich biomass. For example, elephants have been observed (Skarpe unpublished) to feed on the green parts of grasses like *Panicum repens* and *Cynodon dactylon* in the growing season, discarding underground parts, but in the dry season to kick up and eat the nutrient rich rhizomes while discarding the wilted shoots.

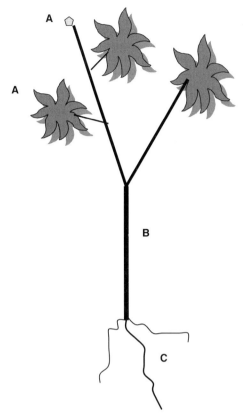

Storage locations: **Vulnerability to mammalian herbivory**

A: leaves and buds: High (unless out of reach)
B: woody stem: Low
C: roots and underground organs: Generally low (except for digging/uprooting)

Fig. 9.1 Plant storage locations and consequences for herbivory and its impacts

9.4 Plant Resistance

9.4.1 Introduction

Plants have 'evolved' various strategies which minimise the negative effect of herbivory on plant fitness (Rosenthal and Kotanen 1994; Strauss and Agrawal 1999). Rosenthal and Kotanen (1994) called all these plant strategies 'resistance' and within that concept distinguished 'avoidance strategies', including escape and defence strategies, and 'tolerance strategies'.

Central to most theories on plant resistance is the assumption that all resistance traits imply a cost for the plant, directly in terms of resource allocation and/or indirectly in the form of ecological costs (Koricheva 2002b; Strauss et al. 2002; Hester et al. 2006b). Only if the fitness cost of resistance is less than the cost for the herbivory it prevents or ameliorates, will plants with resistance traits have higher fitness and a competitive advantage compared to plants without such traits in the same situation (Jokela et al. 2000). Thus, the interactions between resource availability or uptake, resource loss to herbivores, and plant resistance strategies are dynamic in space and time, complex and poorly understood (Stamp 2003; Hester et al. 2006b). However, there is no doubt that differences in resistance strategies in relation to prevalent herbivory regimes are a powerful means by which large browsing and grazing herbivores influence competitive hierarchies between plants both within and across species (Strauss et al. 2002; Cipollini et al. 2003; Hester et al. 2006b), and, hence, vegetation composition.

Avoidance and tolerance can represent alternative evolutionary responses to herbivory (van der Meijden et al. 1988; Juenger and Lennartsson 2000; Fig. 9.2), and are often negatively related in comparisons among species. It was long assumed that the two strategies are mutually exclusive, but recent studies have shown that this is not necessarily the case (Tiffin and Rausher 1999; Mauricio 2000).

9.4.2 Escape Strategies

Plants can avoid herbivory by spatial or temporal escape. Milchunas and Noy-Meir (2002) distinguish between internal and external avoidance, including escape, mechanisms. Internal escape mechanisms encompass morphological traits, as discussed above, such as short stature, maintaining a large proportion of the biomass

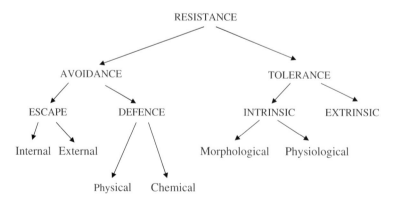

Fig. 9.2 Plant resistance to herbivores: conceptual strategies (derived from Rosenthal and Kotanen 1994; Strauss and Agrawal 1999; Kotanen and Rosenthal 2000; Milchunas and Noy-Meir 2002), modified from Hester et al. (2006)

below herbivory level (depending on the size of the herbivore) or keeping most edible biomass above reach for terrestrial herbivores. The latter strategy, however, requires that young plants survive herbivory to attain tall growth (Vila et al. 2002). Phenological traits making the plant less apparent (sensu Feeny 1976) or attractive to large herbivores include deciduousness, the reduction of the whole plant to underground organs, and survival only as seed. Storage of resources in woody stems or underground is also a kind of escape strategy. Ward and Saltz (1994) report that gazelles digging for lily bulbs in the Negev drive a selection of the lilies to grow to even greater depths to escape herbivory.

In environments with high herbivory pressure, sensitive plant species mainly occur in inaccessible places such as rock outcrops or steep slopes, thus using external/locational escape from herbivory. Grazing-sensitive species may have survived in such refugees for thousands of years in heavily livestock grazed areas, for example in the Middle East (Noy-Meir and Seligman 1979; Milchunas and Noy-Meir 2002) and probably in areas with intense natural herbivory (Skarpe et al. 2004). Such populations may serve as sources of propagules for surrounding sink populations that may not be able to reproduce themselves except in narrow 'windows of opportunity' when herbivore densities are reduced following, for example, disease or drought (Milchunas and Noy-Meir 2002; Skarpe et al. 2004). Plants may also escape herbivory by association with either less palatable or more palatable species, depending on the foraging pattern of the herbivore (see 9.6.4; Hjältén et al. 1993a; Olff et al. 1999; Hester et al. 2006b).

9.4.3 Structural and Chemical Defences

Structural defences, mainly spines, prickles and thorns of different types and origin, are hypothesised to have evolved as a defence against foraging by large herbivores in many parts of the world (Milewski et al. 1991; Myers and Bazely 1991; Grubb 1992; Karban et al. 1999). Spinescence occurs in a wide variety of taxa and habitats, and is particularly abundant in arid and semi-arid regions and in nutrient-rich areas of small extent (Grubb 1992; Ward and Young 2002). Spinescence has been shown to reduce short-term intake rate of many browsers (Cooper and Owen-Smith 1986, 2001) by reducing bite rate and/or bite size (Belovsky et al. 1991, Skarpe et al. 2006). Particularly, spines tend to change foraging mode from twig biting and leaf stripping to the less detrimental picking of leaves from between the spines (Cooper and Owen-Smith 1986; Gowda 1996). Thus, while not actually preventing browsing, spinescence may reduce plant losses to large herbivores.

A large number of chemical compounds in plants have been shown to have deterrent effects on large herbivore foraging. Indeed, variation in plant defences is a major factor governing 'palatability' and animal forage choice (Stamp 2003). A plethora of hypotheses have been presented to explain the evolution of defences, under what conditions plants should invest in defences and in what types of defences (Feeny 1976; Bryant et al. 1983; Coley et al. 1985; Milewski et al. 1991;

Myers and Bazely 1991; Herms and Mattson 1992; Grubb 1992; Karban et al. 1999; Jokela et al. 2000; Koricheva 2002a). According to their effects on animals, Feeny (1976) divided chemical defences into; quantitative defences', where the deterrent effect increases with the concentration of the defence compound, and 'qualitative defences', where the deterrent effect is already high at very low concentrations. However, Foley et al. (1999) questioned the validity of this division. Chemical defences may be constitutive or induced (Rosenthal and Janzen 1979; Palo and Robbins 1991; Foley et al. 1999). It is hypothesised that constitutive defences are an evolutionary response to intense herbivory and act as deterrents, reducing the likelihood and severity of attack. Induced chemical defences are hypothesised to be adaptive, phenotypically plastic, responses which reduce herbivory only when needed, in contrast to the 'continuous' nature of constitutive defences. Particularly in environments with stochastic or periodic herbivory, induced defences may be less costly for the plant than constitutive defences (Åström and Lundberg 1994; Cippolini et al. 2003; Hester et al. 2006b).

Feeney's (1976) definition of quantitative defences includes many of the carbon-based compounds, e.g., different types of phenolics and related substances (Palo and Robbins 1991). Compounds of the phenolic family are widespread in plants of widely different taxonomic and geographic origin, including trees, forbs, and grasses (Cooper et al. 1988; Sunnerheim et al. 1988; du Toit et al. 1991). Some phenolics undoubtedly reduce large herbivore foraging (Bryant et al. 1989; Cooper and Owen-Smith 1985; Woodward and Coppock 1995), and there is evidence that they are selected for by plants under heavy herbivory. These types of chemical defences are mainly found in plants evolved in nutrient-poor environments (Coley et al. 1985; Bryant et al. 1983). Such plants are inherently slow growing and have little ability to compensate for lost biomass, and are therefore predicted to invest in defence (Coley et al. 1985). It has been suggested that the concentration of carbon-based defences fluctuates with the difference between carbon gained by the plant in photosynthesis and carbon used for growth, and hence may be sensitive to differences in growth rate in space and time (Herms and Mattson 1992; Stamp 2003). It is hypothesised that plants in resource-rich environments often do not avoid herbivory but develop tolerance traits (see Sect. 9.4.4). Defence compounds, if they occur in such environments, are often effective in low concentration (Feeney's 'qualitative defences'), such as many alkaloids (Coley et al. 1985). These types of defence are generally not solely carbon based, but contain substances such as nitrogen. For fast-growing plants needing all the carbon they acquire for growth, this is 'cheaper' than defences which require high concentrations to be efficient. A problem with chemical plant defences is the ability of herbivores to develop behavioural and physiological counter adaptations (Iason 2005; Iason and Villalba 2006). For example, many browsers have tannin binding proteins in their saliva, lessening the digestibility-reducing effect of ingested tannins (Robbins et al. 1987; Austin et al. 1989; Juntheikki 1996). More specific adaptations also exist; for example, Pass et al. (2002) found liver enzyme adaptations to eucalypt constituents by various specialist-feeding marsupials.

9.4.4 Plant Tolerance

Plants in resource-rich environments often do not avoid herbivory, but develop tolerance traits to minimise the harmful effects of herbivory. Also in situations where the plant defence is ineffective in preventing loss of biomass, for example, if the herbivore(s) is(are) insensitive to the defence, or if trampling is an important source of damage, tolerance may be a more advantageous strategy than avoidance (Strauss and Agrawal 1999; Jokela et al. 2000). Tolerance is defined here as 'the strategy by plants to minimise the reduction in fitness following herbivory' (Rosenthal and Kotanen 1994; Briske 1996; Strauss and Agrawal 1999; Kotanen and Rosenthal 2000). Tolerance traits include high and flexible rates of nutrient absorption, photosynthesis and growth, and numerous protected meristems (Bradshaw 1965; Rosenthal and Kotanen 1994; Hester et al. 2006b). Like avoidance, tolerance to herbivory involves costs for the plant, e.g., the building and maintenance of stores of energy and nutrients as well as of dormant buds that can be activated following herbivory (McNaughton 1983; van der Meijden et al. 1988; Bilbrough and Richards 1993). Herbivory-tolerant plants are generally palatable to large herbivores, and have in some cases been found to increase in resource-rich environments following herbivory or intense trampling (Olofsson et al. 2001) and may form preferentially grazed patches where nutrient cycling may also be enhanced by the greater herbivore presence creating a positive feedback loop (section 9.5.5; Olofsson et al. 2001; Harrison and Bardgett this volume). Tolerance in plants is assumed to have little direct effect on herbivore fitness, thus is considered unlikely to trigger counter-adaptations in herbivores (Rosenthal and Kotanen 1994; Roy and Kirchener 2000).

9.5 Effects of Herbivory on Plants

9.5.1 Introduction

Foraging by large herbivores can have a wide variety of effects on plants, from instantaneous death to increased growth and competitive ability (Crawley 1983). As discussed above, herbivory interacts with plant architecture, photosynthesis, dynamics of energy and nutrients, and expressions of defence and tolerance (Crawley 1983; Hester et al. 2006b). Effects of herbivory on plants are also influenced by external factors, including previous and subsequent disturbance such as fire or drought, inter- and intraspecific competition and resource availability. The ultimate effect on plant performance depends on the impact by the herbivore on plant fitness, which is difficult to quantify, and is often substituted by measures of biomass, production or reproductive output (Crawley 1997). Herbivore differential foraging among plants and populations of plants and the variation in plant responses to being eaten can have a strong effect on the competitive relations among plants. Thus, by increasing the inter- and intraspecific variation in fitness of plants (Danell et al. 2003), large herbivores can have a profound influence both on vegetation composition and plant evolution (Vourc'h et al. 2001, 2002; Danell et al. 2003).

9.5.2 Reserve Dynamics

Plants growing in seasonal environments do not generally respond to herbivory occurring during the dormant season until the beginning of the following growing season, unless the impact is severe enough to cause desiccation and death, but may respond immediately to grazing/browsing taking place during the growing season (Senn and Haukioja 1994; Hester et al. 2004; Gill 2006). Sprouting of new shoots and leaves following dormancy and damage inflicted during the dormant period is generally initiated using stored resources (Chapin et al. 1990; Millard et al. 2001). Many studies have shown that the response of trees to damage by herbivores depends greatly on the ability of the trees to use stored nutrient resources for growth (Haukioja et al. 1990; Honkanen et al. 1999; Millard et al. 2001; Sect. 9.3.2 this chapter). Millard et al. (2001) stressed the importance of the location of resources in the plant for the preservation of resources for future use, linking this to the considerable difference between evergreen and deciduous species in their ability to resprout following herbivory (Bryant et al. 1983; Danell et al. 1991b, 1997). As many evergreens do not store nutrients, but resources remain in situ in the latest generation of leaves or needles (Sect. 9.3.2), such resources are often lost to herbivores, preventing their recycling to new growth. Millard et al. (2001) suggest that the often high level of defences in evergreen leaves and needles may have evolved more to protect the nutrient resources than the leaves themselves.

9.5.3 Sprouting and Resprouting

In addition to its dependence on availability of stored resources, resprouting following herbivory requires buds that escape the herbivore and can be activated for new growth. Regrowth following herbivory includes refoliation, sprouting of new shoots, or the production of whole new ramets (Miquelle 1983; Danell et al. 1997; Bond and Midgley 2001). Plants vary in their ability to develop new leaves on the defoliated positions (Miquelle 1983). For refoliation to take place, a certain minimum degree of defoliation is often required, and refoliation is more likely to occur early in the growing season, when the expected productive lifetime of the leaf is relatively long, than late in the season (Miquelle 1983; Danell et al. 1994; Skarpe and van der Wal 2002; Hester et al. 2004, 2005). Likewise, resprouting of shoots is generally most intense following herbivory early in the growing season or during dormancy, as plants have more time to compensate before the end of the growing season (Gill 1992; Bergström and Danell 1995; Hester et al. 2005; Guillet and Bergström 2006). When browsing removes leading shoots the apical dominance is reduced (Aarssen 1995), resulting in the development of many lateral shoots, which reduces the height of the plant but often increases its lateral spread compared to unbrowsed individuals (Belovsky 1984; Hester et al. 2004; Makhabu et al. 2006a). High or medium intensity twig browsing in woody species generally removes significant proportions of meristems present, in many species resulting in fewer shoots in the following growing season (Danell et al. 1994; Bergström et al. 2000; Hester et al. 2004). As a result there is less competition for resources among individual shoots, and shoots can become larger and have higher nutrient concentrations than those on undamaged trees (Danell et al.1994; Bergström et al. 2000; Rooke et al. 2004). This may also be influenced by increased root:shoot ratio following the browsing of shoots (McNaughton 1984; Danell and Bergström 1989). The reduction in defence compounds sometimes observed in such shoots may be a result of resources being allocated for fast growth at the expense of defence, and/or the breakdown of existing defence compounds and their components subsequently used for growth (Coley et al. 1985). Conversely, leaf stripping of trees during the growing season has been shown to result in an increase in number of shoots the following season, but a decrease in shoot size (Danell et al. 1994). Following severe damage, e.g., by fire or very intense browsing, some trees, particularly young ones, and many herbs may respond by sprouting basal shoots from the lower part of the stem (Bond and Midgley 2001). Such shoots have the capability to develop into new ramets and enhance the persistence of the plant (Bond and Midgley 2001). Thus, large herbivores may significantly change abundance, morphology, and chemistry of shoots, and hence interfere with their own future food resource as well as with plant fitness and thereby with vegetation composition.

In grasses the apical meristems of vegetative tillers often survive grazing, as described in Sect. 9.2.4, and can produce new leaves following a grazing event (Lemaire and Chapman 1996; Crawley 1997; Fig. 9.3). Many grasses develop new tillers from lateral buds on the basal part of the stem following grazing, whether or not the apical meristem is consumed (Briske 1996). The activation of such buds is influenced by light quality, which varies with the density of the grass canopy above the meristem (Deregibus et al. 1985). The ability to sprout new tillers from lateral buds also varies with species, growth form, and time of grazing (Briske 1996). It is generally strong during much of the growing period in rhizomatous carpet-forming species, often occupying grazed environments, and poorer in erect tussock species (Briske 1996). In clonal species the resprouting modules may develop into leaf bearing tillers, stolons, or rhizomes. The formation of new tillers following grazing, in combination with often reduced internode length, may make a grazed sward lower but denser, with higher biomass per volume and higher leaf-to-stem ratio compared to an ungrazed sward (Hodgson 1981). Such grazed swards provide a larger bite size and higher energy and nutrient intake rate for grazers than ungrazed swards, and therefore may be maintained by herbivores (Hodgson 1981; Bircham and Hodgson. 1983; Jaramillo and Detling 1988).

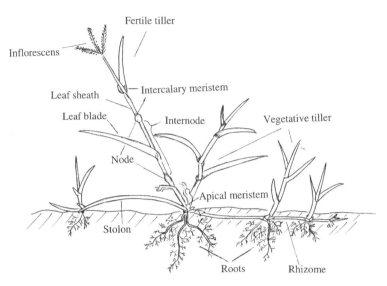

Fig. 9.3 Schematic picture of clonal grass plant with stolons, rhizomes, fertile and vegetative tillers, and positions of meristems

9.5.4 Repeated Herbivory

The reduced height of browsed trees (Sect. 9.5.4 this chapter) leads to a larger proportion of shoots and leaves remaining available within browsing height for terrestrial herbivores (Hester et al. 2000; Makhabu et al. 2006a). Improved accessibility, together with the larger size and generally higher nutritional value of regrowth shoots, has often been observed to increase the probability of browsing for a previously browsed plant (Löyttyniemi 1985; Danell and Bergström 1987; Welch et al. 1991; Bergström et al. 2000; Moore et al. 2000; Makhabu and Skarpe 2006). However, reduced preferences have been recorded in other studies (Danell et al. 1997; Duncan et al. 1998). It is not always clear why these differences are found, but the occurrence of induced defences may offer one explanation (Bryant et al. 1983). Increased browsing of previously browsed plants may result in a 'feeding loop', where browsing-induced changes in the plant lead to further browsing and further changes in the plant (du Toit et al. 1990). The impact of browsing megaherbivores, such as the African elephant (*Loxodonta africana*) on trees has been shown to sometimes lead to the formation of low, intensely coppiced trees or stands of trees with high production of preferred browse, forming what might be called a 'browsing lawn', analogous to 'grazing lawns' (see below; Jachmann and Bell 1985; Makhabu et al. 2006a). Despite the differences in observed preference or avoidance of previously browsed plants, browsing generally increases variation among plants in food quality for herbivores (Danell et al. 2003). Rebrowsing implies that the targeted trees suffer repeated damage and may eventually die, but that a smaller proportion of the population is browsed than would be expected if browsing was randomly distributed. Conversely, a larger proportion of the population than expected should be browsed if previously browsed trees were avoided. Obviously, both scenarios would impact vegetation structure and composition. If the differences between trees leading to browsing in the first place are heritable, rebrowsing could lead to a genetic change in the tree populations, with potential effects for future vegetation composition and herbivory (Vourc'h et al. 2001, 2002). However, surprisingly little is known about what internal and external factors govern the decision by the browser to feed from a certain plant in the first place (Bergström and Danell 1987; Welch et al. 1991; Duncan et al. 1998).

In grasslands, the repeated utilisation by grazers of previously grazed patches with vigorously resprouting, grazing-tolerant grasses, i.e., 'grazing lawns', has long been documented (Lock 1972; McNaughton 1984; Hodgson 1981; Coppock et al. 1983; McIntyre and Tongway 2005). The short-cropped sward of such repeatedly grazed patches may have less biomass per area but higher biomass density (biomass per volume), lower proportion of standing dead, higher leaf to stem ratio, and higher concentration of nutrients and minerals than grasses in less grazed patches (Hodgson 1981; Bircham and Hodgson 1983; McNaughton 1984). Repeated grazing in nutrient rich environments has also been shown to enhance soil nutrient availability, presumably due to increased cycling of nutrients caused by higher quality litter and the deposition of urine and faeces (Hodgson 1981; Ruess and McNaughton 1984; Harrison and Bardgett this volume). While the 'grazing lawn' concept is patch based, and a grazing lawn may include many individual plants, genets and ramets, and also different species, the 'browsing lawn' concept for woody species generally includes a single tree or a stand of trees of the same species (Jachmann and Bell 1985; du Toit et al. 1990; Makhabu et al. 2006a). In contrast to grazing lawns, most studies of repeated browsing have found reduced litter quantity and quality and lower soil N concentrations in the presence of browsing compared to the situation without browsing (Pastor et al. 1993; Ritchie et al. 1998; Persson et al. 2005). Thus, the nutrient dynamics and persistence of a 'browsing lawn' may be fundamentally different from that of a 'grazing lawn' (Naiman et al. 2003; Fig. 9.4).

9.5.5 Compensatory Growth

Plants are defined as compensating for herbivory when the fitness of browsed or grazed plants is the same as of undamaged plants, undercompensating or overcompensating is defined as when damaged plants have lower and higher fitness, respectively, than undamaged plants (McNaughton 1983; Belsky 1986; Maschinski and Whitham 1989; Whitham et al. 1991; Noy-Meir 1993). In most studies biomass is discussed instead of fitness, as this is easier to measure (Strauss and Agrawal 1999). There are many records of 'overcompensation' in terms of above-ground biomass by grasses and sedges in nutrient rich environments (McNaughton 1983; Paige and Whitham 1987; Hik and Jefferies 1990), and rather fewer for woody species (Dangerfield and Modukanele 1996). However, it has been questioned

Fig. 9.4 "Browsing lawn" intensely resprouting mopane trees (*Coelophospermum mopane*) repeatedly browsed by elephants (this species is not much browsed by other browsers).

whether the observed regrowth should strictly be defined as compensatory growth (Järemo et al. 1996); in fact whether overcompensation really occurs or not may be a matter of definition, and this is still the subject of much controversy (Belsky 1986; Crawley 1997; Hester et al. 2006b).

Tolerance traits, including fast (re)growth, stored resources, and numerous meristems, are to a large extent species specific, and are prerequisites for resprouting and compensatory growth (Strauss and Agrawal 1999; Juenger and Lennartsson 2000). Generally, fast growing species, particularly short-lived herbs and grasses, have a greater ability to regrow after biomass loss than do relatively slow growing woody species (Whitham et al. 1991; Bergström and Danell 1987), although trees may compensate over a longer time scale (Haukioja and Koricheva 2000). Total biomass of current season's shoots on trees browsed (or clipped) the previous year

Fig. 9.4 b (continued) 'Grazing lawn' short-cropped repeatedly grazed sward in the foreground with wildebeest (*Connochaetes taurinus*) and Grant's gazelle (*Gazella granti*); Serengeti, Tanzania

is usually smaller or similar to that on unaffected trees (Hjältén et al. 1993b; Danell et al. 1997; Bergström et al. 2000; Hester et al. 2004), rarely larger (Dangerfield and Modukanele 1996, Hester et al. 2004). Generally, the more severe the damage the poorer is the ability of the plant to compensate. However, severity of browsing interacts with plant age and physiological (phenological) stage and with resource availability to affect compensatory ability (Bullock et al. 2001; Gill 1992; Guillet and Bergström 2006). Damage during the dormant period in seasonal ecosystems generally has least effect on plant growth, as many plant resources are stored in stems and roots at that time, although this depends on species (Gill 1992; Millard et al. 2001). Browsing during the growing season may have greater effect on the plant, but early herbivory may cause the smallest reduction in growth, as plants have a longer time to compensate before the end of the growing season (Bergström and Danell 1995; Hester et al. 2005; Guillet and Bergström 2006). In perennial plants, older individuals, with more developed root systems and nutrient stores, are considered more likely to compensate well for herbivory than younger plants (Guillet and Bergström 2006). Plants with high resource availability and 'tolerance' traits are likely to have greater flexibility in growth and nutrient uptake than plants in resource-poor environments, thus regrowing faster after herbivore damage (Coley et al. 1985; Rosenthal and Kotanen 1994; Danell et al. 1997). Furthermore, for trees, plants with high resource availability are likely to grow above browsing height for most herbivores faster than trees in nutrient-poor environments, which will suffer browsing for a longer period (Danell et al. 1991b, 1997). However, plants with greater resource availability and faster growth are likely to suffer greater losses to herbivores whilst still within browsing height (Price 1991; but see Makhabu et al. 2006b). Danell et al. (1991b), for example, found that Scots pine (*Pinus sylvestris*) along a resource gradient lost most biomass to browsers in the resource-rich part of the gradient, but suffered most reduction in growth as a result of herbivory in the resource-poor area.

9.5.6 Regeneration and Persistence

Herbivory may interfere with sexual reproduction in plants, either indirectly by changing physiology and allocation of resources, or directly by consumption of flower buds during the dormant season and flowers and fruits during the growing season.

Large herbivores may also facilitate reproduction in plants by pollinating flowers (du Toit 2003; Skarpe personal observations) and disperse seeds attached to their bodies or in their digestive tracts (Bodmer and Ward 2006). Most often, herbivory causes plants to allocate resources to vegetative growth at the expense of sexual reproduction, which favours species that are able to reproduce vegetatively (Briske 1996; Crawley 1997). Many herbivory-tolerant grasses, herbs, and dwarf shrubs are clonal, and spread vegetatively with rhizomes and/or stolons (Briske 1996; Crawley 1997) which gives them an advantage in vegetation under herbivory. Few trees are clonal, in the sense that they spread vegetatively, but many trees are able to sprout new ramets from meristems on the lower part of the stem or upper parts of the roots following severe damage, e.g., from fire or browsing (Bond and Midgley 2001; Bergström et al. 2000). While sprouting ability in juveniles can be regarded as a recruitment strategy, maintained sprouting behaviour in mature trees is termed a persistence trait (Bond and Midgley 2001), promoting survival in environments where sexual regeneration may be a rare event. Thus, many herbivory-tolerant plants can persist for a considerable time, even up to thousands of years (Kay 1997; Piggott 1993), without sexual reproduction, either by reproducing vegetatively, persisting from suckers or simply by having a long lifespan. Populations of large herbivores generally fluctuate, and may decline temporarily, e.g., from disease, temporary drought, or icing events, providing a 'window of opportunity' for plants to reproduce, although the success depends on the length of the vulnerable stage of the plant in relation to the duration of the 'window'. The result may be distinct cohorts of plants having established during such periods of low herbivore density (Prins and van der Jeugd 1993).

9.6 Herbivore Foraging Behaviour

9.6.1 Introduction

Modes of foraging as well as degree of selectivity vary between preferential browsers (eating woody plants and forbs) and preferential grazers (feeding on graminoids), between ruminants requiring comparatively nutrient rich food and hind gut fermenters, which can sustain themselves on poorer food, and among herbivores with different body sizes (Bell 1971; Jarman and Sinclair 1979; Demment and van Soest 1985; Gordon 2003; Makhabu 2005; Pastor et al. 2006a). This influences the distribution and abundance of herbivore species at different spatial scales (Olff et al. 2002) and, hence, the type and degree of impact on the vegetation. For example, one may expect hindgut fermenting megaherbivores to dominate in a nutrient-poor savanna, such as the Chobe in northern Botswana, whereas smaller ruminants would play a more important role in nutrient-rich systems like the Serengeti. Sensitivity to different plant defences may also vary with animal body size and digestive system (Palo 1987; Stokke 1999). Grazers and browsers tend to encounter very different food resources, and, subsequently, have developed different foraging modes. It has been hypothesised that understanding bite size and its implications is fundamental for understanding the foraging behaviour of large herbivores (Gordon 2003). Grasses constitute a food resource of comparatively homogeneous quality, and bite size is closely associated with sward density. Wilmshurst et al. (1999) and Fryxell et al. (2004, 2005) found that antelopes in the Serengeti closely track the grass height, thus maximising their energy intake rate. In contrast, food for browsers varies strongly in 'available' bite size, nutrient concentration, and defences (Spalinger and Hobbs 1992; Gross et al. 1993; Bergman et al. 2000; Gordon 2003). Thus, food selection in browsers tends to be more complex than for grazers and is less well understood (Spalinger and Hobbs 1992; Gross et al. 1993; Gordon 2003).

9.6.2 Hierarchical Foraging

The direct effects of herbivores on vegetation composition are caused by the interplay between selectivity of animal foraging and plant competitive ability. Selective foraging has often been described as a nested hierarchy of decisions taken by the animal at different spatial scales. This implies that palatable plants in selected feeding sites are more heavily used by the herbivore than plants of the same species outside such sites (Senft et al. 1987; Skarpe et al. 2000, 2007; Boyce et al. 2003; Fryxell et al. 2004). Selected feeding sites may vary over time, as animals track food quality or availability, e.g., along an altitudinal gradient (Albon and Langvatn 1992; Mysterud et al. 2001), or in large-scale migrations (Murray 1995). Selection criteria may be similar at different scales (Schaefer and Messier 1995; Skarpe et al. 2006), for example when areas with highest availability of the preferred food are selected, whereas other studies have found that physical properties of the landscape are more important for habitat selection at larger spatial scales (Senft et al. 1987; Boyce et al. 2003). Selection may be related to occurrence of, for instance, key resources, exposure, snow conditions, or insects. In areas adjacent to permanent water in a dry climate or salt- or clay licks, herbivory may be intense both on preferred and less preferred plants (Senft et al. 1987). Burnt areas may also suffer intense herbivory due to young and nutrient-rich resprouting vegetation and, in grazing systems, low proportions of dead material (Hobbs and Gimingham 1987; Moe et al. 1990). There may also be intense interactions between fire in grasslands and large grazing herbivores (Archibald et al. 2005).

9.6.3 Large Herbivores and Predators

In much of Europe and North America wild ungulates have increased dramatically in recent decades with changing environments, limited hunting, and few large predators (Fuller and Gill 2001; Côté et al. 2004). The re-introduction or natural increase of predators such as wolves in some areas has led to changes in patterns of herbivory, dependent more on non-lethal behavioural effects than to predation, but still causing a decrease in ungulate populations (Lima 1998; Brown et al. 1999, 2001; Ripple and Beschta 2004). Such behavioural changes relate to trade-off decisions between forage optimisation and the high cost of vigilance in a predator rich habitat (Illius and Fitzgibbon 1994). Large herbivores have been found to avoid areas with high predation risk, e.g., core areas of wolf territories or main wolf track routes, and to select either open habitats to facilitate detection of predators, areas with good cover to hide from predators, and/or areas with good escape possibilities (Ripple and Beschta 2003, 2004; Fortin and Beyer 2005). There is evidence that the presence of carnivores profoundly changes spatial habitat use by large herbivores, with concomitant reduced pressure on small areas with preferred vegetation. In North America such behavioural changes in elk (*Cervus elaphus*) and moose (*Alces alces*) have been reported to reduce the browsing pressure on some heavily preferred species, such as *Populus tremuloides* and *Salix* spp. growing in dense vegetation with low visibility and/or constituting an attraction for predators seeking herbivore aggregations (Fortin and Beyer 2005; Ripple and Beschta 2005). Such behavioural differences in large herbivores between habitats with and without predators should be present also in, for example, the predator-rich African savanna ecosystem where small, fenced reserves often are devoid of large predators, but have to our knowledge not been studied there.

9.6.4 The Importance of Neighbours

At smaller scales the position of a plant relative to other plants of higher or lower palatability can influence the probability of being eaten positively or negatively (see above) depending upon what scale the herbivore makes foraging decisions. When the herbivore makes decisions at the stand level, a palatable or less palatable plant may escape herbivory when growing among unpalatable neighbours, but may suffer increased risk of herbivory when growing among tasty neighbours (Hjältén et al.1993a; Palmer et al. 2003). When the herbivore makes decisions at a finer scale, highly palatable plants are sought out and eaten, which may reduce herbivory on less palatable neighbours (Hjältén et al. 1993a). McNaughton (1978) found that a palatable species was protected from grazing by African buffalo (*Syncerus caffer*) and wildebeest (*Connochaetus taurinus*) when growing together with unpalatable neighbours, whereas more selective Thompson gazelles (*Gazella thomsonii*) and zebra (*Equus burchelli*) actively picked out the palatable species.

9.7 Large Herbivore Effects on Vegetation

9.7.1 Introduction

The importance of large herbivores in modifying vegetation and ecosystems depends less on the average proportion of plant biomass removed and more on animal selectivity and exploitation of heterogeneity in plant populations, at many scales of space and time (Hobbs 1996; Augustine and McNaughton 1998). Generally, we suggest that the key factor governing direct impacts by large herbivores on vegetation composition and processes is the interplay between animal differential foraging and competitive relations between plants. Generally, herbivory reduces the competitive strength of a plant, but positive effects and no effects have also been recorded (Grime 1979; Tilman 1988; Stuart-Hill and Tainton 1989; Crawley 1997; Olofsson et al. 2002; Hester et al. 2006b). Although herbivory may kill small plants, e.g., seedlings, instantly, the more common effect is a reduction in growth and resource uptake. This affects competitive hierarchies between plants, which can result in the replacement of strongly impacted plants by species which are less affected (Hobbs 1996; Augustine and McNaughton 1998; Hester et al. 2006b). Indirectly, large herbivores also strongly influence vegetation and ecosystems by interfering with nutrient cycling via changes in above- and below-ground litter quantity and quality and by the deposition of faeces and urine (Pastor and Naiman 1992; Pastor et al. 1993, 2006b; Ritchie et al. 1998; van der Wal et al. 2004;.Persson et al. 2005) Such effects are dealt with elsewhere (Harrison and Bardgett this volume).

9.7.2 Herbivory and Composition of Plant Populations

Large herbivores feed selectively among plants within a population, exploiting phenotypic differences, for example, in plant vigour and concentration of nutrients and defence compounds (Price 1991; Duncan et al. 1998; Vourc'h et al. 2001; Danell et al. 2003). The selective competitive advantage of resistance traits within a population has attracted considerable recent interest (Weis and Hochberg 2000). Generally, genets with resistance traits gain a competitive advantage when costs of herbivory (in plants without resistance traits) are high, costs of resistance are low, and the growth rate of the species is relatively low (Hjältén et al. 1993b; Verwijst 1993; Shabel and Peart 1994; Kelly 1996; Augner et al. 1997). Using a modelling approach, Weis and Hochberg (2000) found that the advantage of resistance traits was more obvious with asymmetric than with symmetric competition. Thus, as herbivory often increases variation in plant size and other plant qualities (Danell et al. 2003), it may strongly influence the composition of plant populations. Over longer time scales, this can also influence the genetic properties of the population. Vourc'h et al. (2001), for example, found that black tailed deer (*Odocoileus hemionus sitkensis*) selectively browsed plants of *Thuja plicata* with low concentrations of defence compounds, which favoured the persistence of plants with high concentrations of defences. However, Vila et al. (2002) found no differences in chemistry or preference by deer (*Odocoileus hemionus sitkensis*) between individual trees in a population of *Picea sitchensis*, and concluded that escape from browsing is only a matter of age.

In dioecious plants, differences in reproductive investment may lead to differences between sexes in growth rate and chemistry, causing herbivores to distinguish between sexes. There is much evidence for preferential selection of male plants over female plants (Alliende 1989; Danell et al. 1991a; Hjältén 1992; Ågren et al. 1999). This may lead to skewed sex ratios in favour of females, as often observed (Alliende 1986; Crawford and Balfour 1990; Danell et al. 1991a; Dormann and Skarpe 2002). However, the correlation between sex-related foraging preferences and plant productivity or concentration of defences is not always strong (Danell et al. 1991a; Hjältén 1992; Dormann and Skarpe 2002), and factors such as phenological differences between sexes, resource availability, latitude or altitude may influence the results strongly (Danell et al. 1991a, Ågren et al. 1999).

9.7.3 *Herbivory and the Composition of Plant Communities*

Although the general 'rules' are the same, the interplay between herbivory and interactions among plant species is often more complex than that within species. This is because of the generally larger variation in animal selectivity and plant response to herbivory among species than within. There may also be more complex interactions among plants of different species than among plants of the same species, including associations among plants leading to increased or reduced susceptibility to herbivory, as discussed above (9.6.4).

The sensitivity of plant community composition, in terms of taxa or combinations of plant traits ('functional types') to herbivory, has been suggested to vary with potential primary production or resource availability (Milchunas and Lauenroth 1993;

Grime et al. 1997). Milchunas and Lauenroth (1993) compared data on grazed and ungrazed vegetation in a worldwide data set from 236 sites of grass-dominated vegetation. They found that the degree of change in species composition with grazing regime depended first of all on resource (water) availability or potential net primary productivity, thereafter on evolutionary history of grazing and only in third place on current herbivory intensity. Many plant traits related to shortage of resources or 'stress' (sensu Grime 1979) may also function as 'neutral resistance' to herbivory (Edwards 1989). For example, in dry grasslands the adaptation to drought is largely convergent with adaptation to grazing (e.g., short stature, clonality, narrow hard leaves, and high root/shoot ratio for better competition for soil resources). In contrast, in a sub-humid grassland the high biomass production and competition for light favours tall, broad-leaved species with a low root/shoot ratio, which makes the plants more vulnerable to herbivory (Milchunas et al. 1988; Milchunas and Lauenroth 1993; Grime et al. 1997). Under such conditions tall tussock or bunch grasses frequently decrease under herbivory when in competition with shorter, often clonal, species (Mitchley 1988; Belsky 1992; Briske 1996; Skarpe 1991a).

Not only the degree but also the type of vegetation change in relation to herbivory varies with resource availability, as that influences the relative competitive advantage of different plant resistance traits (Coley et al. 1985; Herms and Mattson 1992; Grime et al. 1997; Stamp 2003, Sect. 4 in this chapter). In resource-poor environments in particular, herbivory may lead to a decrease in much-eaten plant species and an increase in defended unpalatable plants (Coley et al. 1985). In natural systems this would imply a reduction in future grazing or browsing, as animals would select other foraging areas or suffer reduced densities (Senft et al. 1987; Hobbs 1996). However, this depends on tolerance of the herbivore to a reduction in food quality, and may operate with considerable time lag. In domestic systems with high animal densities and low mobility such vegetation change may be obvious, leading to a reduction in forage quality and/or quantity (Stoddart and Smith 1955; Skarpe 1991a). Similar development can take place in natural systems where animals are attracted to resources other than food, e.g., water or licks (Senft et al.1987). In nutrient-rich systems grazing may lead to a competitive advantage to plants with tolerance traits, and may create a feedback loop whereby grazing promotes nutrient cycling and enhanced forage quality leading to repeated grazing (McNaughton 1984, Hobbs 1996; Olofsson 2001, Sect. 9.5.4 of this chapter; Fig. 9.5). To what extent and under what conditions browsing may lead to a corresponding increase in trees with tolerance traits is not well understood (Jachmann and Bell 1985; Naimann et al. 2003).

The variation in vegetation composition in response to herbivory is suggested to be smaller in plant communities and floristic regions with a long evolutionary and historical exposure to herbivory than in communities and floristic regions without such a history (Milchunas and Lauenroth 1993; Milchunas et al. 1988; Ward 2004, 2006). Grubb (1992), for example, showed that functionally similar vegetation types with a different history of herbivory differed in the relative frequency of thorny (a herbivory-related trait) species, and that even the proportion of thorny

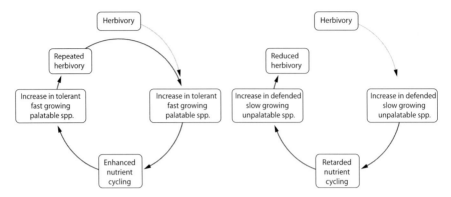

Fig. 9.5 *Left*: Feedback loop enhancing repeated herbivory in resource-rich environment. *Right*: Herbivory leading to reduced forage quality and a decrease in herbivory in resource-poor environment, modified from Skarpe et al. (2004).

species within a plant genus or family differed with different history of herbivory. In African savannas and North American prairies with a long history of grazing, species turnover and changes in primary production following recent changes in grazing regime have been shown to be moderate (O'Connor 1993; Milchunas and Lauenroth 1993). In western North America and the South American pampas, which have had few large herbivores for a considerable time before the introduction of livestock, grazing has more often caused a reduction in primary production and ground cover of the vegetation and an increase in ruderal and alien species (Mack and Thompson 1982; Milton et al. 1994; Rusch and Oesterheld 1997).

As discussed in the first section of this chapter, plant architecture is of key importance for herbivore impacts. Contrary to grasses and forbs, trees and shrubs often grow out of reach for browsers. Thus, it is mainly juveniles, seedlings, and saplings that are vulnerable to browsing impacts. Many studies have shown how ungulates can completely eliminate seedlings or saplings of certain species making regeneration possible only in limited periods, 'windows of opportunity', when herbivore populations are low for one reason or another (Gill 1992; Prins and van der Jeugd 1993; Kay 1997; Engelmark et al. 1998; Skarpe et al. 2004; Côté et al. 2004). Repeatedly browsed young trees in a forest may either get killed by the herbivore or suffer reduced competitive ability relative to other woody or herbaceous plants (Gill 1992; Hester et al. 1996; Hulme 1996). Browser impacts on competitive relations at the seedling and sapling stage are fundamentally important in determining the species composition of the mature tree layer, which, in turn, generally dominates the ecosystem processes in the forest for a considerable period of time (Hulme 1996; Weisberg and Bugmann 2003; Senn and Suter 2003; Gill 2006).

The recent high densities of ungulate browsers in much of Europe and North America are considered to be the main cause of large-scale shifts in tree species composition, population structure, and dynamics in many areas (Allison 1990; Thompson et al. 1992; Thompson and Curran 1993; Kay 1997; Kielland and Bryant 1998; Weisberg and Bugmann 2003; Pastor et al. 2006b; Tremblay et al. 2007).

For some species, re-expansion can be rapid after reduction or removal of herbivores, but for other species it is likely that beyond certain thresholds, a return to a more dominant state will not occur without some other perturbation or intervention (Tremblay et al. 2006). As evergreen species have much of their nutrient resources accessible for large herbivores throughout the year (see above), they may be particularly vulnerable to browsing. This may be a reason for the widespread replacement of evergreen *Tsuga canadensis* with *Acer saccarum* in the USA (Frelich and Lorimer 1985; Rooney et al. 2000). However, changes in the opposite direction have also been recorded, for example, where the evergreen *Picea abies* was found to gain dominance over deciduous *Betula* spp. as a result of much heavier browsing pressure by moose on *Betula* than on the heavily defended *Picea abies* (Engelmark et al. 1998). Tree species regenerating from a 'seedling bank', i.e., a population of suppressed seedlings and saplings that are released after a specific event such as wind-throw, can be particularly exposed to repeated browsing (Tremblay et al. 2006). However, changes in forest composition are also affected by many factors other than browsing, working at different spatial and temporal scales, including silvicultural practices, suppression of fires, and cascading effects of the decline or recovery of large predators (Kay 1997; Senn and Suter 2003; Hobbs 2003; Ripple and Beschta 2004). There is still a lack of predictive understanding of the relative importance of browsers in forest dynamics, for example, the relationship between population dynamics of ungulates and of trees, and processes determining browsing patterns, plant responses at the population and landscape level, and herbivore–plant–soil interactions driving vegetation change (Tremblay et al. 2004, 2006; Gill 2006; Pastor et al. 2006b).

Herbivore mediated interactions among species may also lead to changes in vegetation physiognomy and spatial diversity. To what extent large herbivores once modified primeval forest into more open parkland is debatable (Bradshaw and Mitchell 1999; Birks 2005; Mitchell 2005; Vera et al. 2006). However, in transition zones between forest and open vegetation such as tundra, steppe, or savanna, large herbivores alone or in interaction with fire may increase the openness of forest or woodland (Laws 1970; Caughley 1976; Dublin et al. 1990; van Langevelde et al. 2003). Megaherbivores, by their sheer size and strength, may be of particular importance in reducing the cover or biomass of trees (Caughley 1976; van de Koppel and Prins 1998). However, recent research has emphasised the importance of interactions among herbivores, rather than single-species effects, in driving vegetation dynamics (van de Koppel and Prins 1998). There is also evidence that smaller herbivore species are often responsible for the lack of tree regeneration in many areas (Belsky 1984; van de Koppel and Prins 1998; Prins and van de Jeugd 1993; Rutina et al. 2004). How profound such dynamics may be is illustrated by an example from northern Botswana (Skarpe et al. 2004), where there is evidence that a raised alluvial area close to permanent water shifted from open flats to tall woodland following the dramatic decline in large herbivores around 1900 caused by rinderpest and excessive hunting for ivory. It then changed to thicket vegetation from about the 1950s following the recovery of the herbivore populations, and is now covered by increasingly open shrub vegetation. A number of studies have suggested that interactions among elephants, vegetation, and, to varying extent, fires can lead to cyclic transitions between grassland and woodland (Caughley 1976; Dublin et al. 1990). However, van de Koppel and Prins (1998), modelling the outcome of interactions between vegetation and a large and a small herbivore, found that oscillations between woodland and grassland were only likely under certain conditions involving an efficient large herbivore and a smaller herbivore.

In contrast to large herbivores causing increasingly open vegetation, heavy livestock grazing in arid and semi-arid rangelands with summer rain frequently leads to an increase in woody vegetation—'bush encroachment' or 'shrub encroachment' (van Vegten 1983; Skarpe 1990b). Many factors may contribute to the increase in woody vegetation (Archer et al.1995), but the significance of heavy grazing seems undeniable (van Vegten 1983; Skarpe 1990b). It is important in this context to separate between the often dense regeneration stands of some *Acacias*, for example, which will self-thin over time, and the bush encroachment stands which do so only to a limited extent (Skarpe 1991b). The classical explanation of the expansion of woody plants once the grass layer is severely damaged is the 'the two soil layer model' (Walter 1939; Walker et al. 1981; Walker and Noy-Meir 1982). This is based on the assumption that woody species and grasses largely use different soil layers, and that the decrease in competitive power of shallow rooted grasses promotes deeper rooted woody species. However, this model has been challenged, and it is obvious that the spatial separation of tree and grass roots is far from complete (Stuart-Hill and Tainton 1989). Woody species often have deep roots enabling them to exploit resources unavailable for grasses. However, many encroacher tree/ scrub species also have a remarkably large proportion of shallow roots (van Vegten 1983; Knoop and Walker 1985; Skarpe 1990b), possibly competing with grasses for quickly recycled nutrients from dung and urine (Tolsma et al. 1987a, b). A new model for bush encroachment, with important management implications, has been presented by Ward (2006) and its main departure from the above models relates to the conditions promoting initiation of bush encroachment.

9.8 The Theory of Vegetation Change

9.8.1 Introduction

Classical succession theory (Clements 1916) predicts deterministic, unidirectional change between different states along a continuum from early successional to 'climax' vegetation. Increasing herbivory has been assumed to push the equilibrium towards the early successional end of the gradient (retrogressive succession), whereas a release of herbivory would lead to a progressive succession towards the 'climax' end. There have been a wide variety of developments of, and challenges to, this basic theory. Here we outline just a few of the major ones relating to grazing management.

9.8.2 Succession Theories

The range succession model is one derivative of classical succession theory and this has been widely used by range managers across the globe since about the 1940s (Stoddart and Smith 1943, 1955; Ellison 1960; Briske et al. 2003). The basic assumption of the model is that grazing slows down or prevents natural successional change from one state to another. Heavy grazing and/or drought is assumed to return communities to early successional states, and late successional states being achieved by removal of grazing, with a whole continuum of intermediate successional states in-between. Therefore the underlying assumption is that grazing levels can be manipulated to maintain or create 'desirable equilibrium' range conditions, as required (Westoby et al 1989; Briske et al 2003).

Classical succession theory, although widely and usefully applied in a range of systems, was found to have many predictive weaknesses, with many examples where the assumptions made did not hold and the predicted patterns of change did not occur (e.g., Turner 1971; Noble and Slatyer 1980, West et al 1984; Walker et al 1986; Grime et al 1988; Ellis and Swift 1988; Westoby et al. 1989). One major weakness of classical successional theories is that they have, as an underlying concept, the assumption that 'equilibrium' is a normal or desirable state, and systems are moved between different states by various predictable processes. Much ecosystem description is also based on this assumption (Sullivan 1996). In systems where control variables are relatively consistent and predictable, then this conceptual approach of equilibrium dynamics works pretty well, but in more unpredictable systems, such as semi-arid areas, where rainfall in particular is highly variable and unpredictable, non-equilibrium is in fact more the 'norm' and attempts to manage for 'desirable equilibrium states' can be disastrous (Westoby et al. 1989; Behnke and Scoones 1993; Sullivan 1996; Richardson et al 2005). Various definitions of equilibrium and non-equilibrium exist in ecological literature (e.g. Briske et al. 2003; Illius and O'Connor 2004), the former using the term 'persistent non-equilibrium' to describe systems with high climatic variability. Two widely used non-equilibrium models are state-and-transition and state-and-threshold, as described below.

9.8.3 The State-and-Transition Model

The state-and-transition (ST) model was proposed as an alternative to classical succession theory, since it accommodates non-linear and non-equilibrium theories of plant community dynamics (Westoby et al. 1989). In brief, this model defines stable states and transition probabilities between those states, thus enabling evaluation of different hypotheses about vegetation processes and multidirectional responses to different management actions, for example. ST models have been widely used for grazing management research and application in rangeland systems, often combined with ordination techniques (Gauch 1982; Foran et al. 1986; Friedel 1991; Filet 1994; Allen-Diaz and Bartolome 1998; Stringham et al. 2003; Hill et al.

2005). Allen-Diaz and Bartolome (1998) provide a good summary of comparison with classical succession theory, both for their own sagebrush–grass rangeland dataset and for a range of other studies. This paper highlights how the main limitation of classical succession theory is simply the emphasis on broad, linear transitions. The ST models, on the other hand, give better predictive power but, to achieve this, they require much more detailed data (both time- and site-specific) on transitions in all directions. Such detailed data is rarely fully available for any system and so this limits the usefulness of this modelling approach (Briske et al. 2003; Hill et al. 2005; Bestelmeyer et al. 2006), although it is still widely used and actively applied by both managers and modellers (Kurz et al. 2000; McIntosh et al. 2003; Letnic et al. 2004).

9.8.4 State-and-Threshold and Catastrophe Theories

Transition models fail in systems where discontinuous, irreversible changes take place (Briske et al. 2003; Warren 2005). For this reason Laycock (1991) developed the state-and-threshold model, which allowed for the possibility of systems failing to return to a previous state, even when the cause of change from that state had been removed. Discontinuous, irreversible changes are, by their nature, difficult to predict, unlike the transition model approach. However, the main criticism of this approach was that the state-and-threshold model failed to incorporate the underlying continuous vegetation processes inherent in most systems (i.e., the strength of the previous models). It is seen that none of the models thus far gave a single approach for continuous and discontinuous dynamics (Lockwood and Lockwood 1993; Richardson et al. 2005) The search for a unified approach resulted in the application of catastrophe theory to ecological systems (Loehle 1989; Lockwood and Lockwood 1991; Rietkerk et al. 1996), which was developed by Thom (1975) as a mathematical framework to model sudden, discontinuous phenomena, but one which mathematically could also encompass more continuous dynamics, as per the more classical transition models (Saunders 1980; Hooley and Cohn 2003). In brief, catastrophe theory describes the 'qualitatively different discontinuities of a system' (Lockwood and Lockwood 1993), e.g., rangeland condition classes, which are driven by specific combinations of 'control variables', such as rainfall or temperature, or woodland understorey communities, driven by combinations of light and soil factors (Hooley and Cohn 2003). Lockwood and Lockwood (1993) give a good summary of two expressions of catastrophe theory: 'fold-catastrophe' and 'cusp-catastrophe'; the latter has been more widely and usefully applied in biological sciences. The appeal of catastrophe theory lies in its simplification of control variables to five or less, rather than the multiple combinations of 'state variables' required to model such systems mechanistically—catastrophe theory does not require a mechanistic understanding of these state interactions. However, this has also led to criticism of this approach as providing little insight into the complexity of mechanisms driving a system, which is arguably important in the development of a sound, predictive understanding of how a system works. However, the necessity of such a detailed understanding is debatable (particularly for management purposes), since many systems can be well modelled with relatively few main drivers (Berryman and Stanseth 1984; Lockwood and Lockwood 1993; Hooley and Cohn 2003; Tremblay et al. 2004).

9.9 Conclusions

In this chapter we have outlined a range of factors which are important in defining vegetation dynamics in response to foraging by large mammalian herbivores. We have discussed properties of plants that make them differ in attractiveness and availability to large herbivores, and how animals exploit the variation in such properties among plant modules and genets. We have also seen how plants vary in their ability to deal with herbivory. We conclude that the impacts by large herbivores on vegetation composition are determined by the interplay between the differential foraging by large herbivores and differences in the relative competitive strength of plants. This interplay between plants and large herbivore communities is highly dynamic and interactive, affecting and being affected by cascading effects involving other trophic levels and abiotic factors. While the research on interactions between plants and large herbivores has made considerable progress during the last decades, the function of these interactions in an ecosystem context is still little understood, particularly for browser–woody vegetation systems.

References

Aarssen LW (1995) Hypothesis for the evolution of apical dominance in plants: implications for the interpretation of overcompensation. Oikos 74:149–156
Ågren J, Danell K, Elmqvist T, Ericson L, Hjältén J (1999) Sexual dimorphism and biotic interactions. In: Geber MA, Dawson TE, Delph LE (eds) Gender and sexual dimorphism in flowering plants. Springer, Berlin Heidelberg New York
Albon SD, Langvatn R (1992) Plant phenology and the benefits of migration in a temperate ungulate. Oikos 65:502–513
Allen-Diaz B, Bartolome J (1998) Sagebrush–grass vegetation dynamics: comparing classical and state–transition models. Ecol Applic 8:795–804
Alliende MC (1986) Growth and reproduction in a dioecious tree, *Salix cinerea*. Thesis, Univ of Wales
Alliende MC (1989) Demographic studies of a dioecious tree. II. The distribution of leaf predation within and between trees. J Ecol 77:1048–1058
Allison TD (1990) The influence of deer browsing on the reproductive biology of Canada yew (*Taxus canadensis* Marsh.). 1. Direct effects on pollen, ovule, and seed production. Oecologia 83:523–529
Archer S, Schimel DS, Holland EA (1995) Mechanisms of shrubland expansion – land-use, climate or CO_2. Clim Change 29:91–99
Archibald S, Bond WJ, Stock WD, Fairbanks DHK (2005) Shaping the landscape: fire-grazer interactions in an African savanna. Ecol Appl 15:96–109

Åström M, Lundberg P (1994) Plant defence and stochastic risk of herbivory. Evol Ecol 8:288–298
Atkinson MD (1992) Biological flora of the British Isles. No. 175. *Betula pendula* Roth (*B. verrucosa* Ehrh.) and *B. pubescens* Ehrh. J Ecol 80: 837–870
Augner M, Tuomi J, Rout M (1997) Effects of defoliation on competitive interactions in European white birch. Ecology 78:2369–2377
Augustine DJ, McNaughton SJ (1998) Ungulate effects of the functional species composition of plant communities: herbivore selectivity and plant tolerance. J Wildl Manage 62:1165–1183
Austin PJ, Suchar LA, Robbins CT, Hagerman AE (1989) Tannin-binding proteins in saliva of deer and their absence in saliva of sheep and cattle. J Chem Ecol 15:1335–1347
Bardgett RD, Streeter TC, Bol R. (2003) Soil microbes compete effectively with plants for organic-nitrogen inputs to temperate grasslands Ecology 84:1277–1287
Beck CB, Schmid R, Rothwell GW (1982) Stelar morphology and the primary vascular system of seed plants. Bot Rev 48:691–815
Begon M, Harper JL, Townsend CR (1996) Ecology: individuals, populations and communities, 3rd edn. Blackwell, Oxford
Behnke RH, Scoones I (1993) Rethinking range ecology: implications for range management in Africa. In: Behnke RH, Scoones I, Kerven C (eds). Range ecology at disequilibrium: new models of natural variability and pastoral adaptation in African savannas. Overseas Development Institute, London, pp 1–30
Bell RHV (1971) A grazing ecosystem in the Serengeti. Sci Am 225:86–93
Belovsky GE (1984) Moose and snowshoe hare competition and a mechanistic explanation from foraging theory. Oecologia 61:150–159
Belovsky GE, Schmitz OJ, Slade JB, Dawson TJ (1991) Effects of spines and thorns on Australian arid zone herbivores of different body masses. Oecologia 88:521–528
Belsky AJ (1984) Role of small browsing mammals in preventing woodland regeneration in the Serengeti National Park, Tanzania. Afr J Ecol 22:271–279
Belsky AJ (1986) Does herbivory benefit plants? A review of the evidence. Am Nat 127:870–892
Belsky AJ (1992) Effects of grazing, disturbance and fire on species composition and diversity in grassland communities. J Veg Sci 3:187–200
Bergman CM, Fryxell JM, Gates CG (2000) The effect of tissue complexity and sward height on the functional response of wood bison. Func Ecol 14:61–69
Bergström R, Danell K (1987) Moose winter feeding in relation to morphology and chemistry of six tree species. Alces 22:91–112
Bergström R, Danell K (1995) Effects of simulated summer browsing by moose on leaf and shoot biomass of birch, *Betula pendula*. Oikos 72:132–138
Bergström R, Skarpe C, Danell K (2000) Plant responses and herbivory following simulated browsing and stem cutting of *Combretum apiculatum*. J Veg Sci 11:409–414
Berryman AA, Stanseth NC (1984) Behavioural catastrophes in biological systems. Behav Sci 29:127–137
Bestelmeyer BT, Trujillo DA, Tugel AJ, Havstad KM (2006) A multi-scale classification of vegetation dynamics in arid lands: what is the right scale for models, monitoring and restoration? J Arid Env 65:296–318
Bilbrough CJ, Richards JH (1993) growth of sagebrush and bitterbrush following simulated winter browsing: mechanisms of tolerance. Ecology 74:481–492
Bircham JS, Hodgson J (1983) The influence of sward condition on rates of herbage growth and senescence under continuous stocking management. Grass Forage Sci 38:323–331
Birks HJB (2005) Mind the gap: How open were European primeval forests. Trends Ecol Evol 20:154–156
Bloom AJ, Chapin FS III, Mooney HA (1985) Resource limitation in plants – an economic analogy. Ann Rev Ecol Syst 16:363–392
Bodmer R, Ward D (2006) Frugivory in large mammalian herbivores. In: Danell K, Bergström R, Duncan P, Pastor J (eds) Large herbivore ecology and ecosystem dynamics. Cambridge Univ Press, Cambridge, pp 232–260

Bond WJ, Midgley JJ (2001) Ecology of sprouting in woody plants: the persistence niche. Trends Ecol Evol 16:45–51
Boyce MS, Mao JS, Merrill EH, Fortin D, Turner MG, Fryxell J, Turchin P (2003) Scale and heterogeneity in habitat selection by elk in Yellowstone National Park. Ecoscience 10:421–431
Bradshaw AD (1965) Evolutionary significance of phenotypic plasticity in plants. Adv Genet 13:115–155
Bradshaw RHW, Mitchell FJG (1999) The palaeoecological approach to reconstructing former grazing–vegetation interactions. Forest Ecol Manag 120:3–12
Briske DD (1996) Strategies of plant survival in grazed systems: a functional interpretation. In: Hodgson J, Illius AW (eds) The ecology and management of grazing ecosystems. CAB International, Wallingford, UK, pp 37–68
Briske DD, Fuhlendorf SD, Smeins FE (2003) Vegetation dynamics on rangelands: a critique of the current paradigms. J Appl Ecol 40:601–614
Brown JS, Laundré JW, Gurung M (1999) The ecology of fear: optimal foraging, game theory, and trophic interactions J Mammal 80:385–399
Brown JS, Kotler BP, Bouskila A (2001) Ecology of fear: Foraging games between predators and prey with pulsed resources. Ann Zool Fenn 38:71–87
Bryant JP, Chapin FS, Klein DR (1983) Carbon/nutrient balance of boreal plants in relation to vertebrate herbivory. Oikos 40:357–368
Bryant JP, Kuropat PJ, Cooper SM, Frisby K, Owen-Smith N (1989) Resource availability hypothesis of plant antiherbivore defence tested in a South African savanna ecosystem. Nature 340:227–229
Bullock JM, Franklin J, Stevenson MJ, Silvertown J, Coulson SJ, Gregory SJ, Tofts R (2001) A plant trait analysis of responses to grazing in a long-term experiment. J Appl Ecol 38:253–267
Chapin FS III, Schulze ED, Mooney HA (1990) The ecology and economics of storage in plants. Annu Rev Ecol Syst 21:423–447
Caughley G (1976). The elephant problem - an alternative hypothesis. E Afr Wildl J 14:265–283
Cipollini D, Purringnton CB, Bergelson J (2003) Costs of induced responses in plants. Basic Appl Ecol 4:79–89
Clements FE (1916) Plant succession. Carnegie Institute, Washington, DC
Coleman GD, Chen THH, Ernst SG, Fuchigami H (1991) Photoperiod control of poplar bark storage protein accumulation. Plant Physiol 96:686–692
Coley PD, Bryant JP, Chapin FS (1985) Resource availability and plant antiherbivore defence. Science 230:895–899
Cooper SM, Owen-Smith N (1985) Condensed tannins deter feeding by browsing ungulates in a South African savanna. Oecologia 67:142–146
Cooper SM, Owen-Smith N (1986) Effects of plant spinescence on large mammalian herbivores. Oecologia 68:446–455
Cooper SM, Owen-Smith N (2001) Effects of plant spinescence on large mammalian herbivores. Oecologia 68:446–455
Cooper SM, Owen-Smith N, Bryant JP (1988) Foliage acceptability to browsing ruminants in relation to seasonal changes in leaf chemistry of woody plants in a South African savanna. Oecologia 75:336–342
Coppock DL, Detling JK, Ellis JE, Dyer MI (1983) Plant–herbivore interactions in a North American mixed-grass prairie. I. Effects of black-tailed prairie dogs on intraseasonal aboveground plant biomass and nutrient dynamics and plant species diversity. Oecologia 56:1–9
Côté SD, Rooney TP, Tremblay JP, Dussault C, Waller DM (2004) Ecological impacts of deer overabundance Annu Rev Ecol Evol 35:113–147
Crawford RMM, Balfour J (1990) Female-biased sex ratios and differential growth in arctic willows. Flora 184:291–302
Crawley MJ (1983) Herbivory - the dynamics of animal–plant interactions. Studies in ecology, vol 10. Blackwell, Oxford

Crawley MJ (ed) (1997) Plant ecology. Blackwell, Oxford
Danckwerts JE, Gordon AJ (1987) Long-term partitioning, storage and remobilisation of ^{14}C assimilation by *Lolium perenne* (cv. Melle). Ann Bot 59:55–66
Danell K, Bergström R (1987) Effects of simulated winter browsing by moose on morphology and biomass of two birch species. J Ecol 75:533–544
Danell K, Bergström R (1989) Winter browsing by moose on two birch species: impact on food resources. Oikos 54:11–18
Danell K, Huss-Danell K, Bergström R (1985) Interactions between browsing moose and two species of birch in Sweden. Ecology 66:1867–1878
Danell K, Hjältén J, Ericson L, Elmqvist T (1991a) Vole feeding on male and female willow shoots along a gradient of plant productivity. Oikos 62:145–152
Danell K, Niemela P, Varvikko T, Vuorisalo T (1991b) Moose browsing on Scots pine along a gradient of plant productivity. Ecology 72:1624–1633
Danell K, Bergström R, Edenius L (1994) Effects of large mammalian browsers on architecture, biomass, and nutrients of woody plants. J Mammal 75:833–844
Danell K, Haukioja E, Huss-Danell K (1997) Morphological and chemical responses of mountain birch leaves and shoots to winter browsing along a gradient of plant productivity. Ecoscience 4:296–303
Danell K, Bergström R, Edenius L, Ericsson G (2003) Ungulates as drivers of tree population dynamics at module and genet levels. Forest Ecol Manag 181:67–76
Dangerfield JM, Modukanele M (1996) Overcompensation by *Acacia erubescens* in response to simulated browsing. J Trop Ecol 12: 1–4
Demment MW, van Soest PJ (1985) A nutritional explanation for body-size patterns of ruminant and nonruminant herbivores. Am Nat 125:641–672
Deregibus VA, Sanchez RA, Casal JJ, Trlica MJ (1985) Tillering responses to enrichment of red light beneath the canopy in a humid natural grassland. J Appl Ecol 22:199–206
Devlin RM (1966) Plant physiology. Reinhold, New York
Diaz S, Acosta A, Cabido M (1992) Morphological analysis of herbaceous communities under different grazing regimes. J Veg Sci 3:689–696
Dormann CF, Skarpe C (2002) Flowering, growth and defence in the two sexes: consequences of herbivore exclusion for *Salix polaris*. Funct Ecol 16:649–656
Dublin HT, Sinclair ARE, McGlade J (1990). Elephants and fire as causes of multiple stable states in the Serengeti-Mara woodlands. J Anim Ecol 59:1147–1164
Du Toit JT (2003) Large herbivores and savanna heterogeneity. In: du Toit JT, Rogers KH, Biggs HC (eds) The Kruger experience: ecology and management of savanna heterogeneity. Island Press, Washington, DC, pp 292–309
Du Toit JT, Bryant JP, Frisby K (1990) Regrowth and palatability of *Acacia* shoots following pruning by African savanna browsers. Ecology 71:149–154
Du Toit E.W, Wolfson MM, Ellis RP (1991) The presence of condensed tannins in the leaves of *Eulalia villosa*. J Grassland Soc S Afr 8:74–76
Duncan AJ, Hartley SE, Iason GR (1998) The effect of previous browsing damage on the morphology and chemical composition of Sitka spruce (*Picea sitchensis*) saplings and on their subsequent susceptibility to browsing red deer (*Cervus elaphu*s). Forest Ecol Manag 103:57–67
Eckstein RL, Karlsson PS, Weih M (1998) The significance of resorption of leaf resources for shoot growth in evergreen and deciduous woody plants from a subarctic environment. Oikos 81:567–575
Edwards PJ (1989) Insect herbivory and plant defence theory. In: Grubb PJ, Whittaker JB (eds) Toward a more exact ecology. Blackwell, Oxford, pp 275–297
Ellis JE, Swift DM (1988) Stability of African pastoral ecosystems: alternate paradigms and implications for development. J Range Manage 41:450–459
Ellis RP, Vogel JC, Fuls A (1980) Photosynthetic pathways and the geographical distribution of grasses in South West Africa/Namibia. S Afr J Sci 76:307–314
Ellison L (1960) Influence of grazing on plant succession of rangelands. Bot Rev 26:1–78

Engelmark O, Hofgaard A, Arnborg T (1998) Successional trends 219 years after fire in an old *Pinus sylvestris* stand in northern Sweden. J Veg Sci 9:583–592

Falkengren-Grerup U, Mansson KF, Olsson MO (2000) Uptake capacity of amino acids by ten grasses and forbs in relation to soil acidity and nitrogen availability. Environ Exp Bot 44:207–219

Feeny P (1976) Plant apparency and chemical defence. Rec Adv Phytochem 10:1–40

Filet PG (1994) State and transition models for rangelands. 3. The impact of the state and transition model on grazing lands research, management and extension: a review. Trop Grasslands 28:214–222

Fitter AH (1997) Acquisition and utilization of resources. In: Crawley MJ (ed) Plant ecology, 2nd edn. Blackwell, Oxford

Flower-Ellis JGK (1971) Age structure and dynamics in stands of bilberry (*Vaccinium myrtillus* L.). Royal Coll Forestry, Stockholm

Foley WJ, Iason GR, MacArthur C (1999) Role of plant secondary metabolites in the nutritional ecology of mammalian herbivores. How far have we come in 25 years? In: Young HJG, Fahey GC (eds) Nutritional ecology of herbivores, Am Soc Anim Sci, Savoy, IL, pp 130–209

Foran BD, Bastin G, Shaw KA (1986) Range assessment and monitoring in arid lands: the derivation of functional groups to simplify vegetation data. J Environ Manage 27:85–97

Fortin D, Beyer HL (2005) Wolves influence elk movements: behavior shapes a trophic cascade in Yellowstone National Park. Ecology 86:1320–1330

Frelich LE, Lorimer CG (1985) Current and predicted long-term effects of deer browsing in hemlock forests in Michigan, USA. Biol Conserv 34:99–120

Friedel MH (1991) Range condition assessment and the concept of thresholds: a viewpoint. J Range Manage 44:422–426

Fryxell JM, Wilmshurst JF, Sinclair ARE (2004) Predictive models of movement by Serengeti grazers. Ecology 85:2429–2435

Fryxell JM, Wilmshurst JF, Sinclair ARE, Haydon DT, Holt RD, Abrams PA (2005). Landscape scale, heterogeneity, and the viability of Serengeti grazers. Ecol Lett 8:328–335

Fuller RJ, Gill RMA (2001) Ecological impacts of increasing numbers of deer in British woodland. Forestry 74:193–199

Gauch HG (1982) Multivariate analysis in community ecology. Cambridge Univ Press, Cambridge

Georgiadis NJ, McNaughton SJ (1988) Interactions between grazers and a cyanogenic grass, *Cynodon plectostachyus*. Oikos 51:343–350

Gifford RM, Marshall C (1973) Photosynthesis and assimilate distribution in *Lolium multiflorum* following differential tiller defoliation. Aust J Biol Sci 26:517–526

Gill RMA (1992) A review of damage by mammals on north temperate forests: 1 Deer. Forestry 65:145–169

Gill RMA (2006) The influence of large herbivores on tree recruitment and forest dynamics. In: Danell K, Bergstrom R, Duncan P, Pastor J (eds) Large herbivore ecology, ecosystem dynamics and conservation. Cambridge Univ Press, Cambridge, pp 170–202

Gordon IJ (2003) Browsing and grazing ruminants: are they different beasts? Forest Ecol Manag 181:13–21

Gowda JH (1996) Spines of *Acacia tortilis*: what do they defend and how? Oikos 77:279–284

Grime JP (1979) Plant strategies and vegetation processes. Wiley, Chichester

Grime JP, Hodgson JG, Hunt R (1988) Comparative plant ecology: a functional approach to common British species. Unwin Hyman, London.

Grime JP, Thompson K, Hunt R et al (1997) Integrated screening validates primary axes of specialisation in plants Oikos 79:259–281

Gross JE, Shipley LA, Hobbs NT, Spalinger DE, Wunder BA (1993) Functional response of herbivores in food-concentrated patches; tests of a mechanistic model. Ecology 74:778–791

Grubb P (1992) A positive distrust in simplicity. J Ecol 80:585–610

Guillet C, Bergström R (2006) Compensatory growth of fast-growing willow (*Salix*) coppice in response to simulated large herbivore browsing. Oikos

Harper JL (1977) Population biology of plants. Academic Press, London

Harrison AK, Bardgett RD (2007) Impacts of grazing and browsing by large mammals on soils and soil biological properties. (this volume)

Haukioja E, Koricheva J (2000) Tolerance to herbivory in woody vs. herbaceous plants. Evol Ecol 14:551–562

Haukioja E, Ruohomäki K, Senn J, Suomela J, Walls M (1990) Consequences of herbivory on the mountain birch (*Betulas pubescens* ssp *tortuosa*): importance of the functional organisation of the tree. Oecologia 82:238–247

Heilmeier H, Schulze E-D, Whale DM (1986) Carbon and nitrogen partitioning in the biennial monocarp *Arctium tomentosum* Mill. Oecologia 70:466–467

Herms DA, Mattson WJ (1992) The dilemma of plants – to grow or defend. Quart Rev Biol 67:283–335

Hester AJ, Mitchell FJG, Kirby KJ (1996). Effects of season and intensity of sheep grazing on tree regeneration in a British upland woodland. Forest Ecol Manag 88:99–106

Hester AJ, Edenius L, Buttenshøn RM, Kuiters AT (2000). Interactions between forests and herbivores: the role of controlled grazing experiments. Forestry 73:381–391

Hester AJ, Millard P, Baillie GJ, Wendler R (2004). How does timing of browsing affect above- and below-ground growth of *Betula pendula*, *Pinus sylvestris* and *Sorbus aucuparia*? Oikos 105:536–550

Hester AJ, Lempa K, Neuvonen S, Høegh K, Feilberg J, Arnþórsdóttir S, Iason GR (2005) Birch sapling responses to severity and timing of domestic herbivore browsing – implications for management. In: Wielgolaski FE (ed) Plant ecology, herbivory and human impact in Nordic mountain birch forests. Ecological studies, vol 180. Springer, Berlin Heidelberg New York, pp 139–155

Hester AJ, Scogings PF, Trollope WSW (2006a) Long-term impacts of goat browsing on bush-clump dynamics in a semi-arid subtropical savanna. Plant Ecol 183:277–290

Hester AJ, Bergman M, Iason GR, Moen R (2006b). Impacts of large herbivores on plant community structure and dynamics. In: Danell K, Bergström R, Duncan P, Pastor J (eds) Large herbivore ecology and ecosystem dynamics. Cambridge Univ Press, Cambridge, pp 97–141

Hik DS, Jefferies RL (1990) Increases in the net above-ground primary production of salt-marsh forage grass: a test of the predictions of the herbivore-optimization model. J Ecol 78:180–195

Hill MJ, Roxburgh SH, Carter JO, McKeon GM (2005) Vegetation state change and consequent carbon dynamics in savanna woodlands of Australia in response to grazing, drought and fire: a scenario approach using 113 years of synthetic annual fire and grassland growth. Aust J Bot 53:715–739

Hjältén J (1992) Plant sex and hare feeding preferences. Oecologia 89:253–256

Hjältén J, Danell K, Lundberg P (1993a) Herbivore avoidance by association – vole and hare utilization of woody plants Oikos 68:125–131

Hjältén J, Danell K, Ericson L (1993b) Effects of simulated herbivory and intraspecific competition on the compensatory ability of birches. Ecology 74:1136–1142

Hobbs NT (1996) Modification of ecosystems by ungulates. J Wildl Manage 60:695–713

Hobbs NT (2003) Challenges and opportunities in integrating ecological knowledge across scales. Forest Ecol Manage 181:223–238

Hobbs RJ, Gimingham CH (1987) Vegetation, fire and herbivore interactions in heathland. Adv Ecol Res 16:88–173

Hodgkinson KC (1974) Influence of partial defoliation on photosynthesis, photorespiration and transpiration by Lucerne leaves at different ages. Aust J Plant Physiol 1:561–578

Hodgson J (ed) (1981) Sward measurement handbook. British Grassland Society, Cirencester, UK

Holopainen J, Rikala R, Kainulainen P Oksanen J (1995) Resource partitioning to growth, storage and defence in nitrogen-fertilized Scots pine and susceptibility of the seedlings to the tarnished plant bug *Lygus rugulipennis*. New Phytol 131:521–532

Honkanen T, Haukioja E, Suomela J (1994) Effects of simulated defoliation and debudding on needle and shoot growth in Scots pine (*Pinus sylvestris*): implications of plant source/sink relationships for plant–herbivore studies. Funct Ecol 8:631–639

Honkanen T, Haukioja E, Kitunen V (1999) Responses of *Pinus sylvestris* branches to simulated herbivory are modified by tree sink/source dynamics and by external resources. Funct Ecol 13: 126–140

Hooley JL, Cohn EVJ (2003) Models of filed layer interactions in an experimental secondary woodland. Ecol Model 169:89–102

Hulme PE (1996) Natural regeneration of yew (*Taxus baccata* L.): microsite, seed or herbivore limitation? J Ecol 84:853–861

Hytteborn H (1975) Deciduous woodland at Andersby, eastern Sweden. Above-ground tree and shrub production. Acta Phytogeograph Suecica 61:96

Iason GR (2005) The role of plant secondary metabolites in mammalian herbivory: ecological perspectives. Proc Nutr Soc 64:123–131

Iason GR, Villalba JJ (2006) Behavioral strategies of mammal herbivores against plant secondary metabolites: the avoidance–tolerance continuum. J Chem Ecol 32:1115–1132

Illius AW, Fitzgibbon C (1994) Costs of vigilance in foraging ungulates Anim Behav 47:481–484

Illius AW, Gordon IA (1987) The allometry of food intake in grazing ruminants. J Anim Ecol 56:989–999

Illius A, O'Connor TG (2004) The definition of non-equilibrium and the role of key resource – an ecological perspective. In: Vetter S (ed) Rangelands at equilibrium and non-equilibrium. Univ of Western Cape, Cape Town, p 16

Inghe O, Tamm CO (1985) Survival and flowering of perennial herbs IV. Oikos 45:400–420

Jachmann H, Bell RHV (1985) Utilization by elephants of the Brachystegia woodlands of the Kasungu National Park, Malawi. J Afr Ecol 23:245–258

Jaramillo VJ, Detling JK (1988) Grazing history, defoliation and competition: effects on short grass production and nitrogen accumulation. Ecology 69:1599–1608

Järemo J, Nilsson P, Tuomi J (1996) Plant compensatory growth: herbivory or competition? Oikos 77:238–247

Jarman PJ, Sinclair ARE (1979) Feeding strategy and the pattern of resource partitioning in ungulates. In: Sinclair ARE, Norton-Griffiths M (eds) Serengeti – dynamics of an ecosystem. Univ of Chicago Press, Chicago, pp 130–163

Jefferies RL, Klein DR, Shaver GR (1994) Vertebrate herbivores and northern plant communities – reciprocal influences and response. Oikos 71:193–206

Jia J, Niemelä P, Danell K (1995) Moose *Alces alces* bite diameter selection in relation to twig quality on four phenotypes of Scots pine *Pinus sylvestris*. Wildl Biol 1:47–55

Jokela J, Schmid-Hempel P, Rigby MC (2000) Dr. Pangloss restrained by the Red Queen – steps towards a unified defence theory. Oikos 89:267–274

Juenger T, Lennartsson T (2000) Tolerance in plant ecology and evolution: toward a more unified theory of plant–herbivore interaction Evol Ecol 14:283–287

Juntheikki MR (1996) Comparison of tannin-binding proteins in saliva of Scandinavian and North American moose (*Alces alces*). Biochem Syst Ecol 24:595–601

Karban R, Agrawal AA, Thaler JS, Adler LS (1999) Induced plant responses and information content about risk for herbivory. Trends Ecol Evol 14:443–447

Kay CE (1997) Is aspen doomed? J Forest 95:434

Kelly JK (1996) Kin selection in the annual plant *Impatiens capensis*. Am Nat 147:899–918

Kielland K, Bryant JP (1998) Moose herbivory in taiga: effects on biogeochemistry and vegetation dynamics in primary succession. Oikos 82:377–383

Knoop WT, Walker BH (1985) Interactions of woody and herbaceous vegetation in a southern African savanna. J Ecol 73:235–253

Koricheva J (2002a) The carbon–nutrient balance hypothesis is dead; long live the carbon–nutrient balance hypothesis? Oikos 98:537–539

Koricheva J (2002b) Meta-analysis of sources of variation in fitness costs of plant antiherbivore defences. Ecology 83:176–190

Körner C (1994) Biomass fractionation in plants: a reconsideration of definitions based on plant functions. In: Roy J, Garnier E (eds) A whole plant perspective on carbon–nitrogen interactions. SPB Academic Publishing, The Hague, pp 175–185

Kotanen PM, Rosenthal JP (2000) Tolerating herbivory: does the plant care if the herbivore has a backbone? Evol Ecol 14:537–549

Kozlowski TT (1966) Shoot growth in woody plants. Botan Rev 30:335–392

Kruger EL, Reich PB (1997) Responses of hardwood regeneration to fire in mesic forest openings. III. Whole plant growth, biomass distribution, and nitrogen and carbohydrate relations. Can J Forest Res 27:1841–1850

Kurz WA, Beukema SJ, Klenner W, Greenough JA, Robinson DCE, Sharpe AD, Webb TM (2000) TELSA: the tool for exploratory landscape scenario analyses. Comput Electron Agr 27:227–242

Langer RHM (1972) How grasses grow. Studies in biology 34. Edward Arnold, London

Langstrom B, Tenow O, Ericsson A, Hellgvist C, Larsson S (1990) Effects of shoot pruning on stem growth, needle biomass, and dynamics of carbohydrates and nitrogen in Scots pine as related to season and tree age. Can J Forest Res 20:514–523

Laws RWJ (1970) Elephants as agents of habitat and landscape change in East Africa. Oikos 21:1–15

Laycock WA (1991) Stable states and thresholds of range condition in North America rangeland: a viewpoint. J Range Manage 44:427–435

Lemaire G, Chapman D (1996) Tissue flows in grazed plant communities. In: Hodgson J, Illius AW (eds) The ecology and management of grazing systems. CAB International, Wallingford, UK, pp 3–36

Letnic M, Dickman CR, Tischler MK, Tamayo B, Beh C-L (2004) The responses of small mammals and lizards to post-fire succession and rainfall in arid Australia. J Arid Environ 59:85–114

Letourneau DK, Dyer LA et al. (2004) Indirect effects of a top predator on a rain forest understory plant community. Ecology 85:2144–2152

Lima SL (1998) Non-lethal effects of the ecology of predator–prey interactions. Bioscience 48:25–34

Lock JM (1972) The effects of hippopotamus grazing on grasslands. J Ecol 60:445–467

Lockwood JA, Lockwood DR (1993) Catastrophe theory: a unified paradigm for rangeland ecosystem dynamics. J Range Manage 46:282–288

Loehle C (1989) Catastrophe theory in ecology: a critical review and an example of the butterfly catastrophe. Ecol Model 49:125–152

Ludlow MM, Wilson GL (1971) Photosynthesis of tropical pasture plants III. Leaf age. Aust J Biol Sci 24:1077–1087

Löyttyniemi K (1985) On repeated browsing of Scots pine saplings by moose (*Alces alces*). Silva Fenn 19:387–391

MacFadden BJ (1997) Origin and evolution of the grazing guild in New World terrestrial mammals. Trends Ecol Evol 12:182–187

Mack RN, Thompson JN (1982) Evolution in steppe with few large hooved mammals. Am Nat 119:757–773

Makhabu SW (2005) Resource partitioning within a browsing guild in a key habitat, the Chobe Riverfront, Botswana. J Trop Ecol 21:641–649

Makhabu SW, Skarpe C (2006) Rebrowsing by elephants three years after simulated browsing on five woody plant species in northern Botswana. S Afr J Wildl Res 36:99–102

Makhabu SW, Skarpe C, Hytteborn H (2006a) Elephant impact on shoot distribution on trees and on rebrowsing by smaller browsers. Acta Oecol 30:136–146

Makhabu SW, Skarpe C, Hytteborn H, Mpufu ZD (2006b) The plant vigour hypothesis revisited – how is browsing by ungulates and elephant related to woody species growth rate? Plant Ecol 184:163–172

Marshall C, Sagar GR (1968) The distribution of assimilates in *Lolium multiflorum* Lam following differential defoliation. Ann Bot 32:715–719

Maschinski J, Whitham TG (1989) The continuum of plant responses to herbivory: the influence of plant association, nutrient availability and timing. Am Nat 134:1–19

Mauricio R (2000) Natural selection and the joint evolution of tolerance and resistance as plant defences. Evol Ecol 14:491–507

McIntosh BS, Muetzelfeldt RI, Legg CJ, Mazzoleni S, Csontos P (2003) Reasoning with direction and rate of change in vegetation state transition modelling. Environ Modell Softw 18:915–927

McIntyre S, Tongway D (2005) Grassland structure in native pastures: links to soil surface condition. Ecol Manage Restor 6:43–50

McNaughton SJ (1978) Serengeti ungulates: feeding selectivity influences the effectiveness of plant defense guilds. Science 199:806–807

McNaughton SJ (1983) Compensatory growth as a response to herbivory. Oikos 40:329–336

McNaughton SJ (1984) Grazing lawns, animals in herds, plant form, and coevolution. Am Nat 124:863–886

McNaughton SJ (1985) Ecology of a grazing ecosystem: the Serengeti. Ecol Monogr 55:259–294

Milchunas DG, Lauenroth WK (1993) Quantitative effects of grazing on vegetation and soils over a global ranger of environments. Ecol Monogr 63:327–366

Milchunas DG, Noy-Meir I. (2002). Grazing refuges, external avoidance of herbivory and plant diversity. Oikos 99:113–130

Milchunas DG, Sala OE, Lauenroth WK (1988) A generalised model of the effects of grazing by large herbivores on grassland community structure. Am Nat 132:87–106

Milewski AV, Young TP, Madden D (1991) Thorns as induced defences: experimental evidence. Oecologia 86:70–75

Millard P (1996) Ecophysiology of the internal cycling of nitrogen for tree growth. J Plant Nutr Soil Sci 159:1–10

Millard P, Proe MF (1991) Leaf demography and the seasonal internal cycling of nitrogen in sycamore (*Acer pseudoplatanus* L.) seedlings in relation to nitrogen supply. New Phytol 117:587–596

Millard P, Hester AJ, Wendler R, Baillie G (2001) Remobilization of nitrogen and the recovery of *Betula pendula*, *Pinus sylvestris*, and *Sorbus aucuparia* saplings after simulated browsing damage. Funct Ecol 15:535–543

Milton SJ, Dean WRJ, du Plessis MA, Siegfried WRA (1994) A conceptual model of arid rangeland degradation. The escalating cost of declining productivity. Bioscience 44:70–76

Miquelle DG (1983) Browse regrowth and consumption following summer defoliation by moose J Wildl Manage 47:17–24

Mitchell FJG (2005) How open were European primaeval forests? Hypothesis testing using palaeoecological data. J Ecol 93:168–177

Mitchley J (1988) Control of relative abundance of perennials in chalk grassland in southern England. II. Vertical canopy structure. J Ecol 76:341–350

Moe SR, Wegge P, Kapela EB (1990) The influence of man-made fires on large wild herbivores in Lake Burungi area in northern Tanzania. Afr J Ecol 28:35–43

Mooney HA (1997) Photosynthesis. In: Crawley MJ (ed) Plant ecology, Blackwell, Oxford, pp 345–374

Moore NP, Hart JD, Kelly PF, Langton SD (2000) Browsing by fallow deer (*Dama dama*) in young broadleaved plantations: seasonality, and the effects of previous browsing and bud eruption. Forestry 73:437–445

Murray MG (1995) Specific nutrient requirements and migration of wildebeest In: Sinclair ARE, Arcese P (eds) Serengeti II - Dynamics, management and conservation of an ecosystem. Univ of Chicago Press, Chicago, pp 231–256

Myers JH, Bazeley D (1991) Thorns, spines, prickles and hairs: are they stimulated by herbivory and do they deter herbivores? In: Tallamyr DJ, Raup MJ (eds) Phytochemical induction by herbivores. Academic Press, New York, pp 326–343

Mysterud A, Langvatn R, Yoccoz NG, Stenseth NC (2001). Plant phenology, migration and geographical variation in body weight of a large herbivore: the effect of a variable topography. J Anim Ecol 70:915–923

Naiman RJ, Braack L, Grant R, Kemp AC, du Toit JT, Venter FJ (2003) Interactions between species and ecosystem characteristics. In: du Toit JT, Rogers KH, Biggs HC (eds) The Kruger experience – ecology and management of savanna heterogeneity. Island Press, Washington, DC, pp 221–241

Nambiar EKS (1987) Do nutrients retranslocate from fine roots? Can J For Res 17:913–918

Nambiar EKS, Fife DN (1991) Nutrient retranslocation in temperate conifers. Tree Physiol 9:185–207

Noble IR, Slatyer RO (1980) The use of vital attributes to predict successional changes in plant communities subject to recurrent disturbances. Vegetatio 43:5–21

Noy-Meir I (1993) Compensating growth of grazed plants and its relevance to the use of rangelands. Ecol Appl 3:32–34

Noy-Meir I, Seligman NG (1979) Management of semi-arid ecosystems in Israel. In: Walker BH (ed) Management of semi-arid ecosystems. Elsevier, Amsterdam, pp 398–424

O'Connor TG (1993) The influence of rainfall and grazing on the demography of some African savanna grasses: a matrix modelling approach. J Appl Ecol 30:119–132

Olff H, Vera FWM, Bokdam J (1999) Shifting mosaics in grazed woodland driven by the alternation of plant facilitation and competition. Plant Biol 1:127–137

Olff H, Ritchie ME, Prins HH. (2002). Global environmental controls of diversity in large herbivores. Nature 415:901–904

Olofsson J, Kitti H, Rautiainen P, Stark S, Oksanen L (2001) Effects of summer grazing by reindeer on composition of vegetation, productivity and nitrogen cycling. Ecography 24:13–24

Olofsson J, Moen J, Oksanen L (2002) Effects of herbivory on competition intensity in two arctic–alpine tundra communities with different productivity. Oikos 96:265–272

Paige KN, Whitham RG (1987) Overcompensation in response to mammalian herbivory: the advantage of being eaten. Am Nat 129:407–416

Painter EL, Detling JK (1981) Effects of defoliation on net photosynthesis and regrowth of western wheatgrass. J Range Manage 34:68–71

Painter EL, Detling JK. Steingraeber DA (1989) Grazing history, defoliation, and frequency-dependent competition: effects on two North American grasses. Am J Bot 76:1368–1379

Palmer SCF, Hester AJ, Elston DA, Gordon IJ, Hartley SE (2003) The perils of having tasty neighbors: grazing impacts of large herbivores at vegetation boundaries. Ecology 84:2877–2890

Palo RT (1987) Chemical defence in woody plants and the role of digestive system of herbivores. In: Provenza FD, Flinders JF, McArthur ED (eds) Proceedings from a symposium on plant–herbivore interactions. US Dept Agr Forest Serv, Gen Tech Report INT-222

Palo RT, Robbins CT (eds 1991) Plant defences against mammalian herbivory. CRC Press, Boca Raton, FL

Pass GJ, McLean S, Stupans I, Davies NW (2002) Microsomal metabolism and enzyme kinetics of the terpene p-cymene in the common brushtail possum (*Trichosurus vulpecula*), koala (*Phascolarctos cinereus*) and rat. Xenobiotica 32:383–397

Pastor J, Naiman RJ (1992) Selective foraging and ecosystem processes in boreal forests. Am Nat 139:690–705

Pastor J, Dewey B, Naiman RJ, McInnes PF, Cohen Y (1993) Moose browsing and soil fertility in the boreal forests of Isle Royale National Park. Ecology 74:467–480

Pastor J, Cohen J, Hobbs NT (2006a) The role of large herbivores in ecosystem nutrient cycles. In: Danell K, Bergström R, Duncan P, Pastor J (eds) Large herbivore ecology and ecosystem dynamics. Cambridge University Press, Cambridge, pp 289–325

Pastor J, Danell K, Bergström R, Duncan P (2006b). Themes and future directions in herbivore–ecosystem interactions and conservation. In: Danell K, Bergström R, Duncan P, Pastor J (eds) Large herbivore ecology and ecosystem dynamics. Cambridge University Press, Cambridge, pp 468–478

Pérez-Barbería FJ, Elston DA, Gordon IJ, Illius AW (2004) The evolution of phylogenetic differences in the efficiency of digestion in ruminants. Pro Royal Soc Lond B Biol Sci 271:1081–1090

Persson I-L, Pastor J, Danell K, Bergström R (2005) Impact of moose population density on the production and composition of litter in boreal forests. Oikos 108:297–306

Peterson CJ Jones RH (1997) Clonality in woody plants: a review and comparison with clonal herbs. In: De Kroon H, Groenendael J (eds) The ecology and evolution of clonal growth in plants. Backhuys, Leiden, pp 263–289

Piggott CD (1993) Are the distribution of species determined by failure to set seed? In: Marshall C, Grace J (eds) Fruit and seed production. Cambridge University Press, Cambridge, pp 203–216

Porter H, Nagel O (2000) The role of biomass allocation in the growth response of plants to different levels of light, CO_2, nutrients and water: a quantitative review. Aust J Plant Physiol 27:595–607

Price PW (1991) The plant vigour hypothesis and herbivore attack. Oikos 62:244–251

Prins HHT, van der Jeugd HP (1993) Herbivore population crashes and woodland structure in East Africa. J Ecol 81:305–314

Pugnaire FI, Chapin FS III (1993) Controls over nutrient resorption from leaves of evergreen Mediterranean species. Ecology 74:124–129

Raunkiær C (1937) Plant life forms. Clarendon, Oxford

Raspe O, Finlay C, Jacquemart A-L, (2000) Biological flora of the British Isles, no. 214. *Sorbus aucuparia* L. J Ecol 88: 910–930

Renaud PC, Verheyden-Tixier H, Dumont B (2003) Damage to saplings by red deer (*Cerevus elaphus*): effect of foliage height and structure. Forest Ecol Manage 181:31–37

Richardson FD, Hahn BD, Hoffman MT (2005) On the dynamics of grazing systems in the semi-arid succulent Karoo: the relevance of equilibrium and non-equilibrium concepts to the sustainability of semi-arid pastoral systems. Ecol Model 187:491–512

Rietkerk M, Ketner P, Stroosnijder L, Herbert HT (1996) Sahelian rangeland development: a catastrophe? J Range Manage 49:512–519

Ripple WJ, Beschta RL (2003) Wolf reintroduction, predation risk, and cottonwood recovery in Yellowstone National Park. Forest Ecol Manage 184:299–313

Ripple WJ, Beschta RL (2004) Wolves and the ecology of fear: can predation risk structure ecosystems? BioScience 54:755–766

Ripple WJ, Beschta RL (2005) Willow thickets protect young aspen from elk browsing after wolf reintroduction. West N Am Nat 65:118–122

Ritchie ME, Tilman D, Knops JMH (1998) Herbivore effects on plant and nutrient dynamics in oak savannah. Ecology 79:165–177

Robbins CT, Mole S, Hagerman AE, Hanley TA (1987) Role of tannins in defending plants against ruminants; reduction in dry matter digestion? Ecology 68:1606–1615

Rooke T, Bergström R, Skarpe C, Danell K (2004) Morphological responses of woody species to simulated twig-browsing in Botswana. J Trop Ecol 20:281–289

Rooney TP, McCormick RJ, Solheim SL, Waller DM (2000) Regional variation in recruitment of hemlock seedlings and saplings in the upper Great Lakes, USA. Ecol Applic 10:1119–1132

Rosenthal GA, Janzen DH (1979) Herbivores: their interaction with plant secondary metabolites. Academic press, New York

Rosenthal JP, Kotanen PM (1994) Terrestrial plant tolerance to herbivory. Trends Ecol Evol 9:145–148

Roy BA, Kirchener JW (2000) Evolutionary dynamics of pathogen resistance and tolerance. Evolution 54:51–63

Ruess RW, McNaughton SJ (1984) Urea as a promotive coupler of plant–herbivore interactions. Oecologia 63:331–337

Rusch GM, Oesterheld M (1997) Relationship between productivity, and species and functional group diversity in grazed and non-grazed Pampas grassland. Oikos 78:519–526

Rutina LP, du Toit JT, Moe SR, Hytteborn H (2004) Browsing ungulates limit tree seedling survival in the Chobe riparian zone, northern Botswana. In: Rutina LP (ed) Impalas in an elephant-impacted woodland: browser-driven dynamics of the Chobe riparian zone, northern Botswana. Thesis, Agricultural Univ of Norway

Sachs T, Hassidim M (1996) Mutual support and selection between branches of damaged plants. Vegetatio 127: 25–30

Saunders PT (1980) An introduction to catastrophe theory. Cambridge University Press, Cambridge

Schaefer JA, Messier F (1995) Habitat selection as a hierarchy: the spatial scales of winter foraging by muskoxen. Ecography 18:333–344

Schmitz OJ (2003) Top predator control of plant biodiversity and productivity in an old-field ecosystem. Ecol Lett 6:156–163

Senft RL, Coughenour MB, Baily DW, Rittenhouse LR, Sala OE, Swift DM (1987) Large herbivore foraging and ecological hierarchies. Landscape ecology can enhance traditional foraging theory. BioScience 37:789–799

Senn J, Haukioja E (1994) Reactions of the mountain birch to bud removal: effects of severity and timing, and implications for herbivores. Funct Ecol 8:494–501

Senn J, Suter W (2003) Ungulate browsing on silver fir (*Abies alba*) in the Swiss Alps: beliefs in search of supporting data. Forest Ecol Manage 181:151–164

Senock RS, Sisson WB, Donart GB (1991) Compensatory photosynthesis of *Sporobolus flexuosus* (Thurb.) rydb following simulated herbivory in the northern Chihuahua desert. Bot Gaz 152:275–281

Shabel AB, Peart DR (1994) Effects of competition, herbivory and substrate disturbance on growth and size structure in pin cherry (*Prunus pensylvanica*) seedlings. Oecologia 98:150–158

Shipley LA, Illius AW, Danell K, Hobbs NT, Spalinger DE (1999) Predicting bite size selection of mammalian herbivores: a test of a general model of diet optimization. Oikos 84:55–68

Skarpe C (1990) Shrub layer dynamics under different herbivore densities in an arid savanna, Botswana. J Appl Ecol 27:873–885

Skarpe C (1991a) Impact of grazing in savanna ecosystems. Ambio 20:351–356

Skarpe C (1991b) Spatial patterns and dynamics of woody vegetation in an arid savanna. J Veg Sci 2:565–572

Skarpe C, van der Wal R (2002) Effects of simulated browsing and length of growing season on leaf characteristics and flowering in a deciduous arctic shrub, *Salix polaris*. Arct Antarct Alpine Res 34:282–286

Skarpe C, Bergström R, Bråten A-L, Danell K (2000) Browsing in a heterogeneous savanna. Ecography 23:632–640

Skarpe C Aarrestad PA, Andreassen, HP et al (2004) The return of the giants: ecological effects of an increasing elephant population. Ambio 33:276–282

Skarpe C, Jansson I, Seljeli L, Bergström R, Eivin Røskaft E (2007) Browsing by goats on three spatial scales in a semi-arid savanna. J Arid Environ 68:480–491

Spalinger DE, Hobbs NT (1992) Mechanisms of foraging in mammalian herbivores: new models of functional response. Am Nat 140:325–348

Stamp N (2003) Out of the quagmire of plant defence hypotheses. Quart Rev Biol 78:23–55

Steinlein T, Heilmeier H, Schulze E-D (1993) Nitrogen and carbohydrate storage in biennials originating from habitats of different resource availability. Oecologia 93:374–382

Stoddart LA, Smith AD (1943) Range management. McGraw-Hill, New York

Stoddart LA, Smith AD (1955) Range management, 2nd edn. McGraw-Hill, New York

Stokke S (1999) Sex differences in feeding-patch choice in a megaherbivore: elephants in Chobe National Park, Botswana. Can J Zool 77:1723–1732

Strauss SY, Agrawal AA (1999) The ecology and evolution of plant tolerance to herbivory. Trends Ecol Evol 14:179–185

Strauss SY, Rudgers JA, Lau JA, Irwin RE (2002) Direct and ecological costs of resistance to herbivory. Trends Ecol Evol 17:278–281

Stringham TK, Krueger WC, Shaver PL (2003) State and transition modelling: an ecological process approach. J Range Manage 56:106–113

Stromberg CA (2004) Using phytolith assemblages to reconstruct the origin and spread of grass-dominated habitats in the great plains of North America during the late Eocene to early Miocene. Palaeogeogr Palaeocl 207:239–275

Stuart-Hill GC, Tainton NM (1989) The competitive interaction between Acacia karroo and the herbaceous layer and how this is influenced by defoliation. J Appl Ecol 26:285–298

Sullivan S (1996) Towards a non-equilibrium ecology perspectives from an arid land. J Biogeog 23:1–5
Sunnerheim K, Palo RT, Theander O, Knutsson PG (1988) Chemical defence in birch. Platyphylloside: a phenol from *Betula pendula* inhibiting digestibility. J Chem Ecol 14:549–560
Thom R (1975) Structural stability and morphogenesis. An outline of a general theory of models. Benjamin, Reading, MA
Thomas H, Stoddart JL (1980) Leaf senescence. Ann Rev Plant Physiol 31:83–111
Thompson ID, Curran WJ (1993) A reexamination of moose damage to balsam fir – white spruce forests in Newfoundland: 27 years later. Can J Zool 23:1388–1395
Thompson ID, Curran WJ, Hancock JA, Butler CE (1992) Influence of moose browsing on successional forest growth on black spruce sites in Newfoundland. For Ecol Manage 47:29–37
Tiffin P, Rausher MD (1999) Genetic constraints and selection acting on tolerance to herbivory in the common morning glory *Ipomoea purpurea*. Am Nat 154:700–716
Tilman D (1988) Plant strategies and the dynamics and structure of plant communities. Princeton University Press, Princeton, NJ
Tolsma DJ, Ernst WHO, Verwei RA, Vooijs R (1987a) Seasonal variation of nutrient concentrations in a semi-arid savanna ecosystem in Botswana J Ecol 75:755–770
Tolsma DJ, Ernst WHO, Verwei RA (1987b) Nutrients in soil and vegetation around two artificial waterpoints in eastern Botswana J Appl Ecol 24:991–1000
Tremblay JP, Hester AJ, McLeod J, Huot J (2004) Choice and development of decision support tools for the sustainable management of deer–forest systems. Forest Ecol Mgmt 191:1–6
Tremblay JP, Huot J, Potvin F (2007) Density-related effects of deer browsing on the regeneration dynamics of boreal forests. J Appl Ecol 44:552–562
Tremblay JP, Huot J, Potvin F (2006) Divergent nonlinear responses of the boreal forest field layer along an experimental gradient of deer densities. Oecologia 150:78–88
Turner GT (1971) Soil and grazing influences on a salt-desert shrub range in western Colorado. J Range Manage 24:397–400
Vanderklein DW, Reich PB (1999) The effect of defoliation intensity and history of photosynthesis, growth and carbon reserves of two conifers with contrasting life spans and growth habits. New Phytol 144:121–132
van de Koppel J, Prins HHT (1998) The importance of herbivore interactions for the dynamics of African savanna woodlands: an hypothesis J Trop Ecol 14:565–576
van der Meijden E, Wijn M, Verkaar HJ (1988) Defence and regrowth, alternative plant strategies in the struggle against herbivores. Oikos 51:355–363
van der Wal R, Bardgett RD, Harrison KA, Stien A. (2004) Vertebrate herbivores and ecosystem control: cascading effects of faeces on tundra ecosystems. Ecography 27:245–252
van Langevelde F, van de Vijver CADM, Kumar L, van de Koppel J et al (2003) Effects of fire and herbivory on the stability of savanna ecosystems. Ecology 84:337–350
van Vegten JA, (1983) Thornbush invasion in eastern Botswana. Vegetatio 56:3–7
Vera FWM, Bakker ES, Olff H (2006) Large herbivores: missing partners of western-European light-demanding tree and shrub species? In: Danell K, Bergström R, Duncan P, Pastor J (eds) Large herbivore ecology and ecosystem dynamics. Cambridge University Press, Cambridge, pp 203–231
Verwijst T (1993) Influence of the pathogen *Melampsora epitea* on intraspecific competition in a mixture of *Salix viminalis* clones. J Veg Sci 4:717–722
Vila B, Vourc'h G, Gillon D, Martin JL, Guibal F (2002) Is escaping deer browse just a matter of time in *Picea sitchensis*? Trees 16:488–496
Vourc'h G, Martin JL, Duncan P, Escarré J, Clausen T (2001) Defensive adaptations of *Thuja plicata* to ungulate browsing: a comparative study between mainland and island populations. Oecologia 126:84–93
Vourc'h G, Vila B, Gillon D, Escarré J, Guibal F, Fritz H, Clausen T, Martin JL (2002) Disentangling the causes of damage variation by deer browsing on young *Thuja plicata*. Oikos 98:271–283

Walker BH, Noy-Meir I (1982) Aspects of the stability and resilience of savanna ecosystems. In: Huntley BJ, Walker BH (eds) Ecology of tropical savannas. Springer, Berlin Heidelberg New York

Walker BH, Ludwig D, Holling CS, Peterman RM (1981) Stability of semi-arid savanna grazing systems. J Ecol 69:473–498

Walker BH, Matthews DA, Dye PJ (1986) Management of grazing systems – existing versus an event-orientated approach. S Afr J Sci 82:172

Wallace IL (1990) Comparative photosynthetic responses of big bluestem to clipping versus grazing J Range Manage 43: 58–61

Waller DM (1997) The dynamics of growth and form. In: Crawley MJ (ed) Plant ecology, 2nd edn. Blackwell, Oxford

Walter H (1939) Grassland, Savanne und Busch der arideren Teile Afrikas in ihrer ökologischen Bedingtheit. Jahrb Wissenschaftl Bot 87:750–860

Wang RZ (2002) Photosynthetic pathway type of forage species along grazing gradient from the Sognen grassland, northeastern China. Photosynthetica 40:57–61

Ward D, Young TP (2002) Effects of large mammalian herbivores and ant symbionts on condensed tannins of *Acacia drepanolobium* in Kenya. J Chem Ecol 28:913–929

Ward D (2004) The effects of grazing on plant biodiversity in arid ecosystems. In: Shachak M, Picket STA, Gosz JR, Perevolotsky A (eds) Biodiversity in drylands: towards a unified framework. Oxford Univ Press, Oxford, pp 233–249

Ward D (2006) Long term effects of herbivory on plant diversity and functional types in arid ecosystems. In: Danell K, Bergström R, Duncan P, Pastor J (eds) Large herbivore ecology and ecosystem dynamics. Cambridge Univ Press, Cambridge, pp 142–169

Ward D, Saltz D (1994). Foraging at different spatial scales: Dorcas gazelles foraging for lilies in the Negev Desert. Ecology 75:48–58

Wareing P, Patrick J (1975) Source–sink relationship and partition of assimilates in the plant. In: Cooper JP (ed) Photosynthesis and productivity in different environments. Cambridge Univ Press, Cambridge, pp 481–499

Wareing P, Khalifa MM, Trehane KJ (1968) Rate-limiting processes in photosynthesis at saturating light intensities. Nature 220:453–457

Warren A (2005) The policy implications of Sahelian change. J Arid Environ 63:660–670

Weis AE, Hochberg ME (2000) The diverse effects of intraspecific competition on the selective advantage to resistance: a model and its predictions. Am Nat 156:276–292

Weisberg PJ, Bugmann H. (2003) Forest dynamics and ungulate herbivory: from leaf to landscape. Forest Ecol Manage 181:1–12

Welch D, Staines BW, Scott D, French DD, Catt DC (1991) Leader browsing by red and roe deer on young Sitka spruce trees in western Scotland I. Damage rates and the influence of habitat factors. Forestry 64:61–82

West NE, Provenza FD, Johnson PS, Owens MK (1984) Vegetation change after 13 years of livestock grazing exclusion on sagebrush semidesert in west central Utah. J Range Manage 37:262–264

Westoby M, Walker B, Noy-Meir I (1989) Opportunistic management for rangelands not at equilibrium. J Range Manage 42:266–274

Whitham TG, Maschinski J, Larson KC, Paige KN (1991) Plant responses to herbivory: the continuum from negative to positive and underlying physiological mechanisms. In: Price P, Lewinsohn T, Fernandes W, Benson W (eds) Plant–animal interactions: evolutionary ecology in tropical and temperate regions. Wiley New York, pp 227–256

Wilmshurst JF, Fryxell JM, Farm BP, Sinclair ARE, Henschel CP (1999) Spatial distribution of Serengeti wildebeest in relation to resources. Can J Zool 77:1223–1232

Wilsey B (2002) Clonal plants in a spatially heterogeneous environment: effects of integration on Serengeti grassland response to defoliation and urine-hits from grazing animals. Plant Ecol 159:15–22

Wilson JR, Hacker JB (1987) Comparative digestibility and anatomy of some sympatric C_3 and C_4 arid zone grasses. Aust J Agric Res 38:287–295

Wolfson MM, Tainton NM (1999) The morphology and physiology of the major forage plants. In: Tainton N (ed) Veld management in South Africa, Univ of Natal Press, Pietermauritzburg, pp 54–90

Woodward A, Coppock DL (1995) Role of plant defence in the utilisation of native browse in southern Ethiopia. Agroforest Syst 32:147–161

Chapter 10
The Impact of Browsing and Grazing Herbivores on Biodiversity

Spike E. van Wieren and Jan P. Bakker

10.1 Biodiversity and Large Mammalian Herbivores

Biodiversity definable as the variety of life in all its forms. Biodiversity encompasses the entire biological hierarchy. Two different hierarchical schemes are frequently used to classify biological entities (Sarkar and Margules 2002): a spatial hierarchy and a taxonomic hierarchy. The spatial hierarchy runs from biological molecules and macromolecules, through cell organelles, cells, individuals, populations and meta-populations, communities, ecosystems ultimately to the biosphere; and the taxonomic hierarchy from alleles through loci, linkage groups, genotypes, subspecies, species, genera, families, orders, classes, phyla, and kingdoms. Each level is highly heterogeneous and, even more important, most entities are poorly defined. For example, there is still much confusion about the species concept, to mention one of the key entities in biology (Sarkar and Margules 2002). Furthermore biodiversity also comprises the relationships between entities within a hierarchy. Through all the confusion of the concept of biodiversity it is very difficult to operationalise and, therefore, various proximates have been developed which allow the estimation of (subsets of) biodiversity in the field. According to Ricotta (2005): 'biodiversity may be defined simply as a set of multivariate summary statistics for quantifying different characteristics of community structure'.

The most widely used biodiversity indicators are character or trait diversity, species diversity and species assemblage diversity. The rationale of trait diversity is that evolutionary mechanisms usually impinge directly on traits of individuals (Vane-Wright et al. 1991; Sarkar and Margules 2002). Trait, however, is a difficult term and largely determined by pragmatic considerations and, therefore, not very precise. The use of the species assemblage indicates that what is important is the variety of biotic communities with their associated pattern of interactions, sometimes referred to as functional biodiversity. Focusing on communities will automatically take care of species. Despite all the discussion around the concept, species diversity is the most commonly used indicator of biodiversity, even though we know that there is much more to biodiversity than the variety of species. A few reasons why species richness is used, is that it is thought to correlate with many measures of ecological diversity and that there is a positive relationship with trait richness and topographic diversity (Sarkar and Margules 2002).

A simple, much frequently used, index is the number of species in a location (Desrochers and Anand 2005). This is a crude way to estimate diversity and alternative diversity measures have been developed which combine species richness with relative distribution of species abundances. The Shannon entropy (H) and the Simpson diversity (1/D) indices are currently the most widely used (Pielou 1975; Shannon and Weaver 1949; Simpson 1949). These established measures have recently been extended to incorporate additional information related to ecological complexity (see Desrochers and Anand 2005 for review). Within the recognition of the effect of habitat heterogeneity on biodiversity, 'total' diversity can be partitioned into within- and between-habitat diversity (Whittaker 1972). Thus total diversity (γ-diversity) can be estimated by computing H of all sampling units pooled together; within-habitat diversity (α-diversity) is estimated from the H of all sampling units, and between-habitat diversity is the difference between γ-diversity and α-diversity (Desrochers and Anand 2004). From this the quantity β-diversity can be computed as the ratio of γ-diversity and α-diversity. This measure is useful as it quantifies habitat heterogeneity across habitats, and can be used in environmental assessment because it indicates the degree to which habitats have been partitioned by species (Desrochers and Anand 2004; Ricotta 2005).

In the studies with regard to the effects of grazing and browsing on biodiversity, the most widely used indices are almost identical to the most widely used ones. These are trait diversity (phenotypic and genotypic), species diversity (species richness and species composition), and community diversity (structure and composition).

Evolutionary relationships between mammalian herbivores and biodiversity. Large mammalian herbivores have been around for some time (Janis this volume) and we can ask to what extent this group has contributed to biodiversity as described above. As real proof is absent we can only speculate and indicate probabilities. First of all the group itself is a contribution to terrestrial animal diversity; more than 200 species of ungulates exist now, but summed over the Tertiary a multiple of this number has populated the earth, evolving and going extinct in an irregular fashion. (Hernández Fernández and Vrba 2005). On a higher trophic level we can safely assume that the large herbivores have provided exclusive niches for a number of the larger predator species, among them lions and tigers. Large herbivores and large carnivores most likely co-evolved during the Tertiary.

Given the long evolutionary time scale and the wide distribution of large herbivores across the major biomes, they must have exerted strong selective forces on, in particular, the plant world. It is generally acknowledged that herbivory has led to a number of plant responses which can be seen as adaptations to predation. Examples are the development of morphological structures such as thorns and hairs, and chemical defenses, although the latter probably predate the mammal era by a long time (Lindroth 1988). Thus the first, and maybe the largest, effects would have been on traits of plants inducing larger phenotypic variation and also, perhaps, a larger genotypic variation. It is not known if this selection has been strong enough to lead to speciation and thus to new species although angiosperms and mammalian herbivores co-evolved for a long time during the Tertiary (Tiffney 2004). The development of large fleshy fruits which need to be predigested and dispersed by large mammals is an indication that some plant species are dependent, and can maybe only exist, in the presence of large mammals.

Several species of temperate North American and European thorny shrubs may have evolved under browsing by now extinct large Pleistocene herbivores, since the presence of taxa such as *Pinus, Crataegus, Rhamnus, Rhus,* and *Juniperus* goes back through the Pleistocene even as far as the early Tertiary (Tallis 1991). These ecological anachronisms are often hard to prove. Large herbivores such as cattle and horses, domesticated from their extinct ancestors aurochs and tarpan, respectively, make it possible to test hypotheses in the recruitment strategy of more preferred woody species that are spatially associated with thorny shrubs under grazed conditions. A cross-site comparison of four floodplain woodlands in north-western Europe showed that sessile oak (*Quercus robur*) can regenerate in the presence of large herbivores through spatial association with Blackthorn (*Prunus spinosa*), a clonal thorny shrub (Bakker et al. 2004). An experiment with transplanted oak seedlings revealed that oak seedlings grew best in grassland exclosures and on the edge of thorny shrubs that received most light. Oak survival was strongly reduced in oak woodland with low light availability. However, in sites with high rabbit *(Oryctolagus cuniculus)* density no young trees were found. Rabbits graze both on young ramets of Blackthorn and on young oaks. It was concluded that the process of associational resistance did not work against rabbits, as they consume the oak seedlings and the thorny shrubs that have no thorns when they are young. Under low rabbit densities the young ramets of Blackthorn can escape from grazing, and once established, they give shelter for the oak (Bakker et al. 2004).

It is likely that large herbivores have (had) an effect on community diversity. Even if no new species have evolved in the presence of large herbivores, it can be assumed that at least locally they have changed the structure and species composition of communities. As such they have contributed to community diversity. Some species, such as elephant, moose and beaver, have been credited with even more effective power in the sense that they are thought to be able to control vegetation development on a larger landscape scale (Owen-Smith 1988; McInnes et al. 1992; Naiman et al. 1986). If such effects should exist in the natural world, then some large herbivores would be able to affect biodiversity on the ecosystem level.

In our view the effects of large mammalian herbivores on biodiversity are probably relatively modest, mainly acting through increasing trait diversity and through the reshuffling of species from the regional species pool by creating new communities. This does not exclude, however, the enormous potential for change in species composition that can be achieved with grazing, as will be demonstrated below. Most of the remainder of this chapter is about the application of grazing and browsing in man-modified and/or man-controlled systems and, as such, the examples differ from the natural world. Most grazing is carried out with a preconceived human management goal and it is important to realize that the results of most of the grazing research should be viewed in this context.

Table 10.1 Plant traits which can be affected by herbivory, with indication whether they are aimed at defence (d), escape (e), or tolerance (t)

Trait
Thorns/spines (d)
Trichomes (e.g., hairs) (d)
Chemicals (e.g., tannins, alkaloids, silicate) (d)
Low stature (height/biomass) (e)
Lower reproductive potential (< flowers, seeds) (e)
Prostrate growth form (e)
More rosettes (e)
Increased tillering (t)
Increased photosysnthetic rates (t)

10.2 Effects on Traits

The main short-term and direct mechanism by which herbivores affect plants is through defoliation. Plants react to this kind of attack by various kinds of resistance mechanisms. Plants may try to avoid damage via defence or escape in space and time, or they may tolerate herbivore damage, once attacked. Various plant traits may be affected by herbivory and a list of important ones is given in Table 10.1.

An example of induced defense by herbivory is the variation in spinescence in plants (Young et al. 2003; Obeso 1997). Young et al. (2003) studied the effects of large herbivores on spine length of *Acacia drepanolobium* in East Africa by simulated and natural herbivory. New spines in plots from which herbivores were excluded were 35–40% shorter than spines in plots exposed to all browsers (Fig. 10.1). On low

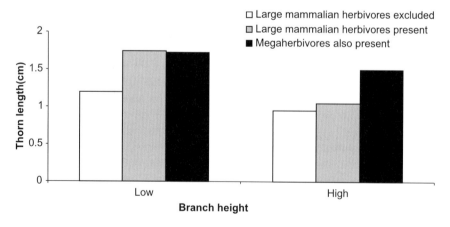

Fig. 10.1 Mean spine (*Thorn*) lengths of low (~1 m) and high (~2 m) branches on *Acacia. drepanolobium* trees accessible to different herbivores in an East African savanna (from Young et al. 2003)

branches, spine lengths were low in total exclusion plots, and longer in plots that allowed wildlife and plots with both megaherbivores (elephants, giraffes). On higher branches, recently produced spines were similar in length in total exclusion plots and plots that allowed wildlife, but were longer in plots in which megaherbivores were allowed. Apart from these induced effects, Young et al. (2003) also showed that simulated large mammal browsing induced greater spine length on trees that had reduced spine length after several years of herbivore exclusion (Fig. 10.2). Interestingly, the induced responses were highly localized and the main variation in natural browsing appeared to be associated with escape in space from browsers by the production of branches outside the reach of browsers, as the trees grow. This escape from herbivory by individual branches high above the ground is seen as an important force in the development of the inducibility of spine length in *Acacia spp.* (Young et al. 2003).

An important aspect of tolerance with respect to herbivory is a change in photosynthetic rates. Widely different effects have been reported: from extreme undercompensation to even overcompensation (Rosenthal and Kotanen 1994; Strauss and Agrawal 1999). This induction of the productivity trait may lead to the development

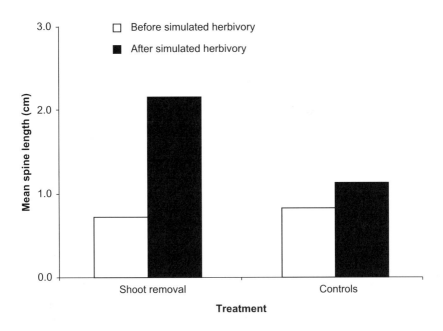

Fig. 10.2 Reinduction of spine length in simulated herbivory experiments on Acacia trees that had been protected from large mammalian herbivory for five years in an East African savanna. Spine lengths before (*open bars*) and one year after (*solid bars*) simulated herbivory. Control branches were either on a paired fork on the same branch or on a branch on the opposite side of the tree crown. Experimental shoot removal resulted in new spine three times longer than before the experiment, but no significant response on other branches on the same tree (after Young et al. 2003)

(or expression) of different ecotypes within species, which behave differently under the same grazing treatment. A nice example is the work at Wind Cave National Park on the physiological responses of prairie grass populations to responses to prairie-dog grazing (Holland and Detling 1990; Fig. 10.3). This work shows that long-term grazing in two ecotypes of *Agropyron smithii* results in a biphasic response curve for above- and below-ground production and for gross and net mineralized N as a function of grazing intensity. Furthermore, the model results suggest that differences in the long-term quasi-equilibrium conditions develop between the two ecotypes, even though both populations show the biphasic response. The populations adapted to grazing contrasted sharply with the less adapted population in terms of grazing-induced potential and N-restricted plant production. The grazing-adapted population shifted its optimal growth conditions at high grazing intensities, whereas the non-grazing-adapted population showed growth limitations at much lower grazing intensities (Fig. 10.3).

Given the above examples, it is clear that herbivory can induce the expression of phenotypic variation of many traits. This may lead to various phenotypes (and genotypes) being simultaneously present in a given area in contrast to a lower phenotypic variation present in ungrazed situations (Loreti et al. 2001). It has also been proposed that resistance to grazing is more varied in species with a long grazing history (Wilsey et al. 1997; McNaughton 1984; Painter et al. 1993), while in species with a short evolutionary history of grazing no, or very little, variation was found (Jaindl et al. 1994; Loreti et al. 2001).

Recently it has been found that herbivory can have effects on plant mating systems. Resource limitation caused by herbivory can negatively affect flower production, flowering phenology, and seed mass and number (Steets and Ashman 2004). Herbivory can also, through leaf damage, reduce the number of flowers on a plant that open synchronously, and reduce flower morphology and nectar reward (Steets and Ashman 2004; Sharaf and Price 2004; Vazquez and Simberloff 2004). Through these effects, phenotypic variation in mating trait expression might change, but effects might also cascade through the system, especially by changing the pollinator pool. This is especially apparent in systems not accustomed to a long history of grazing and where introduced herbivores have strong negative effects on plant species which are important for the community of flower visitors (Vazquez and Simberloff 2004).

Although there is a growing consensus that tolerance and defence are evolving traits under selection from herbivores in natural plant populations (Strauss and Agrawal 1999), and there are many plant systems for which genetic variation in tolerance to herbivory have been reported (Strauss and Agrawal 1999), it is still not clear to what extent trait variation indeed evolved as the result of selective pressures of vertebrate herbivory, let alone that this selection has led to novel genotypes. In fact, many authors stress that the traits affected by herbivory are basic characteristics of plants with important roles in growth and reproduction. Some traits may be by-products of selection for other types of damage: the protected apical meristems, for instance, may have evolved in response to drought and fire (Stowe et al. 2000; Coughenour 1985; Rosenthal and Kotanen 1994). The potential selective pressures exerted by ungulates resulting in speciation seem rather limited.

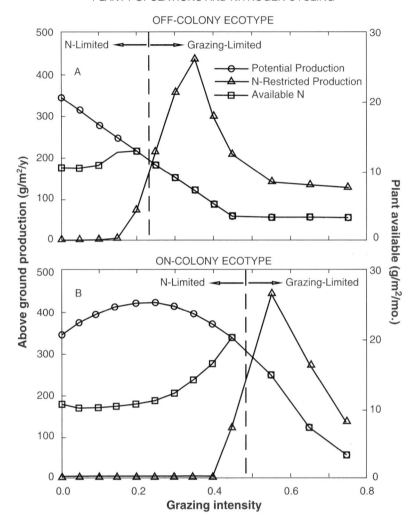

Fig. 10.3 Productivity estimates for two South Dakota western wheatgrass populations growing in the vicinity of prairie dog colonies, as depicted by Century Model studies. *A* shows results from populations not located on prairie dog colonies, *B* from those on such colonies. A pronounced biphasic relationship exists for the *B* population potential production, not so for the *A* population in this regard. Both populations show biphasic responses for N dynamics, but in the interaction between N-limited and grazing-limited responses, the *A* population shows a shift to the left as a function of grazing intensity, while the *B* population shows a shift to the right. This suggests that western wheatgrass has developed ecotypic variation in response to stresses arising during its long-term grazing history (from Holland et al. 1992)

10.3 Effects on Plant Communities

Grasslands. The main mechanisms by which foraging herbivores affect plants are selectivity and trampling. Differences in quality and quantity between plant species lead to preferential grazing selectively affecting plant traits such as tolerance and defence. More intense grazing (and thus trampling) increases the number of vegetation gaps, changing the colonization matrix of the system. All these differential influences may lead to a change in vegetation pattern and species composition (Sternberg et al. 2000).

Because of the above mechanisms, site selection occurs which can be enhanced by feedback mechanisms. Through grazing, plant quality can become enhanced leading to re-utilization of the patch (Bakker et al. 1983; Weber et al. 1998; Adler et al. 2001). Patch size depends on grazing intensity and type of grazer. Sheep create patches of about 60 cm across (Bakker et al. 1984), while cattle patches can be up to several metres (Fuls 1992). The resulting micro patterns can be relatively fixed in time (Bakker 1989). An increase in size of patches can lead to the formation of grazing lawns. Frequently created by (very) large grazers, these can provide good grazing conditions for smaller herbivores (Verweij et al. 2006).

Selective grazing can also lead to changes in plant species composition. Some species are expected to be more affected than others when being grazed, depending on species specific characteristics. Tall annual grasses are frequently negatively affected because their regeneration capacity decreases, while hemicryptic species are less affected. They can withstand heavy grazing as their perennating buds are buried near the soil surface and they frequently have strong physical and chemical defenses (Sternberg et al. 2000). In relation to grazing, plants have been characterized in functional groups such as decreasers and increasers. Figure 10.4 gives an example of how a shift in functional groups can occur under different grazing intensities.

Most of the important grazing studies have been carried out on climatically determined, natural, grasslands and these will be considered first. Climatically determined grasslands (including savanna and shrub steppe) cover about 25% of the earth's land surface (Lauenroth 1979). They receive 250–1000 mm of annual precipitation and have mean annual temperatures between 0° and 26° C.

The effects of grazing in grasslands can be best understood when viewed as superimposed on general relationships between plant species richness and plant biomass. This general relationship is hump-back shaped and based on resource limitation at the lower biomass level and competition for light at the high biomass level with maximum species richness at intermediate biomass levels corresponding with a moderate competition or disturbance level gradient (Grime 1973; Frank 2005). Grazing can be superimposed on this model because herbivores may increase or decrease species richness depending on grazing intensity and the amount of biomass and selective grazing on a dominant species. A number of models have been developed to explain the response of grasslands to large generalist herbivores. Among them are the range succession model (Dyksterhuis

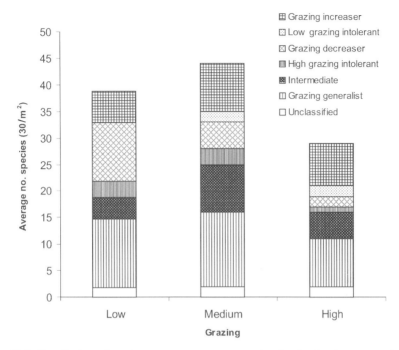

Fig. 10.4 Contribution of grazing response groups to species density for three categories of grazing. Grazing has a significant effect on species density in this design. The study was conducted within the southeast Queensland Bioregion, Australia, with grassy eucalypt woodland as the dominant vegetation type (McIntyre and Martin 2001)

1949), the predation hypothesis (Paine 1966), the intermediate disturbance hypothesis (Grime 1973), the Huston hypothesis, the Milchunas hypothesis (Milchunas et al. 1988), and the state-and-transition model (Westoby et al. 1989; Laycock 1991). Of these hypotheses, only the Milchunas hypothesis will be discussed here because it satisfactorily explains a wider range of real situations than do the other models (Cingolani et al. 2005). In this model, recent insights from the state-and-transition models can be incorporated; this will be discussed in the final paragraph of this chapter.

Milchunas et al. (1988) developed a generalized model of the effects on grassland community structure which proposed that interactions occurring along gradients of evolutionary history of grazing and environmental moisture were primary factors that explained changes in plant species diversity, changes in species composition, and invasion by exotics across gradients in grazing intensity. The direct act of grazing by large herbivores represents a loss of organs to individual plants and an alteration of canopy structure to the assemblage of plants. Plant reaction to the grazing event will depend upon the ability of individuals to compensate for lost organs and the relative impact of the removal on competitive relationships in the canopy. Canopy development and above-to-belowground biomass ratios increase with increasing moisture. Therefore, a given percentage removal from the canopy will have a greater effect in subhumid than in arid environments both in terms of the intensity of plant interactions in the canopy and percentage removal of total plant biomass. Grazing should have greater affect on species composition in more humid areas because adaptations of tall growth forms capable of competing for light in a dense canopy are opposite to those that provide resistance to or avoidance of grazing. In contrast, plant adaptations to frequent loss of organs from drought or herbivory are similar, and under these arid and semiarid conditions competition is primarily for belowground resources. The evolutionary history was brought into the model because it was hypothesized that increasing grazing history, over evolutionary time, results in greater capacities for regrowth following herbivory, favouring prostrate growth forms. In communities of short evolutionary history, and high moisture content, grazing causes rapid shifts in the dynamic balance between suites of species adapted to either grazing avoidance/tolerance or competition in the canopy. Milchunas et al. (1988) presented the model and the experimental evidence for four boundary cases (Fig. 10.5). The model of Milchunas has been confirmed

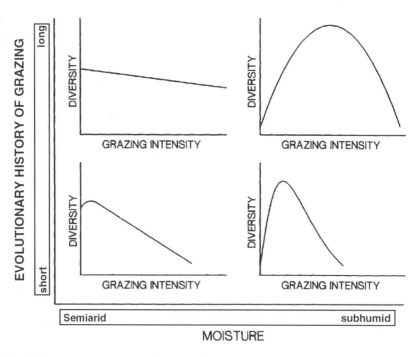

Fig. 10.5 Plant diversity of grassland communities in relation to grazing intensity along gradients of moisture and of evolutionary history of grazing. Increments on the *diversity axis* are equal in all cases, but equal specific values are not implied; that is, relative response, not absolute diversity is implied (from Milchunas et al. 1988)

in a number of studies (Milchunas and Lauenroth 1993, Sternberg et al. 2000; Hunt 2001; see Cingolani et al. 2005 for recent review). The model has been expanded in the sense that for the axis 'moisture' one can substitute 'annual net primary production' or 'productivity' (Milchunas et al. 1988; Cingolani et al. 2005; Frank 2005). It is relevant to note that in all models plant diversity decreases, sometimes strongly, when grazing intensity increases beyond some optimum. It seems that currently most grazing systems of the world (if not all) are situated somewhere at the right hand of any of these curves, which means that most are overgrazed.

Herbivores can affect the rate of succession in grasslands and alter the composition of some successional stages. Salt marshes typically go from an early successional stage with species such as *Puccinellia maritima* and *Plantago maritima* to a late successional stage with a dominance of *Elymus athericus* and *Atriplex portulacoides* (Kuijper and Bakker 2005). Small herbivores such as geese and hares preferentially forage on early successional species which can tolerate grazing well. At the same time these grazers prevent colonization of the later successional species thereby retarding the succession by decades (Van der Wal et al. 2000; Kuijper and Bakker 2005). They also create a mid-successional stage, not present in the absence of herbivores, with a high abundance of grazing-tolerant species (*Festuca rubra, Puccinellia maritima, Plantago maritima, Triglochin maritima*). They thus add, albeit temporarily, an extra species diversity element to the system.

Most of the grasslands in Europe, Japan, eastern North America, and areas of Australia and Asia are of anthropogenic origin. They have been derived from forests and have undergone huge changes (Prins 1998). It is likely that in most cases the species composition no longer bears resemblance to the original situation. Important changes have been a large increase in light reaching the ground layer, with an accompanying increase in the biomass of the field layer, and a decrease in vertical vegetation structure. Although it is almost impossible to estimate the change in biodiversity, we can safely assume that much biodiversity has been lost in the process as so many species are confined to the tree canopy, the various layers in a forest, and to dead wood. The focus of the effects of grazing in anthropogenic grasslands thus has always been on limited subsets of biodiversity (flowering plants, butterflies, dung beetles, ground breeding birds). Within this context, the many grazing studies (summarized by Bakker 1998) have produced an amazing variety of results. A few studies have found an increase in species richness at intermediate grazing intensity (Sternberg et al. 2000; Bakker et al. 1997) but in most studies the results are not clear (Bakker 1998; Bullock et al. 2001; Luoto et al. 2003). Contributing to the inconsistency is the great variety in grazing treatments (winter grazing, summer grazing, year round grazing, rotational grazing, different species of herbivores) imposed on plant communities, the short duration of many experiments, and the probably inherent variability that can be expected in these altered systems. Much work has been done on the effects when grazing is totally removed. Because of changes in agricultural practices, many agricultural grasslands have been abandoned and generally adverse effects (in selected functional groups) have been found. The herb layer frequently becomes dominated by tall grasses and herbs while plant species diversity declines; the aboveground standing crop increases accompanied by litter accumulation (Bakker 1998; Luoto et al. 2003). These effects are also in line with the Milchunas et al. (1988) hypothesis. The frequently occurring development of vegetation communities towards the original state through bush encroachment and tree regeneration are also generally considered as undesirable with respect to biodiversity targets (Bakker 1998; Verdu et al. 2000). It should be noted that restoration attempts can be seriously hampered by seemingly permanent changes in the state of the system (see final section, 10.7).

Woody vegetation. Very little is known of the effects of large herbivores in forests under natural conditions. In the few remnants of natural temperate forests, the density of grazers and browsers is quite low but nevertheless we can expect some effects of browsing—and trampling in particular. The most important ones are modification of the structure and composition of the plant community and effects on the succession rate (Pastor et al. 1997). Forest herbivores will concentrate foraging on available regeneration gaps where they may be able to extend the time to gap closure, enhance ground cover and diversity (Côté et al. 2004). As browsing animals are highly selective, some species in the gaps will be more affected than others, and this is likely to influence the ultimate composition and the vertical structural component of the forest.

It has been hypothesized that, under natural conditions, large herbivores were able to maintain large open areas in temperate forests leading to much more open landscapes than in the absence of grazers (Vera 2000). If so, they should truly be regarded as keystone species. This hypothesis, however, is not very well supported by evidence from pollen analysis (Mitchell 2005) and is difficult to test.

By far the greater part of the forests of the world are heavily modified by man. Tree planting after logging has led to early successional forests on a grand scale in the temperate and boreal regions. This has increased the food base for deer tremendously, which in turn has led to an explosion in deer numbers, both in Europe and North America (Fuller and Gill 2001). It is not surprising that under these changed conditions herbivores will have profound effects on forest structure and composition with cascading effects on other taxa. The results of many effect studies have been reviewed recently (see Côté et al. 2004) and here only a short summary of the main type of changes is given. Heavy browsing reduces twig density, resulting in a more open canopy. Because more light reaches the forest floor, the production of shrubs and herbs increases, decomposition rates change, changing field layer vegetation affecting vertebrate communities (Persson et al. 2000; Suominen et al. 1999). Deer may cause a shift in the canopy composition by selective browsing, generally in the direction of the less preferred, more browsing-tolerant species (Table 10.2). When browsing pressure increases further, the rate and direction of forest succession can be seriously affected and succession can be stalled.

All the above effects are viewed as negative, both from (timber) production and conservation perspectives. Given the poor state of these systems are in, there is no easy solution. The most frequently mentioned counteractive measure is control of deer numbers (Côté et al. 2004). This is not always easy to do and there may be associated negative side effects (changes in habitat use, effects on predators, increasing vigilance; see final section, 10.7). Protection of forest for some time may also help.

Table 10.2 Compositional shifts in dominant tree species induced by deer browsing in boreal and temperate forests (from Côté et al. 2005)

Former dominant	New dominant
Balsam fir (*Abies balsamea*)	White spruce (*Picea glauca*)
Birch (*Betula* spp.)	Norway spruce (*Picea abies*)
Eastern hemlock (*Tsuga canadensis*)	Sugar maple *(Acer saccharum)*
Mixed hardwoods	Black cherry (*Prunus serotina*)
Oak (*Quercus* spp.)	Savanna type system
Scots pine (*Pinus sylvestris*)	Hardwoods and Norway spruce

If the system is allowed to consist again of the mature-age component for a large part, this will slowly turn the tide.

A combination of these measures is an option as well. Simply letting nature run its course can be an option, too, but in some cases this can take a long time, resulting in extinction of taxa of conservation interest in the meantime. It is increasingly recognized, however, that a simple population approach to deer management is not sufficient and that an approach should be taken that considers whole-ecosystem effects (Côté et al. 2004; McShea et al. 1997; Gaillard et al. this volume). Much more work should be done to study relationships between community composition across taxa and various population sizes of deer to understand the full range of deer impacts on biodiversity (Kramer et al. 2003).

10.4 Effects on Invertebrates

Invertebrates comprise a huge group occupying a large number of niches and it would be expected that there are many ways in which large herbivores affect this group. The studies that have been done in this field always have to focus, by necessity and through preference, on a limited subset of taxa, which makes it difficult to judge the overall interactions of invertebrates with large herbivores.

Herbivores may affect invertebrates directly and indirectly through changes in, for example, quality and quantity of litter, dead wood, flowers, seeds, cover, egg laying sites, nesting sites, and vegetation structure. In natural systems we would expect large herbivores to affect insect diversity predominantly through the presence of their living and dead bodies, their excreta, and through local effects on vegetation structure and the creation of gaps. There are many intimate associations between large herbivores and commensals and parasites. These can be species-specific such as the deer bot-fly (*Cephenomya auribarbis*) and the deer ked (*Lipoptena cervi*; Hutson in Stewart 2001). Dead ungulates attract carrion-feeding invertebrates, such as rove beetles as well as assorted flies, some of them being specific to large carcasses. There exists a whole community of coprophilous insects feeding and laying their eggs on the faeces of large herbivores, including *Atomaria peltataeformes* and *Aphodius nemoralis* (Kronblad 1971; Peterson 1998; both cited by Persson et al. 2000). The most obvious taxa in dung are dung beetles (Geotrupidae) and scarab beetles (Scarabidaeidae). Although many dung-associated species are rather unspecific, some species are reported to be associated with only primarily one species of ungulate (Duncan et al. 2001). Large changes in ungulate density will affect the relative abundance of those species associated with them (Putman et al. 1989). Dung-associated invertebrates are an important source of food for a number of predatory animals and, as such, large changes in abundance of the coprophilous community can have cascading effects on, e.g., tawny owls (*Strix aluco*) and different species of woodland bats (in the UK, Duncan et al. 2001).

Through trampling and local intensive grazing, large herbivores can create gaps which are usually favourable for ants, as they provide potential nest sites and adequate microclimatic conditions (see Gonzales-Megia et al. 2004). In forests, herbivores may also create gaps or keep gaps open for a prolonged time. This will benefit insect diversity because the rarest species in woodlands are associated with the opposite ends of the succession spectrum: open clearings and mature or senescent habitats, especially dead and decaying wood (Stewart 2001).

Given the above effects, we can assume that large herbivores, by their presence and activities, do contribute positively to invertebrate diversity. The main mechanism through which this has happened is increase in (mainly habitat) heterogeneity. If grazing pressure increases, however, this positive force can be seriously compromised because vegetation structure (both horizontally and vertically) will be simplified, direct competition for food will increase, and other important niches may disappear (for instance dead wood). There are many studies that report a decline in at least some (but frequently in many) invertebrate groups as a result of heavy grazing pressure (Kearns 1997; Kruess and Tscharntke 2002; Sudgen 1985; Vazquez and Simberloff 2004; Souminen et al. 1999; Gonzalez-Megias et al. 2004; Duncan et al. 2001; Hutchinson and King 1980). Kruess and Tscharntke (2002) studied the insect diversity on intensively and less-intensively grazed pastures and on 5- to 10-year-old ungrazed grasslands in northern Germany and related diverse taxa to vegetation structure. Here we give their results for butterfly adults and lepidopteran caterpillars (Fig. 10.6). The total number of butterfly species and lepidopteran caterpillars increased with decreasing grazing intensity. For both these groups, but also for other groups studied, mean canopy height was the best predictor of both species richness and abundance. It has to be noted that the less-intensive grazing treatment must still be considered quite high from a conservation point of view (1.4 cattle/ha from 1 May to 15 November). From the above example it may seem that cessation of grazing will lead to the highest insect diversity but, in fact, most workers conclude that insect diversity in grasslands (and probably also in forests) is highest at moderate herbivore densities (see Di Giulio et al. 2001; Swengel and Swengel 2001). Indeed, cessation of livestock grazing after intensive grazing, may be beneficial for invertebrates in the short term, when plants can flower. However, long-term cessation of grazing resulted in the disappearance of characteristic halobiontic invertebrates in salt marshes in Germany (Andresen et al. 1990) and France (Pétillon et al. 2005). They were replaced by ruderal species typical for tall inland forb communities. As such, we can link overall insect

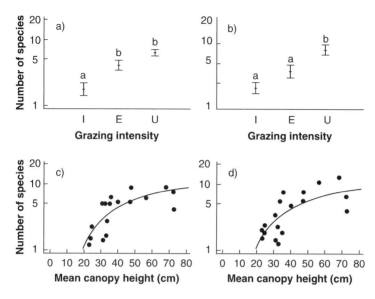

Fig. 10.6 Effects of grazing intensity (*I*. Intensively grazed pastures; *E* extensively grazed pastures; *U* ungrazed grasslands) on butterfly adults and lepidopteran caterpillars: *a*) mean (±1 SE) number of adult species ($V_{2,15} = 13.9$, $p < 0.001$, $n = 18$, and *b*) mean (±1 SE) number of caterpillar species ($F_{2,15} = 11.35$, $p < 0.001$, $n = 18$. Different *letters* above *bars* indicate significant differences (Tukey's honest significant difference). Correlation between mean canopy height and *c*) number of adult butterfly species ($Y = 2.9-55.5/X$, $F_{1,16} = 28.6$, $r^2 = 0.64$, $p < 0.001$, $n = 18$ and *d*) number of caterpillar species ($Y = 2.8 - 54.6/X$, $F_{2,16} = 10.47$, $r^2 = 0.55$, $p < 0.001$, $n = 18$) (from Krauss and Tscharntke 2002)

diversity to the general model of Milchunas et al. (1988; Fig. 10.5) and agree with Hutchinson (1959) and Hunter and Price (1992) that a high α-plant diversity is associated with a high α-diversity in higher trophic levels. We can also conclude that, although some invertebrate groups benefit from heavy grazing, in many systems insect diversity has declined considerably because of present densities of large herbivores.

10.5 Effects on Birds

As with plants and insects, we have little information on the overall effects of large herbivores in natural systems. Large herbivores provide carrion and dung which, through the invertebrates that live in and on them, provide feeding opportunities for a number of bird species, notably corvids and raptors (Duncan et al. 2001). Large herbivores carry parasites and disturb vegetation while moving, providing niches for commensals such as the cattle egret (*Bubulcus ibis*) and cowbirds which feed on insects or parasites carried by the herbivores (Mayfield 1965). On a larger scale, herbivores may have an impact on the composition and structure of the vegetation,

Table 10.3 Distribution of western European species of birds and mammals (ungulates excluded) across preferred habitat types; species can be related to more than one habitat type. 57 species of mammals, 257 species of birds (after Van Wieren 1998)

	Mammals	Birds	
		Nesting site	Feeding site[a]
Bare soil		18	40
Short herbaceous	5	62	57
Tall herbaceous	22	81	47
Shrub	27	20	12
Woodland	41	76	30

[a] Aquatic species excluded

with all the cascading effects on other taxa, but this depends very much on the density of herbivores in a particular system.

As almost all natural grasslands have been modified by man and are now grazed by livestock, most studies have focused on the changes brought about by different grazing intensities. Birds are particularly sensitive to changes in the vegetation structure but also to changes in plant species composition affecting food supplies and nesting opportunities (see also Table 10.3). Many studies have shown that cattle grazing reduces the richness and densities of avian communities in North American prairie habitats (Kantrud 1981; Chamberlain et al. 2000; Gates and Donald 2000; Söderström et al. 2001). It has also been shown that waterfowl nest density and nesting success are significantly reduced in grazed prairies compared to ungrazed ones (Klett et al. 1988; Kirby et al. 1992). Sometimes overall species richness remains the same but a major shift in species composition can be noted. As with invertebrates, heavy grazing leads to a reduction in vegetation heterogeneity with negative consequences for many ground dwelling bird species, e.g., nest losses through trampling (Beintema and Müskens 1987; Barker et al. 1999; Grant et al. 1999), exposure to predation pressure (Brua 1999; Wilson and Hartnett 2001), or a decrease of large insects for the larger insectivores (Söderström et al. 2001).

Most grasslands in Europe and many in North America are man-modified although they frequently have evolved under grazing pressure for centuries. A native flora and fauna has developed in these grasslands, which has adapted to grazing disturbance. Due to changes in agricultural practices the greater part of these grasslands have disappeared, or have become much more intensively used or have been abandoned, with a later decline in typical farmland birds (Newton 1998). When abandoned, the bird species richness in these habitats can increase quickly due to an increase in bushes and trees (Van Wieren 1998) but frequently this direction of change is not much appreciated as the focus is on typical farmland birds (Söderström et al. 2001). When looking at total bird species richness it seems that a low to moderate grazing pressure yields the highest number of species in grasslands. In the Argentine pampa, moderate grazing showed a higher density, species richness, and diversity of species when compared to ungrazed or heavily grazed areas (Zalba and Cozzani 2004). This general finding can be attributed to the higher vegetation heterogeneity in moderately grazed grasslands. Another example of this can be found on salt marshes. Salt marshes are natural grasslands which are generally grazed by livestock in the summer months. In many areas changes in grazing management have led to either abandonment or intensification of the grazing pressure (Norris et al. 1998). Cessation of grazing leads to a tall rank sward while heavy grazing results in a short, uniform dense sward. Both developments are detrimental for some ground breeding birds such as, for instance, the redshank *Tringa totanus*, an important meadow bird in northwestern Europe. When relating breeding density of redshank to grazing intensity, Norris et al. (1998) found the highest breeding density on low to moderately grazed swards. Again, a moderate grazing intensity provided the greatest heterogeneity in the ground vegetation (Norris et al. 1998).

Salt marshes are also important for avian herbivores such as geese. They prefer the high quality low growing plants on the lower salt marshes, while the taller, later successional stages are much less attractive because of the development of tall, less-preferred plant species. When these older salt marshes are grazed they become attractive to geese due to the development of short high quality swards (Van der Graaf et al. 2002; Stock and Hofeditz 2000), a clear example of grazing facilitation. As most wintering geese in western Europe rely on agricultural grasslands adjacent to the coast, livestock grazing also plays an important facilitating role (Bos et al. 2005).

As birds are particularly sensitive to changes in vegetation structure, especially in the understorey (Miyashita et al. 2004; Fuller 2001), we would expect large changes in the bird communities associated with forests when the density of large herbivores increases. In a controlled browsing experiment De Celesta (1994) studied the relationships between deer density and bird communities in northern hardwood forest in the US. A clear non-linear, negative relationship was found between bird diversity and deer abundance. Declines of 2% and 37% in species richness and abundance of intermediate canopy nesters between lowest and highest deer densities were found. The density threshold for the major changes was between 7.9 and 14.9 deer/km^2. Also in other hardwood forest increased deer densities led to lower species richness and abundance of canopy feeders (McShea and Rappole 2000). Sometimes no differences between density treatments were found because of species replacement effects (McShea and Rappole 2000; Moser and Witmer 2000; Fuller 2001). This indicates that, like insects, birds occupy a wide variety of niches and that each landscape has its own particular type of bird community. The pasture woodlands found in western Britain for instance, carry a whole suite of species, very typical for that habitat, including pied flycatcher (*Ficedula hypoleuca*), redstart (*Phoenicurus phoenicurus*), and tree pipit (*Anthus trivialis*; Fuller 2001). For both grasslands and forests we can conclude that heavy grazing and browsing leads to a decrease in bird species richness while low to moderate herbivore pressure has the highest species richness and diversity. Avian herbivores such as geese are facilitated by moderate to heavy grazing pressures in grasslands.

10.6 Effects on Mammals

Large herbivores in grassland systems. Generally more than one species of large herbivore can be found in any system. In Africa especially, species richness can be very high, up to 31 species in the larger national parks. To what extent does the presence of one species affect the presence of others? Species within an ungulate community frequently differ in many respects (size, morphology, digestive physiology) which results in the various species occupying different parts of the ecosystem: resource partitioning. Resource partitioning can simply be the result of adaptation to a particular niche but may also have been (partly) driven by competition which is frequently a selective force shaping the composition and structure of ungulate communities; although evidence for this is still scarce (Arsenault and Owen-Smith 2002). Murray and Illius (2000) studied the foraging behaviour of sympatric topi (*Damaliscus lunatus*) and wildebeest (*Connochaetes taurinus*) in East Africa and found that the very selective grazing style of the smaller species (topi) was able to deplete the green leaf component in the tall sward to the detriment of forage quality for the larger species. Apart from competition, facilitative interactions have also been proposed to have shaped ungulate communities. Facilitation may arise when one species makes more grass accessible to another species, e.g., by reducing canopy height and removing stems (habitat facilitation), or when grazing by one species stimulates grass regrowth, thereby enhancing the nutritional quality for another species (feeding facilitation; Arsenault and Owen-Smith 2002). Although these mechanisms are not mutually exclusive, habitat facilitation has been the most frequently observed in the field. For example, grazing hippopotamus (*Hippopotamus amphibius*) can transform tall grasslands into extensive grazing lawns (Owen-Smith 1988) which may be essential to grazing mesoherbivores such as the western kob (*Kobus kob kob*) in Cameroon (Verweij et al. 2006). The same hippo, however, may exert competitive effects on grazers which prefer tall grass. Elimination of hippos from a part of Queen Elisabeth National Park, Uganda, was followed by a substantial increase in elephant, buffalo, and waterbuck. Following the recovery of the hippos, the numbers of these other species declined again in the region (Arsenault and Owen-Smith 2002). Population responses are thus possible but to what extent both competition and facilitation within natural ungulate communities have shaped the composition of the regional species pool, or are confined only to local effects, is not known (Prins and Olff 1998).

As described above, it is not surprising that the effects of changing herbivore numbers or the introduction of domestic livestock on habitat modification and habitat fragmentation can have far-reaching consequences for the functioning and structure of ecosystems, including effects on native herbivore species and small mammal communities. This is especially apparent when the density of very large species changes. For example, the effects of crowding by elephants in many parts of Africa, can lead to either open parkland or shrubland and positively affect grazers such as oryx (*Oryx gazella*) and zebra (*Equus burchelli*), while browsers such as the lesser kudu (*Tragelaphus imberbis*) and gerenuk (*Litocranius walleri*) decline (Arsenault and Owen-Smith 2002). Where graziers with their livestock have encroached on national grasslands, native large herbivore communities have suffered tremendous losses, both in species richness and numbers. Different mechanisms have been responsible for this decline but two important ones stand out: a direct effect through overhunting and a more indirect effect through competition by livestock for food. Numbers of bison on the Great Plains in North America were estimated at between 30 and 60 million before the 1800s. Overhunting then reduced them to a mere few thousand animals by the early twentieth century (Flores 1991). Competition with livestock is generally through modification and combined use of the food base and it is to be expected that grazer species will be more affected than browsers, especially if the species are of similar size. Mishra et al. (2002) studied ungulate assemblages in the Indian Trans-Himalaya where potentially seven species of wild ungulates had to co-exist with five species of domestic livestock. A number of combinations of two species have shown great overlap in size, feeding habits, while frequently one is the domestic form of the also-present wild type (Table 10.4). It was found that at least four species of the original assemblage of wild species were missing from the region and this could be attributed to competitive exclusion by the similar domestic species. Mishra et al. (2001) also found evidence that the (Trans)-Himalaya is severely overstocked, a condition which has been frequently described for many other parts in the world (Wilson and Macleod 1991; du Toit and Cummings 1999).

Despite the many competitive interactions, facilitation of native species by introduced livestock has also been reported. Interestingly, these were mainly between a larger domestic grazer (notably cattle) and smaller medium sized grazers such as red deer (cattle – red deer *Cervus elaphus*, Gordon 1988; cattle – wapiti *Cervus elaphus*, cattle – red deer and wild boar *Sus scrofa*, Kuiters et al. 2005). We can hypothesize that the large domestic grazer here plays the role of a similar wild

Table 10.4 Pair-wise weight ratios for most similar pairs (smallest ratios) of wild and domestic grazers in the Trans-Himalaya (from: Mishra et al. 2002)

	Species pairs	Weight ratio
1	Tibetan Argali – Donkey	1.03
2	Chiru – Goat	1.08
3	Kiang – Domestic Yak	1.08
4	Chiru – Sheep	1.09
5	Kiang – Horse	1.11
6	Ibex – Donkey	1.18
7	Yak – Domestic Yak	1.39
8	Kiang – Cow	1.44
9	Bharal – Sheep	1.57
10	Bharal – Goat	1.62
11	Bharal – Donkey	1.64
12	Yak – Horse	1.67
13	Yak – *Dzomo*	1.82
14	Yak – Cow	2.16

species originally present in the system, notably the bison in North America and the extinct aurochs in western Europe. Facilitative effects are most apparent at low to moderate grazing intensity, but are quickly replaced by competitive interactions as densities of livestock increase (McCullough 1999).

Small mammals. Large herbivores can affect small mammals by changing vegetation structure and composition thereby affecting the living conditions for this group of species. When large grazers enhance the structure and quality of the grass sward, small herbivores such as rabbits (Oosterveld 1983) and hare (Kuijper 2005) can be facilitated. At higher densities other lagomorph species such as the European brown hare (*Lepus europaeus*; Frylestam 1976) and the Basin pigmy rabbit (*Brachylagus idahoensis*; Siegel et al. 2004) become negatively affected as a result of decreased grass cover, direct disturbance, and destruction of burrows.

Apart from the food base, rodents are particularly sensitive to cover (Leirs et al. 1996; Bowland and Perrin 1989), because lower cover levels increases predation risk (Goheen et al. 2004). The effects of large herbivores on rodents is not always clear. Both positive effects (Keesing and Crawford 2001; Jones and Longland 1999) and negative effects (Keesing 1998, 2000; Jones and Longland 1999) have been reported; the effects being dependent on the type of habitat and species. Granivorous species seem to be more negatively affected than herbivorous species such as the field mouse (*Microtus agrestis*) at low densities (Beever and Brussard 2004), although Steen et al. (2005) could not detect an effect of sheep grazing on the bank vole (*Clethrionomys glareolus*). High grazing intensities generally have negative effects on both species richness and abundance of rodent communities (Steen et al. 2005; Goheen et al. 2004). Although very few studies have been conducted where rodent response has been estimated at various grazing intensities, there is some evidence that species richness and abundance is higher at low grazing intensity than when grazing is completely absent (Steen et al. 2005; Kerley 1992; Smit et al. 2001). At low grazing intensity suitable habitat conditions may be created consisting of alternating short, grazed patches, with higher, ungrazed patches.

Large herbivores in woodlands. Not many studies have been conducted on the effects of large herbivores on mammals in forests (Côté et al. 2004). If, as is assumed here, densities of large herbivores in natural forest ecosystems are relatively low, their impacts on other mammals will most likely be modest. Most forests, however, are heavily modified, to the detriment of many forest mammal species (Lessard et al. 2005; Côté et al. 2004). In most European forests large herbivore species richness has decreased, like that of the large predator community. At the same time numbers of a few deer species have exploded throughout the northern hemisphere as the result of an enormous increase in the early successional stages represented by commercial timber production (Côté et al. 2004). Some late successional species such as the woodland caribou (*Rangifer tarandus caribou*) have decreased as result of changes in the forest. Negative effects of high deer numbers on abundance, diversity, and species richness have been reported for small mammals (Moser and Witmer 2000), while McShea and Rappole (2000) found higher rodent abundance in areas from which deer were excluded. As in grasslands, heavy grazing generally has had a negative effect on small mammals in the UK's

New Forest (Putman et al. 1989). Introduced exotic herbivores can have large negative effects on the native ones. In Argentina, many areas formerly occupied by huemul (*Hippocamelus bisulcus*) are now occupied by red deer (Povilitis 1998). Feral pigs have modified entire communities and ecosystems around the world through their digging and rooting activities (Mack and D'Antonio 1998). If forests become increasingly more open or are converted into grassland, it can be expected that this will be accompanied by a general decrease in mammal species richness because a majority of mammals are related to wooded habitats (Table 10.3).

Effects on mammalian predators. Large herbivores are important prey for a suite of large predators and, as such, they contribute to mammalian diversity. A number of the smaller predators are carrion feeders (fox, badger, jackal) and many of them feed on invertebrates living on large herbivore dung or carcasses. Large changes in large herbivore abundance can thus affect the predator guild as well. As shown above, heavy grazing and browsing have generally negative effects on bird and small mammal species composition and it can be expected that these effects will cascade through the food chain. Heavy grazing and browsing reduced the diversity and abundance of rodents of the New Forest and this had a significant effect on the foraging behaviour, diet, population density, and breeding success of the fox (*Vulpes vulpes*) and the badger (*Meles meles*), as on a number of avian predators (Tubbs and Tubbs 1985; Putman 1986). Conversely, a cessation of grazing had the opposite effect (Petty and Avery 1990).

If large herbivore densities increase substantially, they may affect other species of large herbivores through apparent competition. This occurs when the presence of multiple non-competing prey species elevates predator abundance above levels maintained by single prey species, which increases predation pressure on multiple prey assemblages (Morin 1999). A good example of this is the increase of moose in the northern forests of North America which led to an increase in wolves, which also prey on woodland caribou. The increased wolf population has now a much greater impact on the more vulnerable caribou, because of a much greater abundance of the alternative prey species, the moose, threatening the caribou population with extinction (Lessard et al. 2005). Introduced herbivores may have similar effects when an increase in native predator abundance, because of the increased prey basis, leads to increased predation on native prey species (Vazquez 2002).

10.7 Large Herbivores and Biodiversity

Synthesis. When evaluating the overall effects of large herbivores on diversity, we can conveniently come back to the theoretical model of Milchunas et al. (1988; Fig. 10.5) because this model is still valid (see above and the recent review by Cingolani et al. 2005). In fact, the model can also be extended to include life forms other than the plants for which it was developed; the model seems to hold for birds and mammals as well as for some groups of invertebrates. When looking carefully at the results, it seems, however, that for most animal groups the hump lies a bit more to the left than for plants, and that for animals maximum diversity is found at a lower level of herbivory. This latter has also been suggested for semi-natural grasslands (Vickery and Herbert 2001; Luoto et al. 2003). The model can also be generalized with respect to its X-axis which originally is a moisture gradient but can be indicated as a nutrient or primary productivity gradient (Olff et al. 2002). Therefore, we can conclude that in most systems there will be an optimum level of herbivory whereby biodiversity will be maximal. This level will not be the same in all systems because, apart from abiotic factors, the length of the evolutionary history the system has experienced with large herbivores is of great significance. Systems with a long grazing and browsing history are much more adapted to herbivory than systems with only a short history containing more grazing-tolerant species and species which exhibit compensatory growth (Adler et al. 2004). Biodiversity in these systems changes less with varying herbivore densities while those with only a short history are more vulnerable to change (Fig. 10.5).

Problems. The biodiversity in the world would be highest when each system is managed to be positioned on top of its hump (see Fig. 10.5). However, from the foregoing it may be clear that this generally is not the case and in reality systems can be found on either side of the curve, although too high a grazing pressure is much more common than too low. In some systems, notably in western Europe and the Mediterranean, much of the agricultural land is being abandoned with generally negative consequences for biodiversity (Verdu et al. 2000; Newton 1998). It has to be noted, though, that these evaluations are always made with respect to only a selected subset (e.g., dung beetles, flowering plants, meadow birds) and it can not be ruled out that total biodiversity would actually increase when these systems revert back to the original state, especially in that ancient woodlands and forests were very species rich (Niemala 1997).

The real problems are that in the greater part of the world manipulation of both the species composition and the abundance of the large herbivores (both native and domesticated) has led to high, sometimes very high, densities. Although it has been stated that livestock have done more damage than all chainsaws and bulldozers combined (Noss 1994), it has to be stressed that other man-related activities have contributed (e.g., clearing, lumbering, mowing, burning, fertilizing), especially in the semi-arid systems (De Pietri 1992). Although the negative effects of heavy grazing pressure on biodiversity have been highlighted, there are many other factors that play a part, e.g., decreasing secondary production, soil erosion, increasing surface water runoff, a decrease in landscape appreciation and hence a lower recreational value, and an increase in invasive species (Heitschmidt et al. 2004).

Before trying to formulate some solutions, one particular and nagging problem has to be discussed. In the past few decades, more and more systems are suffering from an increase in grazing-tolerant shrubs which can be either exotics or native species. This can happen when grazing pressure is lowered (Manier and Hobbs 2006; Coughenour 1991; Verdú et al. 2000) but more frequently when herbivore pressure has become very high (Veblen et al. 1989; Archer et al. 1987; Vazquez and Simberloff 2004; George and Bazzaz 1999; Raffaele and Veblen 2001; Weber et al.

1998; Sternberg et al. 2000). The shrub problem is thus a worldwide phenomenon and it has been suggested that a strong increase in shrub development only happens when grazing exceeds a threshold level (Weber et al. 1998). Attempts to counteract these developments have generally failed, and this has led to the notion that systems may have irreversibly changed, the return to the previous ('natural') systems now closed because these systems have moved to another stable state. This finding underscores recent theoretical insights that many systems are not characterized by one single climax state but can produce, as the result of self-organizing and positive feedback processes, an array of stable states, each of which can be highly resilient and resistant to change (Peterson et al. 1999; Fig. 10.7). In this sense large herbivores can also act as switches able to move systems towards alternative successional pathways which can be stable. Such non-linear dynamics have been described in rangeland pastures (Laycock 1991; Lockwood and Lockwood 1993), savanna-woodland systems (Dublin 1995; Rietkerk et al. 2002), and temperate and boreal forests (Pastor et al. 1993). If this assertion is correct then it will be hard indeed to reverse developments causing decreased biodiversity.

Possible solutions for an overstocked world. This much is clear: the major road to improvement of most systems is a substantial decrease in large herbivore pressure. This is true not only when biodiversity is the goal but also when the aim is for development of a more sustainable rangeland agriculture system (Heitschmidt et al. 2004). In systems where production, in some form, is an important objective, more management activities are acceptable than in systems where restoration of biodiversity is the main focus (national parks, reserves). In calculating a more

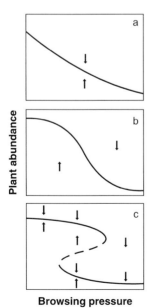

Fig. 10.7 Three hypothetical relationships between the abundance of a forage plant and deer browsing pressure. (*a*) Deer have only modest and monotonic effects on the population. (*b*) A reversible threshold exists beyond which plant abundance drops precipitously. (*c*) Browsing beyond a certain threshold point causes a nonlinear decline that is not simply reversible. The plant population requires a large (or prolonged) reduction in browsing as well as a disturbance factor that promotes an increase of its abundance to recover. This requirement indicates an 'alternate stable state'. *Arrows* indicate dynamic changes at various points (from Côté et al. 2005)

optimal herbivore density in production systems, reference could be made to the (evolutionary) grazing history of the system, sensu Milchunas et al. (1988; Fig. 10.5). Systems with a long history of herbivory can sustain higher densities and a greater variation in density than those with only a short history. As we know that the systems with a long history of grazing may have (had) high densities of large herbivores, it is puzzling that even those systems can become overstocked and degraded. It is suggested here that this may not simply be an effect of too many animals (with additional supplementary feeding and continuous water supply), but could also be the result of constant heavy grazing in almost every corner of an area. There is hardly any part of a grazed area which experiences a temporary relaxation of grazing and for many species this is simply too much. McNaughton (1984) found that, in systems with a long history of grazing, there exist two pools of species: one short and one tall. This suggests that these systems have apparently evolved under fluctuating densities of large herbivores and this could be a key factor in restoration management. Thus, apart from a general decrease in numbers, a greater spread of grazing intensity in time and space would improve biodiversity (Gordon 2006). This calls for larger areas to be grazed by free ranging herds.

When restoration for conservation is the main goal, various management strategies are open, mostly aiming at attacking problems to help nature follow its own course again and thus of a temporal nature. Sometimes a non-intervention policy can be applied; this may be an option even when a transition phase has to be accepted where unwanted developments take place. Restoration generally should not be aimed at zero grazing because there are very few situations where large herbivores would not naturally be present and thus this would lead to a lower biodiversity. In western Europe attempts are being made to restore ecosystems by restoring native communities of large herbivores, referred to as naturalistic grazing, but the lands set aside for these efforts are said to be too small (see report on Naturalistic Grazing by Hodder et al. 2005).

Nevertheless, given the earlier-mentioned possibility of irreversible changes, a reversal to some earlier state may not happen and become impossible. Additional measures such as culling, burning, bush clearance, or predator control may help here but there is no guarantee that the problem will thereby permanently be solved, and furthermore these measures can be costly and produce unwanted side effects. It is not possible to create a situation where there will be only winners. But when the main goal is biodiversity, the best option seems to be herbivore densities that are in close agreement with the historical herbivore density under which the system evolved, with the lowest possible burden of additional management measures.

References

Adler LS, Karban R, Strauss SY (2001) Direct and indirect effects of alkaloids on plant fitness via herbivory and pollination. Ecology 82:2032–2044

Adler P, Milchunas D, Lauenroth W, Sala O, Burke I (2004) Functional traits of graminoids in semi-arid steppes: a test of grazing histories. J Appl Ecol 41:653–663

Andresen H, Bakker JP, Brongers M, Heydemann B, Irmler U (1990) Long-term changes of salt marsh communities by cattle grazing. Vegetatio 89:137–148

Archer S, Garrett MG, Detling JK (1987) Rates of vegetation change associated with prairie dog (*Cynomys ludovicianus*) grazing in North American mixed-grass prairie. Vegetatio 72:159–166

Arsenault R, Owen-Smith N (2002) Facilitation versus competition in grazing herbivore assemblages. Oikos 97:313–318

Bakker JP (1989) Nature management by grazing and cutting. On the ecological significance of grazing and cutting regimes applied to restore former species-rich grassland communities in the Netherlands. Kluwer, Dordrecht

Bakker JP (1998) The impact of grazing to plant communities. In: Wallis de Vries MF, Bakker JP, Van Wieren SE (eds) Grazing and conservation management. Kluwer, Dordrecht, pp 137–184

Bakker JP, De Leeuw J, Van Wieren SE (1984) Micro-patterns in grassland vegetation created and sustained by sheep-grazing. Vegetatio 55:153–161

Bakker JP, Esselink P, Van der Wal R, Dijkema KS (1997) Options for restoration and management of coastal salt marshes in Europe. In: Urbanska KM, Webb NR, Edwards PJ (eds) Restoration ecology and sustainable development. Cambridge Univ Press, Cambridge, pp 286–322

Bakker ES, Olff H, Vandenberghe C, De Maeyer K, Smit R, Gleichman JM, Vera FWM (2004) Ecological anachronisms in the recruitment of temperate light-demanding tree species in wooded pastures. J Appl Ecol 41:571–582

Barker AM, Brown NJ, Reynolds CJM (1999) Do host-plant requirements and mortality from soil cultivation determine the distribution of graminivorous sawflies on farmland? J Appl Ecol 36:271–282

Beever EA, Brussard PF (2004) Community and landscape-level responses of reptiles and small mammals to feral-horse grazing in the Great Basin. J Arid Environ 59:271–297

Beintema AJ, Müskens GJDM (1987) Nesting success of birds in a Dutch agricultural grassland. J Appl Ecol 24:743–758

Bos D, Loonen M, Stock M, Hofeditz F, Van Der Graaf S, Bakker JP (2005) Utilisation of Wadden Sea salt marshes by geese in relation to livestock grazing. J Nat Conserv 15:1–15

Bowland AE, Perrin MR (1989) The effect of overgrazing on the small mammals in Umfolozi Game Reserve. Z Saeugetierkd 54:251–260

Brua RB (1999) Ruddy duck nesting success: do nest characteristics deter nest predation? Condor 101: 867–870

Bullock JM, Franklin J, Stevenson MJ, Silvertown S, Coulson SJ, Gregory SJ, Tofts R (2001) A plant trait analysis of responses to grazing in a long-term experiment. J Appl Ecol 38:253–267

Chamberlain DE, Fuller RJ, Bunce RGH, Duckworth JC, Shrubb M (2000) Changes in the abundance of farmland birds in relation to the timing of agricultural intensification in England and Wales. J Appl Ecol 37:771–788

Cingolani AM, Noy-Meir I, Dýaz S (2005) Grazing effects on rangeland diversity: a synthesis of contemporary models. Ecol Appl 15:757–773

Côté S, Rooney TP, Tremblay JP, Dussault C, Waller DM (2004) Ecological impacts of deer overabundance. Annu Rev Ecol Evol Syst 35:113–147

Coughenour MB (1985) Graminoid responses to grazing by large herbivores: adaptations, exaptations and interacting processes. Ann Mo Bot Gard 72:852–863

Coughenour MB (1991) Spatial components of plant–herbivore interactions in pastoral, ranching, and native ungulate ecosystems. J Range Manage 44:530–542

De Pietri J (1992) Alien shrubs in a national park: can they help in the recovery of natural degraded forest? Biol Conserv 62:127–130

Desrochers RE, Anand M (2004) From traditional diversity indices to taxonomic diversity indices. Int J Ecol Environ Sci 30:93–99

Di Giulio M, Edwards PJ, Meister E (2001) Enhancing insect diversity in agricultural grasslands: the roles of management and landscape structure. J Appl Ecol 38:310–319

Dublin HT (1995) Vegetation dynamics in the Serengeti–Mara ecosystem: the role of elephants, fire and other factors. In: Sinclair ARE, Arcese P (eds) Serengeti II: dynamics, management and conservation of an ecosystem. Univ of Chicago Press, Chicago, pp 71–90

Duncan AJ, Hartley SE, Thurlow M, Young S, Staines BW (2001) Clonal variation in monoterpene concentrations in Sitka spruce (*Picea sitchensis*) saplings and its effect on their susceptibility to browsing damage by red deer (*Cervus elaphus*). Forest Ecol Manage 148:259–269

du Toit JT, Cummings DHM (1999) Functional significance of ungulate diversity in African savannas and the ecological implications of the spread of pastoralists. Biodiv Conserv 8:1643–1661

Dyksterhuis EJ (1949) Condition and management of rangeland based on quantitative ecology. J Range Manage 41:450–459

Flores D (1991) Bison ecology and bison diplomacy: the southern plains from 1800 to 1850. J Am Hist 78:465–485

Frank DA (2005) The interactive effects of grazing ungulates and aboveground production on grassland diversity. Oecologia 143:629–634

Frylestam B (1976) Effects of cattle grazing and harvesting of hay on density and distribution of an European hare population. In: Pielowski Z, Pucek Z (eds) Ecology and management of European hare populations. Polish Hunting Assoc, Warsaw, pp 199–203

Fuller RJ (2001) Responses of woodland birds to increasing numbers of deer: a review of evidence and mechanisms. Forestry 74:289–298

Fuller RJ, Gill RMA (2001) Ecological impacts of increasing numbers of deer in British woodland. Forestry 74:193–199

Gates S, Donald PF (2000) Local extinction of British farmland birds and the prediction of further loss. J Appl Ecol 37:806–820

George LO, Bazzaz FA (1999) The fern understory as an ecological filter: emergence and establishment of canopy-tree seedlings. Ecology 80:833–845

Goheen JR, Keesing F, Allan BF, Ogada DL, Ostfeld RS (2004) Net effects of large mammals on Acacia seedling survival in an African savanna. Ecology 85:1555–1561

Gordon IJ (1988) Facilitation of red deer grazing by cattle and its impact on red deer performance. J Appl Ecol 25:1–10

Gordon IJ (2006) Restoring the function of grazed ecosystems. In: Danell K, Bergström R, Duncan P, Pastor J, Olff H (eds) Large herbivore ecology and ecosystem dynamics. Cambridge Univ Press, Cambridge, pp 449–467

Grant MC, Orsman C, Easton J et al (1999) Breeding success and causes of breeding failure of curlew *Numenius arquata* in Northern Ireland. J Appl Ecol 36:59–74

Grime JP (1973) Control of species diversity in herbaceous vegetation. J Environ Manage 1:151–167

Heinken T, Raudnitschka D (2000) Do wild ungulates contribute to the dispersal of vascular plants in central European forests by epizoochory? Forstwiss Centralbl 121:179–194

Heitschmidt RK, Vermeire LT, Grings EE (2004) Is rangeland agriculture sustainable? J Anim Sci 82(E Suppl):138–146

Hernández Fernández M, Vrba ES (2005) A complete estimate of the phylogenetic relationships in Ruminantia: a dated species-level supertree of the extant ruminants. Biol Rev 80:269–301

Hodder KH, Bullock JM, Buckland, P, Kirby KJ (2005) Large herbivores in the wildwood and modern naturalistic grazing systems. English Nature Res Reports Nr 648, Peterborough, UK

Holland EA, Detling JK (1990) Plant response to herbivory and belowground nitrogen cycling. Ecology 71:1040–1049

Hunt LP (2001) Heterogeneous grazing causes local extinction of edible perennial shrubs: a matrix analysis. J Appl Ecol 38:238–252

Hunter MD, Price PW (1992) Playing chutes and ladders: heterogeneity and the relative roles of bottom-up and top-down forces in natural communities. Ecology 73:724–732

Hutchinson GE (1959) Homage to Santa Rosalia, or why are there so many kinds of animals? Am Nat 93:145–159

Hutchinson KJ, King KL (1980) The effect of sheep stocking level on invertebrate abundance, biomass and energy utilization in a temperate sown grassland. J Appl Ecol 17:369–387

Jaindl RG, Doescher P, Miller RF, Eddleman LE (1994) Persistence of Idaho fescue on degraded rangelands: adaptation to defoliation or tolerance. J Range Manage 47:54–59

Jones AL, Longland WS (1999) Effects of cattle grazing on salt desert rodent communities. Am Midl Nat 141:1–11

Kantrud H (1981) Grazing intensity effects on the breeding avifauna of North Dakota native grasslands. Can Field Nat 95:404–417

Keesing F (1998) Ecology and behaviour of the pouched mouse, *Saccostomus mearnsi*, in central Kenya. J Mammal 79:919–931

Keesing F 2000 Cryptic consumers and the ecology of an african savanna. BioScience 50:205–215

Keesing F, Crawford T (2001) Impacts of density and large mammals on space use by the pouched mouse (*Saccostomus mearnsi*) in central Kenya. J Trop Ecol 17:465–472

Kerley GIH (1992) Ecological correlates of small mammal community structure in the semi-arid Karoo, South Africa. J Zool 227: 17–27

Kirby RE, Ringelman JK, Anderson DA, Sodja RS (1992) Grazing on National Wildlife Refuges: do the needs outweigh the problems? Trans N A Wildl Nat Res Conf 57:611–626

Kiviniemi K (1996) A study of adhesive seed dispersal of three species under natural conditions. Acta Bot Neerl 45: 73–83

Klett AT, Shaffer TL, Johnson DH (1988) Duck nest success in the meadow pothole region. J Wildl Manage 52:431–440

Kramer K, Groen TA, Van Wieren SE (2003) The interacting effects of ungulates and fire on forest dynamics: an analysis using the model FORSPACE. Forest Ecol Manage 181:205–222

Kruess A, Tscharntke T (2002) Contrasting responses of plant diversity to variation in grazing intensity. Biol Conserv 106:293–302

Kuijper DPJ (2005) Small herbivores losing control – plant–herbivore interactions along a natural productivity gradient. Thesis, Univ of Groningen

Kuijper DPJ, Bakker JP (2005) Top-down control of small herbivores on salt-marsh vegetation along a productivity gradient. Ecology 86:914–923

Kuiters AT, Groot B, Geert WTA, Lammertsma DR (2005) Facilitative and competitive interactions between sympatric cattle, red deer and wild boar in Dutch woodland pastures. Acta Theriol 50:241–252

Laycock WA (1991) Stable states and thresholds of range condition on North American rangelands: a viewpoint. J Range Manage 44:427–433

Leirs H, Verheyen W, Verhagen R (1996) Spatial patterns in *Mastomys natalensis* in Tanzania (Rodentia, Muridae). Mammalia 60:545–555

Lessard RB, Martell S, Walters CJ, Essington TE, Kitchell JFK (2005) Should ecosystem management involve active control of species abundances? Ecol Soc 10:1–23

Lindroth RL (1988) Adaptations of mammalian herbivores to plant chemical defenses. In: Spencer KC (ed) Chemical mediation of coevolution. Academic Press, San Diego, pp 425–445

Lockwood JA, Lockwood DR (1993) Catastrophe theory: a unified paradigm for rangeland ecosystem dynamics. J Range Manag 46:282–88

Loreti J, Oesterheld M, Sala O (2001) Lack of intraspecific variation in resistance to defoliation in a grass that evolved under light grazing pressure. Plant Ecol 157:195–202

Luoto M, Pykälä J, Kuussaari M (2003) Decline of landscape-scale habitat and species diversity after the end of cattle grazing. J Nat Conserv 11:171–178

Mack MC, D'Antonio CM (1998) Impacts of biological invasions on disturbance regimes. Trends Ecol Evol 13:195–198

Manier DJ, Hobbs NT (2006) Large herbivores influence the composition and diversity of shrubsteppe communities in the Rocky Mountains, USA. Oecologia 146:641–651

Mayfield HA (1965) The brown-headed cowbird with old and new hosts. Liv Bird 4:13–28

McCullough DR (1999) Density dependence and life-history strategies of ungulates. J Mammal 80:1130–1146

McInnes PF, Naiman RJ, Pastor J, Cohen Y (1992) Effects of moose browsing on vegetation and litter of the boreal forest, Isle Royale, Michigan, USA. Ecology 73:2059–2075

McIntyre S, Martin TG (2001) Biophysical and human influences on plant species richness in grasslands – comparing variegated landscapes in sub-tropical and temperate regions. Austral Ecol 26:233–245

McIntyre S, Heard KM, Martin TG (2003) The relative importance of cattle grazing in subtropical grasslands: does it reduce or enhance plant biodiversity? J Appl Ecol 40:445–457

McNaughton SJ (1984) Grazing lawns: animals in herds, plant form, and coevolution. Am Nat 124:863–886

McShea WJ, Rappole JH 2000 Managing the abundance and diversity of breeding bird populations through manipulation of deer populations. Conserv Biol 14:1161–70

McShea WJ, Underwood HB, Rappole JH (eds; 1997) The science of overabundance: deer ecology and population management. Smithson Inst Press, Washington, DC

Milchunas DG, Lauenroth WK (1993) Quantitative effects of grazing on vegetation and soils over a global range of environments. Ecol Monogr 63:327–366

Milchunas DG, Lauenroth WK, Sala OE (1988) A generalized model of the effects of grazing by large herbivores on grassland community structure. Am Nat 132:87–106

Mishra C, Prins HHT, Van Wieren SE (2001) Overstocking in the Trans-Himalayan rangelands of India. Environ Conserv 28:279–283

Mishra C, Van Wieren SE, Heitkönig IMA, Prins HHT (2002) A theoretical analysis of competitive exclusion in a Trans-Himalayan large-herbivore assemblage. Anim Conserv 5:251–258

Mitchell FJG (2005) How open were European primeval forests? Hypothesis testing using palaeoecological data. J Ecol 93:168–177

Miyashita T, Takada M, Shimazaki A (2004) Indirect effects of herbivory by deer reduce abundance and species richness of web spiders. Ecoscience 11:74–79

Morin PJ (1999) Community ecology. Blackwell, Malden, MA, USA

Moser BW, Witmer GW (2000) The effects of elk and cattle foraging on the vegetation, birds, and small mammals of the Bridge Creek Wildlife Area, Oregon. Int Biodeter Biodegr 45:151–157

Murray MG, Illius AW (2000) Vegetation modification and resource competition in ungulates. Oikos 89:501–508

Naiman RJ, Melilo JM, Hobbie JE (1986) Ecosystem alteration of boreal forest streams by beaver (*Castor canadensis*). Ecology 67:1254–1269

Newton I (1998) Bird conservation problems resulting from agricultural intensification in Europe. Ecol Appl 11:307–322

Niemala J (1997) Invertebrates and boreal forest management. Conserv Biol 11:601–610

Norris K, Brindley E, Cook T, Babbs S, Forster Brown C, Yaxley R (1998) Is the density of redshank *Tringa totanus* nesting on saltmarshes in Great Britain declining due to changes in grazing management? J Appl Ecol 35: 621–634

Noss RF (1994) Cows and conservation biology. Conserv Biol 8:613–616

Obeso JR (1997) The induction of spinescence in European holly leaves by browsing ungulates. Plant Ecol 129:149–156

Olff H, Ritchie ME, Prins HHT (2002) Global environmental controls of diversity in large herbivores. Nature 415: 901–904

Oosterveld P (1983) Eight years of monitoring of rabbits and vegetation development on abandoned arable fields grazed by ponies. Acta Zool Fenn 174:71–74

Owen-Smith N (1988) Megaherbivores. The influence of very large body size on ecology. Cambridge Univ Press, Cambridge

Paine RT (1966) Food web complexity and species diversity. Am Nat 100:850–860

Painter EL, Detling JK, Steingraeber DA (1993) Plant morphology and grazing history: relationships between native grasses and herbivores. Vegetatio 106:37–62

Pastor J, Dewey B, Naiman RJ, McInnes PF, Cohen Y (1993) Moose browsing and soil fertility in the boreal forest of Isle Royale National Park. Ecology 74:467–480

Pastor J, Moen R, Cohen Y (1997) Spatial heterogeneities, carrying capacity, and feedbacks in animal–landscape interactions. J Mammal 78:1040–1052

Persson I-L, Danell K, Bergstrom R (2000) Disturbance by large herbivores in boreal forests with special reference to moose. Ann Zool Fenn 37:251–263

Peterson GD, Allen CR, Holling CS (1999) Ecological resilience, biodiversity and scale. Ecosystems 1:6–18

Pétillon J, Ysnel F, Canard A, Lefeuvre JC (2005) Impact of an invasive plant (*Elymus athericus*) on the conservation value of tidal salt marshes in western France and implications for management: responses of spider populations. Biol Conserv 126:103–117

Petty SJ, Avery MJ (1990) Forest bird communities. Occasional paper 26. Forestry Commission, Edinburgh

Pielou EC (1975) Ecological diversity. Wiley, New York

Povilitis A (1998) Characteristics and conservation of a fragmented population of huemul *Hippocamelus bisulcus* in central Chile. Biol Conserv 86:97–104

Prins HHT (1998) Origins and development of grassland communities in northwestern Europe. In: Wallis DeVries MF, Bakker JP, Van Wieren SE (eds) Grazing and conservation management. Kluwer, Boston, pp 55–106

Prins HHT, Olff H (1998) Species-richness of African grazer assemblages: towards a functional explanation. In: Newberry DM, Prins HHT, Brown N (eds) Dynamics of tropical communities. Blackwell, Oxford, pp 449–490

Putman RJ (1986) Competition and coexistence in a multispecies grazing system. Acta Theriol 31:271–291

Putman RJ, Edwards PJ, Mann JCE, How RC, Hill SD (1989) Vegetational and faunal changes in an area of heavily grazed woodland following relief of grazing. Biol Conserv 47:13–32

Raffaele E, Veblen TT (2001) Effects of cattle grazing on early regeneration of matorral in northwest Patagonia, Argentina. Nat Area J 21:243–249

Ricotta C (2005) Through the jungle of biological diversity. Acta Biotheor 53:29–38

Rietkerk M, Van Boerlijst MC, Van Langevelde F et al (2002) Self-organization of vegetation in arid ecosystems. Am Nat 160:524–530

Rosenthal J, Kotanen PM (1994) Terrestrial plant tolerance to herbivory. Trends Ecol Evol 9:117–157

Sarkar S, Margules C 2002 Operationalizing biodiversity for conservation planning. J Bioscience 27(S2):299–308

Shannon CE, Weaver W (1949) The mathematical theory of communication. Univ of Illinois Press, Urbana, IL

Sharaf KE, Price MV (2004) Does pollination limit tolerance to browsing in *Ipomopsis aggregata*? Oecologia 138:396–404

Simpson EH (1949) Measurement of diversity. Nature 163:688

Smit R, Bokdam J, den Ouden J, Olff H, Schot-Opschoor H, Schrijvers M (2001) Effects of introduction and exclusion of large herbivores on small rodent communities. Plant Ecol 155:119–127

Söderström B, Pärt T, Linnarsson E (2001) Grazing effects on between-year variation of farmland bird communities. Ecol Appl 11:1141–1150

Steen H, Mysterud A, Austrheim G (2005) Sheep grazing and rodent populations: evidence of negative interactions from a landscape scale experiment. Oecologia 143:357–364

Steets JA, Ashman TL (2004) Herbivory alters the expression of a mixed-mating system. Am J Bot 91:1046–1051

Sternberg M, Gutman M, Perevolotsky A, Ungar ED, Kigrl J (2000) Vegetation response to grazing management in a Mediterranean herbaceous community: a functional group approach. J Appl Ecol 37:224–237

Stewart AJA (2001) The impact of deer on lowland woodland invertebrates: a review of the evidence and priorities for future research. Forestry 74:259–70

Stock M, Hofeditz F (2000) Der Einfluss des Salzwiesen-Managements auf die Nutzung des Habitates durch Nonnen- und Ringelgänse. In: Stock M, Kiehl K (eds) Die Salzwiesen der Hamburger Hallig. Landesamt Nationalpark Schleswig-Holsteinisches Wattenmeer, Tönning, DE, pp 43–55

Stowe KA, Marquis RJ, Hochwender CG, Simms EL (2000) The evolutionary ecology of tolerance to consumer damage. Annu Rev Ecol Syst 31:565–595

Strauss SY, Agrawal AA (1999) The ecology and evolution of plant tolerance to herbivory. Trends Ecol Evol 14:179–185

Sudgen EA (1985) Pollinators of *Astragalus monoensis* Berneby (Fabaceae): new host records; potential impact of sheep grazing. Great Basin Nat 45:299–312

Suominen O, Danell, K, Bergström R (1999) Moose, trees, and ground-living invertebrates: indirect interactions in Swedish pine forest. Oikos 84:215–226

Swengel AB, Swengel SR (2001) Effects of prairie and barrens management on butterfly faunal composition. Biodivers Conserv 10:1757–1785

Tallis JH (1991) Plant community history. Long-term changes in plant distribution and diversity. Chapman and Hall, London

Tiffney BH (2004) Vertebrate dispersal of seed plants through time. Annu Rev Ecol Evol S 35:1–29

Tubbs CR, Tubbs JM (1985) Buzzards, *Buteo buteo*, and land use in the New Forest, Hampshire, England. Biol Conserv 31:41–65

Van der Graaf AJ, Bos D, Loonen MJJE, Engelmoer M, Drent RH (2002) Short-term and long-term facilitation of goose grazing by livestock in the Dutch Wadden Sea area. J Coastal Conserv 8:179–188

Van der Wal R, Van Wijnen H, Van Wieren SE, Beucher O, Bos D (2000) On facilitation between herbivores: how Brent Geese profit from Brown Hares. Ecology 81:969–980

Vane-Wright RI, Humphries CJ, Williams PM (1991) What to protect: systematics and the agony of choice. Biol Conserv 55:235–254

Van Wieren SE 1998 Effects of large herbivores upon the animal community. In: Wallis DeVries MF, Bakker JP, Van Wieren SE (eds) Grazing and conservation management. Kluwer, Boston, pp 185–214

Vazquez DP (2002) Multiple effects of introduced mammalian herbivores in a temperate forest. Biol Invasions 4:175–191

Vazquez DP, Simberloff D (2004) Indirect effects of an introduced ungulate on pollination and plant reproduction. Ecol Monogr 74:281–308

Veblen TT, Mermoz M, Martýn C, Ramilo E (1989) Effects of exotic deer on forest regeneration and composition in northern Patagonia. J Appl Ecol 26:711–724

Vera FWM (2000) Grazing ecology and forest history. CABI International, Wallingford, UK

Verdú JR, Crespo MB, Galante E (2000) Conservation strategy of a nature reserve in Mediterranean ecosystems: the effects of protection from grazing on biodiversity. Biodivers Conserv 9:1707–1721

Verweij RJT, Verrelst J, Loth PE, Heitkönig IMA, Brunsting AMH (2006) Grazing lawns contribute to the subsistence of mesoherbivores on dystrophic savannas. Oikos 114:108–116

Vickery PD, Herkert JR (2001) Recent advances in grassland bird research: where do we go from here? Auk 118:11–15

Weber GE, Jeltsch F, Van Rooyen N, Milton SJ (1998) Simulated long-term vegetation response to grazing heterogeneity in semi-arid rangelands. J Appl Ecol 35:687–699

Westoby M, Walker B, Noy-Meir I (1989) Opportunistic management for rangelands not at equilibrium. J Range Manage 42:266–274

Whittaker RH (1972) Evolution and measurement of species diversity. Taxon 21:213–251

Wilsey BJ, Coleman JS, McNaughton SJ (1997) Effects of elevated CO_2 and defoliation on grasses: a comparative ecosystem approach. Ecol Appl 7:844–853

Wilson AD, Macleod ND (1991) Overgrazing—present or absent. J Range Manage 44:475–482

Wilson GWT, Hartnett DC (2001) Effects of ungulate grazers on arbuscular mycorrhizal fungal communities in tallgrass prairie.Annu Rev Ecol Evol S 35:435–466

Young TP, Stanton ML, Christian CE (2003) Effects of natural and simulated herbivory on spine lengths of *Acacia drepanolobium* in Kenya. Oikos 101:171–179

Zalba SM, Cozzani NC 2004 The impact of feral horses on grassland bird communities in Argentina. Anim Conserv 7:35–44

Chapter 11
Managing Large Herbivores in Theory and Practice: Is the Game the Same for Browsing and Grazing Species?

Jean-Michel Gaillard, Patrick Duncan, Sip E. Van Wieren, Anne Loison, François Klein, and Daniel Maillard

11.1 Introduction

Management of large wild herbivores sometimes includes action on the availability of resources and/or on the abundance of predators, but the management of most populations of these animals is limited to the choice of hunting quotas (Williams et al. 2002). The dynamics of the target populations, in interaction with their habitats, is therefore a central process in management. The dynamics of populations is driven by their demographic parameters, i.e., the rates of reproduction and survival. Demographic rates vary according to the sex and age of the individuals, and also show different degrees of temporal and spatial variation among populations (Gaillard et al. 2000; Coulson et al. 2001). In this chapter we first address the question: do browsing and grazing large herbivores have different demographic rates driving the dynamics of their populations?

Management of most populations of large herbivores being, as stated, based on the choice of quotas for offtake by hunting, a reliable assessment of trends in population size is required. Total counts of individuals in the whole population area or sample counts of plots/transects have been the most popular way for assessing population sizes (e.g., Caughley 1977; Seber 1982; for deer see Cederlund et al. 1998 and for mountain ungulates Loison et al. 2006). In the context of game species such as these, hunting-related methods such as drives (McCullough 1979), observations from vantage points (Ratcliffe 1987), and number of animals seen during hunts (Ericsson and Wallin 1999) are often used. However, such counts are often imprecise, and biased (Williams et al. 2002). Caughley (1977) pointed out that it is difficult to achieve a coefficient of variation of less than 20% for counts of animals. The roe deer (*Capreolus capreolus*), the most abundant deer species in Europe (Andersen et al. 1998), is particularly difficult to count and underestimates of density can be as high as 50% (Cederlund et al. 1998). Since grazers and browsers live in different habitats (Jarman 1974), they pose different problems in estimation of trends in abundance, due in particular to differences in visibility between habitats and to differences in variance because of contrasting group sizes.

Another problem in the management of large herbivores is that the response of populations to a given offtake varies according to the relative importance of density-dependent and density-independent processes acting on the population (Caughley 1977). Populations of animals in good condition (because resources are abundant) respond differently from populations of animals in poor condition to the same offtake (Caughley 1977). Monitoring populations of large herbivores should, therefore, be rooted in the concept of density dependence, the functional dependence of a demographic rate on changes in population size (Williams et al. 2002). This provides a measure of the population–environment relationships, as illustrated in Fig. 11.1. Density dependence is a commonly observed process in populations of large herbivores (Saether 1997; Gaillard et al. 2000). It is widely accepted that populations of large herbivores vary in numbers according to a generalized logistic model (also called theta logistic model; Lande et al. 2003) that includes non-linear density-dependent changes in population growth (Gilpin and Ayala 1973; Fowler 1981). At very low density, the population shows a 'colonising' status and the growth rate is close to maximal (r-max sensu Caughley 1977). The population's productivity is, however, quite low because of the small size of the population. As the size of the population increases, small decreases in individual performance lead the growth rate of the population to decrease slowly, while the population productivity increases. When the population size reaches 50–90% (depending on the generation time, Fowler 1988) of the carrying capacity (defined here as the population size at which the population no longer increases), the population productivity peaks (MSY—maximum sustainable yield—in Caughley 1977). After this point, the

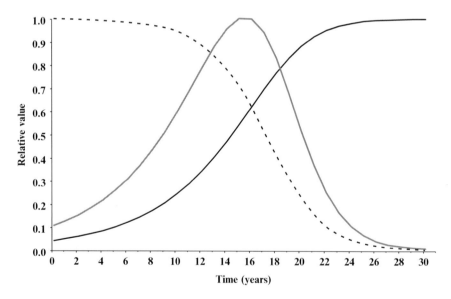

Fig. 11.1 Growth, individual growth and size of a theoretical population according to a theta logistic model. The *black full line* corresponds to changes of the population size over time, the *grey full line* corresponds to the variation of the population productivity according to the population size, and the *dotted line* corresponds to the variation of the individual performance according to the population size

performance of the individuals decreases markedly with declining body condition at high population density, and the productivity of the population declines, too. The demographic status of the population is then close to saturation. The pattern described here can happen at very different absolute densities in different areas because population density relative to habitat quality, rather than to absolute density, influences life-history traits (Van Horne 1983).

Populations of large herbivores have increased tremendously in the temperate countries (Gill 1990) and high-density populations are now widespread (McShea et al. 1997; Warren 1997). Their management, therefore, requires knowledge of the relationships between the animals and their resources as well as an assessment of the size of the population.

We show that a set of ecological indicators of the temporal changes in abundance and of the relationships between the population and the habitat can be a reliable basis for monitoring. This approach has been developed independently in France for the roe deer, and for other large herbivores in other countries (du Toit 2002). We address the issue of whether the same methods of monitoring and management are appropriate for grazers and browsers in Sects. 11.3 Monitoring, and 11.4 Management.

11.2 The Dynamics of Grazer and Browser Populations

Several recent papers have described differences in biological traits among feeding types in large herbivores. From these, Gordon (2003) concluded there is little substantive evidence for differences in morphology and physiology between feeding types once body mass has been accounted for. However, most previous comparative studies of grazers and browsers have dealt with morphology and physiology, and very few have focussed on life-history traits (but see Saether and Gordon 1994 for a notable exception). According to the life history theory (Stearns 1992), life history tactics of vertebrates range along a fast-slow continuum involving the association between relatively large reproductive output, short lifespan, and early age at first reproduction (see Stearns 1983 on mammals, Gaillard et al. 1989 on mammals and birds, Shine and Charnov 1992 on reptiles, Rochet et al. 2000 on fishes). We aim to test whether grazers and browsers have different positions on this fast-slow continuum using published demographic data. We focussed on the reproductive traits (age at first reproduction and litter size), adult survival, and on generation time, which latter provides a direct measure of the speediness of life history (Gaillard et al. 2005). We did not include juvenile survival because juvenile survival varies greatly among both years within populations and populations within a given species (see Gaillard et al. 1998a, 2000 for reviews), so that a large number of possible confounding effects would need to be accounted for to get a reliable comparison among the survival of juveniles of the different feeding types. Although the influence of phylogeny has been discussed in the context of comparisons between grazers and browsers (see, e.g., Perez-Barberia and Gordon 2000 versus

Mysterud 1998), we did not analyse phylogenetic influences. The usefulness of correcting for phylogenetic inertia has been challenged (e.g., Price 1997; Ricklefs and Starck 1996), leading many authors to perform both a standard analysis using the raw data and a comparative analysis accounting for phylogenetic inertia (independent contrasts in most cases; see Fisher and Owens 2000; Nagy and Bradshaw 2000; Perez-Barberia and Gordon 2001 for examples). As the correction for phylogenetic inertia increases the p-values of the statistical tests, but does not affect the parameter estimates, we will consider the size of the effects more than the significance of the statistical tests. To test for life-history differences among feeding types, we split the species of large herbivores into three categories of feeding types: browsers with a diet including less than 10% grass, intermediate feeders with a diet including between 10 and 90% grass, and grazers with a diet of more than 90% of grass; in agreement with most recent analyses focussing on feeding types (e.g., Van Wieren 1996; Perez-Barberia and Gordon 2000). We used available data bases (restricted to ruminant species) to assess the influence of feeding types on life history traits. To account for the marked allometric relationships that occur in most life history traits (Peters 1983; Calder 1984; Brown and West 2000), we performed an ANCOVA with the log-transformed life history trait as the dependent variable, the species-specific body mass (after log-transformation) as the covariate, and the feeding type as the factor. We ran five models in each case: (1) the model including an interaction term between feeding types and body mass that corresponds to different allometric relationships among feeding types; (2) the model including an additive effect between feeding types and body mass that corresponds to a similar allometric relationship, but with differences in the life history trait at a given mass among feeding types; (3) the model including the allometric relationship only (i.e., no influence of feeding types); (4) the model including differences among feeding types only (i.e., no influence of body mass); and (5) the model including a constant value of the life history trait (i.e., no influence of feeding types or body mass). According to the recommendations of Burnham and Anderson (1998), we used the model with the lowest Akaiké Information Criterion (AIC) as the best model (best compromise between precision and accuracy).

Differences in reproductive traits among feeding types. We used the species-specific data available in Van Wieren (1996) to assess the influence of feeding types on age at first breeding (first recorded matings, N=140 species) and on fecundity (N=141), principally bovids and cervids from a wide variety of habitats in Africa, Eurasia, and the Americas. For age at first breeding, the best model included additive effects of body mass and feeding types (Table 11.1). As expected, the age at first breeding was allometrically related to body mass, although the slope was much less than the expected value of 0.25 for biological time (slope 0.147, SE=0.019). For a given body mass browsers bred for the first time earlier than intermediate feeders (difference 0.269, SE=0.058) and grazers (difference 0.349, SE=0.066). The additive effects of body mass and feeding types accounted for 61.8% of the variability observed in age at first breeding. For the fecundity (measured as the number of offspring produced per year), the same model was retained (Table 11.1). As expected, fecundity decreased allometrically with increasing body mass

11 Managing Large Herbivores in Theory and Practice

Table 11.1 Model selection for the analysis of the variation in some life history traits. The table reports the AIC value of each model fitted (1: Interactive effects between feeding type and body mass, 2: additive effects of feeding type and body mass, 3: effect of body mass (allometric effect), 4: effect of feeding type, 5: constant value for the life history trait). The selected model (lowest AIC value) occurs in *italics*

Model Life history trait	1	2	3	4	5
Age at first breeding	30.61	*27.63*	53.99	74.79	156.41
Fecundity	90.43	*86.57*	88.86	111.73	146.45
Adult survival	49.58	45.84	44.27	43.86	*42.27*
Generation time	11.28	*7.98*	8.33	23.36	20.18

although the slope differed from −0.25, the expected allometric exponent for frequency (slope −0.130, SE=0.024). For a given body mass browsers tended to have higher fecundity than intermediate feeders (difference 0.095, SE=0.072) and had higher fecundity than grazers (difference 0.202, SE=0.081). The additive effects of body mass and feeding types accounted for 37.5% of the variability observed in annual fecundity.

Differences in adult survival among feeding types. We used the species-specific data available in Gaillard et al. (2000) to assess the influence of feeding types on adult survival (N=24 species). The model with a constant adult survival (2.42, SE=0.11 on a logit scale corresponding to a survival rate of 0.92) was selected (Table 11.1). The additive effects of body mass and feeding types only accounted for 9.63% of the variability observed in adult survival.

Differences in generation time among feeding types. We used the data in Sinclair (1996) and Gaillard et al. (2000) to assess the influence of feeding types on generation time (N=26 species). The best model included additive effects of body mass and feeding types (Table 11.1). As expected, the generation time was allometrically related to body mass, with a slope very close to the expected value of 0.25 for biological times (slope 0.227, SE=0.050). For a given body mass browsers had shorter generation time than intermediate feeders (difference 0.251, SE=0.132) and grazers (difference 0.223, SE=0.134). The additive effects of body mass and feeding types accounted for 50.4% of the variability observed in generation time (Fig. 11.2).

These analyses show that browsers have faster life histories for a given body mass than intermediate feeders and grazers. A faster life history can involve an earlier age at first breeding, a higher fecundity rate, and/or a shorter lifespan: we show that browsers have faster life histories than grazers and intermediate feeders because they breed earlier and produce more offspring. Thus, a 100 kg browser breeds for the first time at 1.15 years of age and produces an average of 1.37 offspring/year, whereas a grazer of the same mass breeds for the first time at 1.62 years of age and produce an average of 1.12 offspring/year (see Table 11.2). As expected from scaling theory (Brown and West 2000), the reproductive output of large herbivores decreased with increasing body mass at similar rates in all feeding types (Table 11.2). Adult survival of large herbivores, on the other hand, was quite constant at 0.92 and did not depend either on body mass or on feeding type. Such a constancy of adult survival fits with previous analyses of large herbivores that

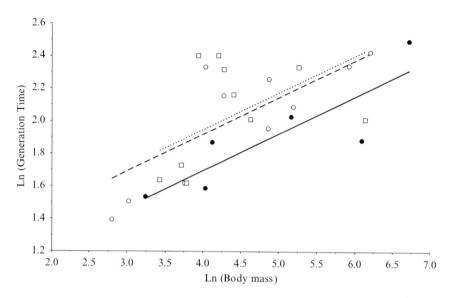

Fig. 11.2 Allometric relationship between generation time and body mass in ruminants in relation to the feeding type. The *lines* (*full line* for browsers, *dashed line* for intermediate feeders, and *dotted line* for grazers) that correspond to the selected model (i.e., additive effects of body mass and feeding types) and the data points (*filled circles* for browsers, *open squares* for intermediate feeders, and *open circles* for grazers) are shown

Table 11.2 Expected age at first breeding (AFB, in years) and annual fecundity (Fec, in number of offspring born) for large herbivores of different body mass and belonging to different feeding types. The estimates correspond to the models selected in each analysis (see text for further details)

	Browsers		Intermediate feeders		Grazers	
	AFB	Fec	AFB	Fec	AFB	Fec
25 kg	0.934	1.636	1.222	1.488	1.324	1.337
50 kg	1.034	1.495	1.353	1.360	1.466	1.222
100 kg	1.145	1.367	1.498	1.243	1.623	1.117
500 kg	1.451	1.109	1.898	1.008	2.056	0.906

reported high survival rates resilient to environmental variations (Gaillard et al. 2000; Gaillard and Yoccoz 2003). The shorter generation time of browsers compared to intermediate feeders and grazers is thus due only to a higher reproductive output. While this shorter generation time we report for browsers increases the relative influence of reproductive traits on population growth rate (Gaillard et al. 2005), our analyses demonstrate the existence of a single general demographic pattern common to all large herbivores. Indeed, a 2 kg browser is the fastest large herbivore, in terms of its demography, that can exist. For such an animal, the expected generation time estimated from our selected model is 2.57 years. According to the direct mathematical link that relates elasticities of demographic parameters (i.e., the relative potential impact of demographic parameters on population growth) and generation time (see Lebreton and Clobert 1991, Gaillard et al. 2005), any population with a generation time longer than 2 years will be more sensitive to a given change in adult survival than to the same change in recruitment. This means that the population growth of all large herbivores, independently of size and feeding type, must be more sensitive to a given change in adult survival than to the same change in a reproductive rate.

The physiological mechanisms behind this difference in generation time remain to be elucidated: the meta-analyses that have so far been conducted do not prove that the digestive ability of browsers is greater than grazers and intermediate feeders, and it is not established that selective feeding by browsers on plants with high levels of cell contents provides them with richer diets (higher protein contents and/or lower cell-wall constituents; Duncan, A. and Poppi, this volume). However, the quality of the data is not good enough to rule out these possible mechanisms.

The life-history differences between browsers and other feeding categories we report here are likely to have general consequences for the status of populations of large herbivores in the future, in the context of global change. The marked increases in CO_2 concentrations (IPCC 2001) may favour woody plants within savannas (Bond and Midgley 2000), so browsers may increase relative to large herbivores with other feeding types in this habitat (see Janis et al. 2000 and Gordon and Prins this volume). However, climate change will not affect the CO_2 concentrations alone, and may decrease the area of savannas, so that the loss of favourable habitat may overcompensate the benefit of increased CO_2 concentrations. On the other hand, the increase of woodland areas in western Europe should benefit browsers. The success of roe deer (Andersen et al. 1998) is consistent with this view. With a faster turnover, browsers should be more resilient to overhunting than grazers and intermediate feeders. A comparative analysis of hunting bags over a large geographical scale in relation to feeding types would be required to test for such a prediction. Lastly, the faster turnover of browsers as compared to large herbivores of other feeding types may confer browsers faster speciation rates. Vrba (1987) has identified some links between ecological features and speciation rates that could indicate that such differences between browsers and grazers could occur; this prediction, however, requires testing explicitly.

11.3 Monitoring

The relationships between the animals and their habitats should be monitored on the basis of 'indicators of ecological change' (sensu Cederlund et al. 1998). As defined by Waller and Alverson (1997), these are 'efficient and reliable indicators capable of serving as "early warnings signs" of impending ecological change' (see also Cairns et al. 1993). Such indicators include measures of trends in abundance of the condition of the animals and of the quality of their habitats. Dale and Beyeler (2001) have reviewed all criteria that an ecological indicator should meet: 'be easily measured, be sensitive to stresses on the system, respond to stress in a predictable manner, be anticipatory, predict changes that can be averted by management actions, be integrative, have a known response to disturbances, anthropogenic stresses, and changes over time, and have low variability in response'.

As no single indicator fulfils all these criteria, a manageably small set of indicators needs to be derived that, together, meet these criteria (Dale and Beyeler, 2001). Moreover, because we cannot possibly monitor every component of an ecological system, we are constrained to identify a small number of indicators that characterise the relevant parts of the system.

Monitoring the abundance of grazers and browsers. The concepts outlined above are equally valid for grazers and browsers. However, large herbivores of different feeding types live in different habitats (Hofmann 1989; Fritz and Loison 2006), where visibility is different. Grazers typically live in open areas that have less variable, and overall greater, visibility than the often quite closed areas such as forests where browsers live. This means that undercounting might be less of a problem, and bias error might be lower in counts of grazers as compared to browsers. However, this does not mean that total counts of grazers are reliable. For instance, both the number of ibex *Capra ibex* (Gaillard et al. 2003) and chamois *Rupicapra rupicapra* (Loison et al. 2006) counted in open mountainous areas, and the number of African ungulates counted from the air in savanna (Jachmann 2002) have been reported to be severely underestimated. However, the rate of underestimation when counting grazers should be less than that when counting browsers. For instance, an underestimation by a factor of 3 or more has been commonly reported for roe deer (Andersen 1953).

As a corollary of differences in habitat openness, group sizes also differ among feeding types (Jarman 1974), with browsers living in smaller groups than grazers (Fritz and Loison 2006). As the variance of abundance estimate increases with increasing group size (Burnham et al. 1980), the precision of counts, all other things being equal, should be better for browsers than for grazers.

As visibility and group size differ markedly among feeding types and strongly influence the sampling design of population counts, the methods used to monitor trends in the abundance of browsers and grazers should be different.

For browsers like roe deer, the Kilometric Index (measured as the number of animals observed per km of foot transect; Vincent et al. 1991) is closely correlated with true changes in the abundance of deer (as assessed by using Capture-Mark-Recapture methods; see Gaillard et al. 1993) in two forests where the exploitation of trees for timber caused considerable changes in visibility over 20 years. The passage of hurricane Lothar in late 1999 led to the breaking or uprooting of a third of the trees in these forests: there may have been an effect of lower visibility on the index in the three years after the hurricane, but this did not cause any bias in the long term. Visibility that varies spatially is not a major problem, provided that the mean value remains reasonably constant over time. This approach is now widely used for browsers (Maillard et al. 1999). If measures of absolute densities are required, 'distance' methods are insensitive to changes in visibility of the animals (Burnham et al. 1980), as the distances at which they are sighted are taken into account.

For grazers, total or sample counts (most often aerial counts; Jachmann 2001) are usually performed. Although they can provide data for detailed demographic analyses (Owen-Smith and Mason 2005; Owen-Smith and Mills 2006), the accuracy of these procedures has not yet been assessed. However, available empirical evidence indicates that size (larger is better) and colour (high contrast between coat colour and ground colour is better) influence the reliability of aerial counts of large herbivores (Redfern et al. 2002; Jachmann 2002).

Monitoring the condition of grazers and browsers. The phenotypic quality of individuals varies strongly with population density (Fowler 1987), and hence offers a potential tool to monitor the population status of large herbivores. For instance, the body mass of fawns in winter (Vincent et al. 1995; Gaillard et al. 1996), the cohort jaw length (Hewison et al. 1996), and the hind foot length of fawns (Toïgo et al. 2006; Zanneïse et al. 2007) have all been reported to decrease markedly with increasing population density of roe deer and are now used as indicators of population changes in a management context (e.g., Blant and Gaillard 2004 in Switzerland). The winter body mass of fawns is widely used as a short-term measure, since this is related to events (such as rainfall) in the previous summer, and the amount of acorn mast in autumn (Kjellander et al. 2006). The length of the hind foot of fawns integrates events over a longer time-scale, and is very convenient to measure. For longer time horizons the jaw length is a useful index of the quality of individuals (Hewison et al. 1996). Resource shortage in early life leads to permanent effects on the size of many mammals, including red deer *Cervus elaphus* (Mysterud et al. 2002) and roe deer (Pettorelli et al. 2002). This causes variation in performance among year classes and thus to delayed density-dependent effects on the dynamics of their populations ('cohort variation'; Albon et al. 1987).

The reproductive performance of individuals in the population can be monitored by the number of offspring per female (Vincent et al. 1995) or as the number of fawns per reproducing female (i.e., females with fawns at heel; Boutin et al. 1987). However, for polytocous species such as roe deer, the pattern of variation of reproductive performance as assessed by sampling females can be blurred by differential family effects in relation to cohort quality, since in poor cohorts fawns born in the same family commonly have similar fates (i.e., both die or both survive in most cases), whereas there is no family effect during good years (Gaillard et al. 1998b). Further work is therefore required to assess the suitability of such indicators for management.

The abundance of resources, and the level of their use, can be measured easily on woody plants (Aldous 1944). However, there are strong biases due to differences among observers in their estimations of plant cover (Morellet 1998). The browsing index has, therefore, been developed (Morellet et al. 2001); this is calculated for each plant species using sample plots of $1\,m^2$ as the proportion of plots in which that species was browsed. Data on the presence/absence of plant species and of signs of browsing on them in quadrants of $1\,m^2$ have been found to be much more repeatable, and therefore better indices of the abundance and use of resources (Morellet et al. 2001; Morellet et al. 2003).

11.4 Management

For the management of populations of large herbivores to succeed, it needs to be based on (Nichols et al. 1995; Cederlund et al. 1998):

- an assessment of the current state of the animal–habitat relationships based on monitoring ecological changes
- an analysis of the range of possible alternatives to population counts, based on the system's current state, on ecological principles (such as density dependence; see Fig. 11.1), and on the benefits and costs of the possible states to human society
- a choice of management actions by the different stake-holders based on a vision of the optimal future state of the animal-habitat relationships, taking into account uncertainty as to the consequences of management actions in complex systems

Such management is 'adaptive'; its essence is the use of a wide range of indices of animal abundance and condition, and ideally of the abundance of resources and the impact of the animals on resources and on human activities. Managers learn about systems as they manage them (Lancia et al. 1996). Adaptive management (Walters 1986) deals with scientific uncertainty by incorporating a set of models representing the competing hypotheses about system responses to management (Runge and Johnson 2002). With adaptive management the information on the system response to management is gathered continuously so that the information is used to improve biological understanding and to inform future decision-making (Nichols et al. 1995; Shea et al. 1998; Williams et al. 2002). Variations on the theme of adaptive management are used in many wildlife management areas (du Toit 2002).

The two essential decisions in the management of populations of large herbivores concern (1) the objective and (2) the quota. The objective is often decided simply on the basis of the objectives of hunters, but should be based on a wide range of considerations of the benefits and costs of high/low densities of large herbivores for society as a whole (taking into account ethical, social and economic dimensions) as proposed in Cederlund et al. (1998).

Having decided on the objective, which could be to increase the population, decrease it, or maintain the current size, the quota needs to be decided. Given that the approach based on indicators of ecological changes proposed above do not provide information on population sizes, the quota will initially have to be chosen on the basis of expert opinion, and it will be necessary to take into account the size of the target species and its feeding type. The potential population productivity of small and large grazers and browsers are given in Fig. 11.3; intermediate feeders will have intermediate productivities. As expected, it appears that the population productivity is higher for a given relative population size of small than of large species and of browsers than of grazers. Small browsers are thus expected to be much less sensitive to hunting and/or poaching impact than large grazers.

The quota should then be applied for three years or more, without change, in order to get estimates of the ecological indicators for a given hunting pressure, so that the success of the management can be judged from the monitoring. In the initial

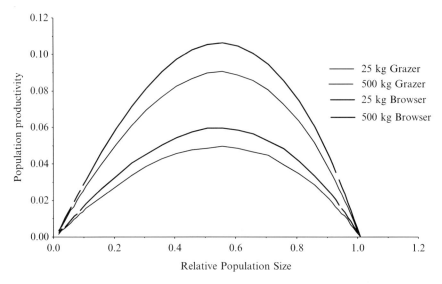

Fig. 11.3 Population productivity (measured as the population increase over two consecutive years) as a function of population size for small (25 kg) and large (500 kg) browser and grazer. To get estimates of population productivities, we used a Leslie demographic model parameterized with the demographic parameters expected for 25 and 500 kg browsers and grazers (as derived from the allometric relationships presented in the text). We thus obtained estimates of population growth (r=0.324 for a 25 kg browser, 0.275 for a 25 kg grazer, 0.275 for a 500 kg browser, and 0.181 for a 500 kg grazer). Then we used the Fowler's equation relating the relative population size at which the maximum sustainable yield occurs in a population (R), its population growth and its generation time (Fowler 1988). We obtained very similar R values among the four case studies (from 0.536 for a large grazer to 0,560 for a small browser). We thus used a constant R of 0.55. We then calculated the population productivity as a function of N, r, and R by using the theta logistic function (Lande et al. 2003)

stages, the management of populations of large herbivores cannot be more accurate than a trial-and-error process dictates, however, as managers gain knowledge of the population–environment system over the years, management will become more finely tuned. Such an approach is thus based on long-term monitoring: the longer the system is monitored, the better it is understood.

11.5 Conclusions

The main source of variation in the life history of large herbivores is the body size of the animal. However, the feeding type, i.e., grazing or browsing, also influences the turnover of the population, being faster for browsers and slower for grazers, but the relatively more important role of adult survival as compared to recruitment parameters on the population growth rate of large herbivores does not change. Feeding type strongly affects the use of habitat and the size of the groups the animals live in. This means that different methods need to be used for the monitoring and management of grazers and browsers. Management should be adaptive, based on the biology of the animals, and appropriate monitoring of ecological and socio-economic changes in the animal–habitat relationships.

References

Albon SD, Clutton-Brock TH, Guinness FE (1987) Early development and population dynamics in Red Deer. II. Density-independent effects and cohort variation. J Anim Ecol 56:69–81
Aldous SE (1944) A deer browse survey method. J Mammal 25:130–136
Andersen J (1953) Analysis of a Danish roe-deer population (*Capreolus capreolus* L) based upon the extermination of the total stock. Dan Rev Game Biol 2:127–155
Andersen R, Duncan P, Linnell JDC (1998) The European roe deer: the biology of success. Scandinavian Univ Press, Oslo
Blant M, Gaillard JM (2004) Use of biometric body variables as indicators of roe deer (*Capreolus capreolus*) population density changes. Game Wildl Sci 21:21–40
Bond WJ, Midgley GF (2000) A proposed CO_2-controlled mechanism of woody plant invasion in grasslands and savannas. Glob Change Biol 6:865–869
Boutin JM, Gaillard JM, Delorme D, Van Laere G (1987) Suivi de l'évolution de la fécondité chez le chevreuil (*Capreolus capreolus*) par l'observation des groupes familiaux. Gibier Faune Sauvage 4:255–265
Brown JH, West GB (2000) Scaling in biology. Oxford Univ Press, New York
Burnham KP, Anderson DR (1998) Model selection and inference: a practical information-theoretic approach. Springer, Berlin Heidelberg New York
Burnham KP, Anderson DR, Laake JL (1980) Estimation of density from line transect sampling of biological populations. Wildlife Monogr No 72
Cairns J, McCormick PV, Niederlehner BR (1993) A proposed framework for developing indicators of ecosystem health. Hydrobiologia 236:1–44
Calder WA (1984) Size, function and life history. Harvard Univ Press, Cambridge, MA
Caughley G (1977) Analysis of vertebrate populations. Wiley, Chichester, NY
Cederlund G, Bergqvist J, Kjellander P, Gill R, Gaillard J-M, Boisaubert B, Ballon P, Duncan P (1998) Managing roe deer and their impact on the environment: maximising the net benefits to society. In: Andersen R, Duncan P, Linnell JDC (eds) The European roe deer: the biology of success. Scandinavian Univ Press, Oslo, pp 337–372
Coulson T, Catchpole EA, Albon SD, Morgan BJT, Pemberton JM, Clutton-Brock TH, Crawley MJ, Grenfell BT (2001) Age, sex, density, winter weather, and population crashes in Soay sheep. Science 292:1528–1531
Dale VH, Beyeler SC (2001) Challenges in the development and use of ecological indicators. Ecol Indic 1:3–10
Duncan AJ, Poppi DP, this volume
du Toit JT (2002) Wildlife harvesting guidelines for community-based wildlife management: a southern African perspective. Biodivers Conserv 11:1403–1416
Ericsson G, Wallin K (1999) Hunter observations as an index of moose *Alces alces* population parameters. Wildlife Biol 5:177–185
Fisher DA, Owens IPF (2000) Female home range size and the evolution of social organization in macropod marsupials. J Anim Ecol 69:1083–1098
Fowler CW (1981) Density dependence as related to life history strategy. Ecology 62:602–610
Fowler CW (1987) A review of density-dependence in populations of large mammals. In: Genoways HH (ed) Current mammalogy. Plenum, New York, pp 401–441
Fowler CW (1988) Population dynamics as related to rate of increase per generation. Evol Ecol 2:197–204

Fritz H, Loison A (2006) Large herbivores across biomes. In: Danell K, Duncan P, Bergström R, Pastor J (eds) Large herbivore ecology, ecosystem dynamics and conservation. Cambridge Univ Press, Cambridge, pp 19–49

Gaillard J-M, Yoccoz NG (2003) Temporal variation in survival of mammals: a case of environmental canalization? Ecology 84:3294–3306

Gaillard J-M, Pontier D, Allainé D, Lebreton JD, Trouvilliez J, Clobert J (1989) An analysis of demographic tactics in birds and mammals. Oikos 56:59–76

Gaillard J-M, Delorme D, Boutin JM, Van Laere G, Boisaubert B, Pradel R (1993) Roe deer survival patterns: a comparative analysis of contrasting populations. J Anim Ecol 62:778–791

Gaillard J-M, Delorme D, Boutin JM, Van Laere G, Boisaubert B (1996) Body mass of roe deer fawns during winter in two contrasting populations. Journal of Wildlife Management 60:29–36

Gaillard J-M, Festa-Bianchet M, Yoccoz NG (1998a) Population dynamics of large herbivores: variable recruitment with constant adult survival. Trends Ecol Evol 13:58–63

Gaillard J-M, Andersen R, Delorme D, Linnell JDC (1998b) Family effects on growth and survival of juvenile roe deer. Ecology 79:2878–2889

Gaillard J-M, Festa-Bianchet M, Yoccoz NG, Loison A, Toïgo C (2000) Temporal variation in fitness components and population dynamics of large herbivores. Annu Rev Ecol Syst 31:367–393

Gaillard J-M, Loison A, Toïgo C (2003) Variation in life history traits and realistic population models for wildlife management. In: Festa-Bianchet M, Apollonio M (eds) Animal behavior and wildlife conservation. Island Press, Washington, DC, pp 115–132

Gaillard J-M, Yoccoz NG, Lebreton JD, Bonenfant C, Devillard S, Loison A, Pontier D, D Allainé (2005) Generation time: a reliable metric to measure life history variation among mammalian populations. Am Nat 166:119–123

Gill R (1990) Monitoring the status of European and North American cervids. GEMS information series global environment monitoring system. UN Environment Programme, Nairobi, Kenya

Gilpin ME, Ayala FJ (1973) Global models of growth and competition. Proc Nat Acad Sci USA, 70:3590–3593

Gordon IJ (2003) Browsing and grazing ruminants: are they different beasts? Forest Ecol Manage 181:13–21

Gordon IJ, Prins HHT, this volume

Hewison AJM, Vincent J-P, Bideau E, Angibault J-M, Putman RJ (1996) Variation in cohort mandible size as an index of roe deer (*Capreolus capreolus*) densities and population trends. J Zool 239:573–581

Hofmann RR (1989) Evolutionary steps of ecophysiological adaptation and diversification of ruminants: a comparative view of their digestive system. Oecologia 78:443–457

IPCC (2001) Working Group I third assessment report, climate change 2001: the scientific basis. http://www.gcrio.org/online.html

Jachmann H (2001) Estimating abundance of African wildlife: an aid to adaptive management. Springer, Berlin Heidelberg New York

Jachmann H (2002) Comparison of aerial counts with ground counts for large African herbivores. J Appl Ecol 39:841–852

Janis CM, Damuth J, Theodor JM (2000) Miocene ungulates and terrestrial primary productivity: where have all the browsers gone? Proc Nat Acad Sci USA 97:7899–7904

Jarman PJ (1974) The social organisation of antelope in relation to their ecology. Behaviour 48:215–265

Kjellander P, Gaillard J-M, Hewison AJM (2006) Density-dependent responses of fawn cohort body mass in two contrasting roe deer populations. Oecologia 146:521–530

Lancia RA, Braun CE, Collopy MW, Dueser RD, Kie JG, Martinka CJ, Nichols JD, Nudds TD, Porath WR, Tilghman NG (1996) ARM! For the future: adaptive resource management in the wildlife profession. Wildlife Soc B 24:436–442

Lande R, Engen S, Saether B-E (2003) Stochastic population dynamics in ecology and conservation. Oxford Univ Press, Oxford

Lebreton JD, Clobert J (1991) Bird population dynamics, management and conservation: the role of mathematical modelling. In: Perrins CM, Lebreton JD, Hirons GJM (eds) Bird population studies: their relevance to conservation and management. Oxford Univ Press, Oxford, pp 105–125

Loison A, Appolinaire J, Jullien J-M, Dubray D (2006) How reliable are total counts to detect trends in population size of chamois *Rupicapra rupicapra* and *R. pyrenaica*. Wildlife Biol 12:77–88

Maillard D, Gaultier P, Boisaubert B (1999) Revue de l'utilisation des différentes méthodes de suivi des populations de chevreuils en France. Le bulletin mensuel de l'office national de la chasse 244:30–37

McCullough DR (1979) The George Reserve deer herd: population ecology of a K-selected species. Univ of Michigan Press, Ann Arbor, MI

McShea WJ, Underwood HB, Rappole JH (1997) The science of overabundance: deer ecology and population management. Smithsonian Institution Press, Washington, DC

Morellet N (1998) Des outils biométriques appliqués aux suivis des populations animales: l'exemple des cervidés. Thesis, Univ of Lyon

Morellet N, Champely S, Gaillard J-M, Ballon P, Boscardin Y (2001) The browsing index: new tool uses browsing pressure to monitor deer populations. Wildlife Soc B 29:1243–1252

Morellet N, Ballon P, Boscardin Y, Champely S (2003) A new index to measure roe deer (*Capreolus capreolus*) browsing pressure on woody flora. Game Wildlife Sci 20:155–173

Mysterud A (1998) The relative roles of body size and feeding type on activity time of temperate ruminants. Oecologia 113:442–446

Mysterud A, Langvatn R, Yoccoz NG, Stenseth NC (2002) Large-scale habitat variability, delayed density effects and red deer populations in Norway. J Anim Ecol 71:569–580

Nagy KA, Bradshaw SD (2000) Scaling of energy and water fluxes in free-living arid-zone Australian marsupials. J Mammal 81:962–970

Nichols JD, Johnson FA, Williams BK (1995) Managing North American waterfowl in the face of uncertainty. Annu Rev Ecol Syst 26:177–199

Owen-Smith N, Mason DR (2005) Comparative changes in adult vs. juvenile survival affecting population trends of African ungulates. J Anim Ecol 74:762–773

Owen-Smith N, Mills MGL (2006) Manifold interactive influences on the population dynamics of a multispecies ungulate assemblage. Ecol Monogr 76:73–92

Perez-Barberia FJ, Gordon IJ (2000) Differences in body mass and oral morphology between the sexes in the Artiodactyla: evolutionary relationships with sexual segregation. Ecol Evol Res 2:667–684

Perez-Barberia FJ, Gordon IJ (2001) Relationships between oral morphology and feeding style in the Ungulata: a phylogenetically controlled evaluation. P Roy Soc Lond B Bio 268:1023–1032

Peters RH (1983) The ecological implication of body size. Cambridge Univ Press, Cambridge

Pettorelli N, Gaillard J-M, Van Laere G, Duncan P, Kjellander P, Liberg O, Delorme D, Maillard D (2002) Variations in adult body mass in roe deer: the effects of population density at birth and of habitat quality. P Roy Soc Lond B Bio 269:747–753

Price T (1997) Correlated evolution and independent contrasts. Philos T Roy Soc B 352:519–529

Ratcliffe PR (1987) The management of red deer in the commercial forests of Scotland related to population dynamics and habitat changes. Thesis, Univ of London

Redfern JV, Viljoen PC, Kruger JM, Getz WM (2002) Biases in estimating population size from an aerial census: a case study in the Kruger National Park, South Africa. S Afr J Sci 98:455–461

Ricklefs RE, Starck JM (1996) Applications of phylogenetically independent contrasts: a mixed progress report. Oikos 77:167–172

Rochet MJ, Cornillon PA, Sabatier R, Pontier D (2000) Comparative analysis of phylogenetic and fishing effects in life history patterns of teleost fishes. Oikos 91:255–270

Runge MC, Johnson FA (2002) The importance of functional form in optimal control solutions of problems in population dynamics. Ecology 83:1357–1371

Saether B-E (1997) Environmental stochasticity and population dynamics of large herbivores: a search for mechanisms. Trends Ecol Evol 12:143–149

Saether, B-E and I J Gordon 1994 The adaptive significance of reproductive strategies in ungulates. P Roy Soc Lond B Bio 256:263–268

Seber GAF (1982) The estimation of animal abundance and related parameters, 2nd edn. Griffin, London

Shea K, NCEAS working group (1998) Management of populations in conservations, harvesting and control. Trends Ecol Evol 13:371–375

Shine R, Charnov EL (1992) Patterns of survival, growth, and maturation in snakes and lizards. Am Nat 139:1257–1269

Sinclair ARE (1996) Mammal populations: fluctuation, regulation, life history theory and their implications for conservation. In: Floyd RB, Sheppard AW, De Barro PJ (eds) Frontiers of population ecology. CSIRO Publishing, Melbourne, pp 127–154

Stearns SC (1983) The influence of size and phylogeny on patterns of covariation among life history traits in the mammals. Oikos 41:173–187

Stearns SC (1992) The evolution of life histories. Oxford Univ Press, Oxford

Toïgo C, Gaillard J-M, Van Laere G, Hewison AJM, Morellet N (2006) How does environmental variation influence body mass, body size and body condition? Roe deer as a case study. Ecography 29:301–308

Van Horne B (1983) Density as a misleading indicator of habitat quality. J Wildlife Manage 47:893–901

Van Wieren SE (1996) Digestive strategies in ruminants and nonruminants. Thesis, Wageningen Agricultural Univ, Wageningen, Netherlands

Vincent J-P, Gaillard J-M, Bideau E (1991) Kilometric index as biological indicator for monitoring forest roe deer populations. Acta Theriol 36:315–328

Vincent J-P, Bideau E, Hewison AJM, Angibault JM (1995) The influence of increasing density on body weight, kid production, home range and winter grouping in roe deer (*Capreolus capreolus*). J Zool 236:371–382

Vrba ES (1987) Ecology in relation to speciation rates: some case histories of Miocene–recent mammal clades. Evol Ecol 1:283–300

Waller DM, Alverson WS (1997) The white-tailed deer: a keystone herbivore. Wildlife Soc B 25:217–226

Walters CJ (1986) Adaptive management of renewable resources. MacMillan, New York

Warren RJ (1997) The challenge of deer overabundance in the 21st century. Wildlife Soc B 25:213–214

Williams BK, Nichols JD, Conroy MJ (2002) Analysis and management of animal populations. Academic Press, San Diego

Zanneïse A, Baïsse A, Gaillard JM, Hewison AJM, Saint-Hilaire K, Toïgo C, Van Laere G, Morellet N (2007) Hind foot length: an indicator for monitoring roe deer populations at a landscape scale. Wildlife Soc B (In press)

Chapter 12
Grazers and Browsers in a Changing World: Conclusions

Iain J. Gordon and Herbert H.T. Prins

12.1 Introduction

The world on which we live is constantly changing, slowly through geological and evolutionary change and rapidly as the result of rainfall events or droughts, storms, hurricanes and cyclones, cold snaps or heat waves, and diurnal changes in light levels. Some of these changes have only local effects but others have global effects. In recent times the world has undergone a dramatic global change in form of the levels of CO_2 in the atmosphere, much of it derived from anthropogenic sources (King 2005). Currently, the level of CO_2, 380 ppm, has not been experienced by the biota on the planet for over 650,000 years (Siegenthaler et al. 2005). This increased CO_2 level is leading to changes in climate with consequences for rainfall and temperature patterns across the globe (IPCC 2001) with knock-on effects for vegetation community composition, and plant chemistry. For example, it is predicted that there will be a shift in the balance from grass to tree dominance in many savanna systems (Higgins et al. 2000; Bond et al. 2003). These changes in tree/grass ratios are happening rapidly (e.g., Lewis 2002) and will provide both opportunities and constraints both for humans in managing their domestic livestock species and also for wild herbivores. Much of this change is also brought about because of changes in land use over the last century with much land being taken out of agriculture, particularly in the temperate regions of the northern hemisphere (e.g., Ford 1971; Kjekshus 1977; Prins and Gordon, Chap. 1 this book). Whatever the exact causes or the interaction between them, an increase in the amount of browse in the landscape clearly will increase the amount of food available for browsers, whilst a reduction in the amount of grass, because of competition with woody species, will reduce the amount of food available for grazers. However, the system responses are unlikely to be that simple, because of such effects as non-linearity in the system and positive and negative feedbacks (e.g., Van de Koppel et al. 2002, van Langevelde *et al.* 2003). As a consequence systems responses will be difficult to predict at the scale at which they are managed and managers, therefore, need to be able to manage systems for unpredictable dynamics; that is, adopting risk averse, adaptive management principles (see Convention on Biological Diversity 2001). Scientists will have to work with managers to develop modelling frameworks that incorporate both hard and soft (expert opinion) knowledge. Scientists will also have to use what knowledge we currently have to make 'best bet' predictions as to what the responses of systems might be to changes in management, and not fall back on the 'we don't yet know enough' argument to continue to collect data before giving advice.

Putting our money where our mouth is, in this chapter we will outline the likely biophysical responses of vegetation systems to CO_2-induced climate change, in terms of the quantity and quality of plant material available to grazers and browsers. We then assess the ways in which browsers and grazer populations are likely to respond to these changes in food quantity and quality and the consequences for the large herbivore community structure. Finally we give some advice as to how browsers and grazers might be managed in changing landscapes with particular emphasis on the collaboration that has to be built between managers and scientists to support the development of adaptive systems levels management approaches.

12.2 Responses in Plant Species Composition

Increases in CO_2 allow shrubs and trees to grow more quickly, and increase their water use efficiency (Drake et al. 1997; Ehleringer et al. 1997), relative to grasses, leading to shrub and tree encroachment into areas that have long been grasslands (Bond et al. 2003). Elevated CO_2 levels may cause higher allocation of carbohydrates to the roots of woody species. This results in, for instance, larger stores of starch resulting in faster regrowth after the dormant season (Bond et al. 2003). Because of this faster growth, woody species have an increased chance to escape a fire trap (Higgins et al. 2000) where fire can kill saplings before their vulnerable growing points have reached a height above the flames. Once saplings and young trees or shrubs are outside this zone, they are not reset by every fire to a prostate post-fire form again, and can start expanding their crown, thus suppressing grass growth. This in turn leads to a reduced fire frequency and fire intensity, which feeds back into expanding woody cover.

Grasslands comprise approximately 30% of the global vegetation communities (Parton 1995), savannas about 20% (Sankaran et al. 2005), whilst forests cover 30% (FAO 2005). With changes in the circulating CO_2 levels in the atmosphere there will be changes in the balance of trees and grass in many terrestrial ecosystems leading to a reduction in the area of the globe covered by grassland and an increase in the area covered by forest. Two types of systems are most prone to change; the first are savannas because in these systems woody species are interspersed with grass at all possible scales, and transition from mixed grass–tree systems to tree-dominated ones can thus happen over large areas very fast (e.g., Stuart-Hill and Tainton 1999). Moreover, many savannas appear to have an unstable balance between woody species and grasses (Sankaran et al. 2005). It is noteworthy that contrary to the generally believed notion that forests are decreasing in the tropics, they are in fact spreading again in West and Central Africa. Two wet years in a row are often enough to allow forest margins to expand into an area that was under savanna before. Only an increased disposal of manpower can keep secondary forests at bay. The second system type prone to change, is man-made agricultural grassland that has been converted from original natural forest. Large parts of the eastern United States, nearly the whole of Europe and also Japan have been cleared of forest in the service of agriculture. De-intensification of agricultural practices, then, can lead to the rapid re-emergence of woody species and forest (see Chap. 1, this book).

It has been predicted that the C4 grasses will not respond as effectively as C3 grasses to changes in atmospheric CO_2 (Bowes 1993; Ainsworth and Long 2005; and see Wand et al. 1999). However, the extent to which C3 and C4 grass species will respond to changes in atmospheric CO_2 levels will depend upon the water availability. Under well-watered conditions, C3 plants show increased photosynthesis and growth. Well-watered C4 plants exhibited increased photosynthesis in response to increasing CO_2, but total mass and leaf area were unaffected (Ward et al. 1999). In response to drought, C3 plants drop a large amount of leaf area and maintain relatively high leaf water potential in the remaining leaves, whereas C4 plants retain greater leaf area, but at a lower leaf water potential, which suggests that C4 species may have an advantage over C3 species in response to increasing atmospheric CO_2 and more frequent and severe droughts.

12.3 Responses in Plant Chemistry

Plant metabolism is centred around carbon, oxygen, and water. Plants get their carbon and oxygen primarily from the air; thus the concentrations of carbon and oxygen in the atmosphere play a major part in the chemistry of plants, most particularly in the photosynthesis rates. Increased CO_2 levels potentially leads to increased levels of photosynthesis (called the CO_2 fertilisation effect) and improve their water-use efficiency (because plants can restrict stomata, reducing transpiration, while fixing the same amount of CO_2) but the extent to which this potential is realised depends on the effects of other limiting factors, such as water and nutrients, including N and P. Therefore, the response of plants to elevated CO_2 will depend on local climatic circumstances, e.g. the degree of aridity and soil nutrient status (Kimball 1983; Nowak et al. 2004).

Increased CO_2 in both grass and browse plants generally leads to increases in C/N ratios and sugar concentrations, and to decreases in nitrogen and phosphorus concentrations (Kinney et al. 1997; Coley et al. 2002; Goverde et al. 2004; Hattas et al. 2005). Thus, whilst nitrogen concentration is reduced, the concentration of soluble carbohydrates is increased. Other responses are less predictable, with some authors describing no changes in plant secondary compounds to increases in CO_2 (Goverde et al. 2004), whereas others find an increase in concentrations of certain secondary compounds (tannins and ellagitannins, Kinney et al. 1997; leaf phenols, Coley et al. 2002) and, depending on the species, even variation in response (Kinney et al. 1997; Coley et al. 2002).

Since concentrate selectors tend to be energy limited while bulk or roughage feeders are nitrogen limited (Sinclair 1977; Prins and Beekman 1989; Prins 1996), the changes in vegetation quality associated with elevated levels of atmospheric CO_2 are likely to lead to improved levels of nutrition for browsers but not for roughage grazers. All other things being equal, increased overall levels of browser populations are to be expected. If the ratio of nutrients/secondary compounds increases, a relative increase in the nutritional value of browse is likely, resulting in an increase in the nutritional ecology of browsers relative to grazers.

Within the class of roughage grazers there is likely to be a difference in response between ruminants and hind-gut fermenters; roughage hindgut-fermenting grazers (equids, white rhinoceros) are perhaps less severely nitrogen limited than ruminant bulk grazers (many bovids and some cervids). Hindgut-fermenters are often seen as a sort of evolutionary dead end; but as Janis pointed out in Chapter 2 of this book, this opinion is not supported by analysis of the fossil record. We believe that within the class of bulk or roughage grazers, elevated CO_2 levels may even tip the balance of herbivore community composition towards hindgut-fermenters and away from the large ruminant roughage grazers.

It does not appear there will be any great differences between browse and grass in their response to the increased CO_2, in terms of quality, and so we might expect similarities in the nutritional responses among browsers and populations of selective grazers. Yet, in areas where C3 and C4 grasses co-exist, we expect a shift towards vegetations dominated by C4 grasses. Such a shift counterbalances a trend towards increased soluble carbohydrates in the plants, making the vegetation as a whole less attractive to both selective grazers and roughage grazers.

12.4 Responses in Terms of Population Dynamics

With changes in vegetation community structure, that is, the evolving dominance of ecosystems by shrubs and trees, there may be a shift towards browsers dominating the majority of herbivore communities. In turn, there is a possibility that browsers may limit the degree to which shrubs and trees survive in systems, counteracting the CO_2-generated propensity for shrubs and trees to increase. Ultimately, the response will depend on the degree to which the large herbivore/plant ecosystems are food or predator limited. To date there is little evidence for large herbivores being predator limited (Mduma et al. 1999), even where they are confined to islands (Vucetich and Peterson 2004), and there is ample evidence for a relationship between vegetation biomass/productivity and herbivore population growth rates/density (Coe et al. 1976; Fritz and Duncan 1994). There is a great deal of evidence that herbivore population size is positively correlated with the quantity of vegetation biomass available (Coe et al. 1976; Fritz and Duncan 1994). There is also much evidence that herbivores are bottom-up controlled (Drent and Prins 1987).

However, whilst herbivore populations may respond positively to increases in vegetation productivity and nutritional quality, it is not clear whether this will result in top-down control of vegetation composition by herbivores (Bond 2005). As such, it is unlikely that any increase in herbivore population size will have an overall effect on vegetation composition other than on a local scale (Palmer et al. 2003). It is most likely, therefore, that vegetation composition will change towards one that is dominated by trees and shrubs and away from grasses. What effect is this likely to have on the structure of the large herbivore community?

This will depend on the degree to which herbivores are coupled to the dynamics of the vegetation. Where the coupling is strong (e.g., in temperate systems) the potential exists for herbivores to control vegetation structure and composition. Arid and semi-arid grazing systems, however, are prone to the effects of highly variable rainfall, with droughts causing frequent episodic mortality in herbivore populations. This has led to the suggestion that these are actually nonequilibrium or decoupled systems, in which animal impacts on plants are strongly attenuated or absent (Behnke and Scoones; 1992; Scoones 1994). Illius and O'Connor (1999) argued that even in arid and semi-arid systems animal numbers are regulated in a density-dependent manner by the limited forage available in key resource areas, which are utilized in the dry season, but this re-emphasizes the point that herbivores and vegetation are decoupled (not linked through density dependent effects) in the biggest part of the population's range.

12.5 Reponses in Herbivore Community Structure

So, worldwide we expect forests to expand in those areas where people do not actively suppress its regrowth. We thus predict that the wetter parts of the tropics (with more than 600 mm annual rainfall: Sankaran et al. 2005), and the wetter parts of the temperate zone (with more than about 400 mm/yr rainfall, which demarcates the natural steppe formation) will experience this tendency towards increased forest cover. Browsers will benefit from this trend; mixed feeders may benefit too, roughage grazers will not.

It appears easier to make predictions about trends in the types of food than about the chemical composition of the different food stuffs for the different classes of herbivores. We predict that for browsing concentrate selectors, food quality will increase. We also predict that in areas that are dominated by C4 grasses (steppes and tropical savannas), roughage grazers will not benefit from elevated CO_2 levels. Because they are nitrogen-limited; for this class of grazers the situation will become even less favourable. This will also apply to grazers in C3-grass-dominated vegetation (in montane tropical areas and mesic temperate areas). It might be that selective grazers, such as hares or oribi, will be favoured by increased CO_2 levels because of elevated soluble carbohydrates. However, in the contact zones where C4-grass-dominated vegetation borders on C3-grass-dominated vegetation, we predict an expansion of C4 grasses to the detriment of all grazing species that are adapted to using C3 grasses (see Table 12.1).

It makes sense to suggest that selective grazing species of the Ruminantia are energy-limited as are their browsing counterparts. Not many herbivores with a

Table 12.1 Predicted changes in the numerical abundance of classes of mammalian large herbivores in response to increased carbon dioxide levels. Two effects are foreseen, namely changes in food quality and changes in the balance between woody species and grasses. Browsers in grassland may appear a contradiction in terms, but browsers may be living on herbs or make use of small pockets of woody species in a generally grass-dominated biome

	Species typical for tropical open forests	Species typical for tropical grasslands	Transition from C4 to C3 grasslands	Species typical for temperate grasslands	Species typical for temperate forests
Roughage grazers	Rare, and will become rarer	Dominant, but will decrease	Rare (since the beginning of the Holocene), and will further decrease but 'southern species' could invade	Rare (since the beginning of the Holocene), and will further decrease	Extinct (European wild 'forest' horse and aurochs are extinct). Remaining species are rare, and will become rarer
Selective grazers	Rare, but could increase	Common, but could increase	Rare, but could increase and 'southern species' could invade	Will increase	Absent
Mixed grazers	Common and can further increase	Common, but could increase	Rare, but could increase and 'southern species' could invade	Will increase	Will increase
Browsers	Common and will further increase	Rare, but could increase	Rare, but could increase and 'southern species' could invade	Will increase	Will increase

mass greater than that of rodents fall into that group, but the oribi is a candidate. Large non-ruminant concentrate selectors do also exist and most of these are either browsers that include fallen fruits in their diets (like the tapiroids) or mixed-feeders with a strong leaning towards omnivory (suids and tayassuidae). In this respect, lagomorphs (hares, picas, and rabbits), warthogs, and hippos deserve special interest as to whether they are energy-limited or nitrogen-limited. The evolution of coprophagy in lagomorphs may indicate that they have 'solved' nitrogen limitation (Hirakawa 2001). If this is true, then with rising CO_2 levels we predict an increase of the densities of lagomorphs and duiker antelopes; we keep our powder dry for the warthog and the hippo. We predict that, everything else being equal, suids, tayassuidae, tapiroids, and browsing rhinoceroses will increase.

As pointed out already, roughage-eating hindgut-fermenting grazers (equids, white rhinoceros) are less severely nitrogen-limited than ruminant bulk grazers (many bovids and some cervids). On the basis of that, we cautiously predict a shift in the class of roughage grazers from artiodactyls towards equids and rhinos. However, such a shift in numerical abundance will possibly be hampered by lack of sufficient biological variety within the group of hindgut-fermenters remaining for a future renaissance of this group of animals. Indeed, the northern white rhino (*Ceratotherium simum*) went extinct in 2005, and all wild equids except for the Burchell's zebra (*Equus burchellii*) are vulnerable or threatened. But other grass-eating rhinos went extinct before, like the steppe rhinoceros (*Dicerorhinus kirchbergensis*) and the woolly rhino (*Coelodontata antiquitatis*). North American horses went extinct recently, and so did *E. mauritianum* in North Africa and *E. hydruntinus* in Europe. There is thus much less variety left than in the artiodactyl group.

The same lack of potential may apply to the rebound we predict for browsers or for that most versatile group, the mixed feeders. In the temperate zone, there is a definite lack of biological variation to capitalise on the more favourable circumstances. Not only did the North American cameloids go extinct, but perhaps the animal whose all-round adaptations are missed most sorely is the extinct straight-tusked elephant (*Loxodonta atlanticum* / *Palaeoloxodon antiquus*) of the last interglacial of the temperate zone. In the tropics there are two species that typically may benefit from both trends (increase of woody cover and soluble carbohydrates). These are the Asian elephant (*Elephas maximus*) and the African elephant (*Loxodonta africana*). Ironically, through their desire for ivory and their need to protect their crops, humans keep elephant population numbers low, but through their use of fossil fuels, people probably create better conditions for elephants. We do not believe that in the modern world we will quickly see new elephant populations from either African or Asian stock taking an abode in the new forests of Europe, Siberia, or North America. The northern forests thus will lack bulldozer herbivores to facilitate a secure livelihood for smaller species, necessitating perhaps continued management. In Table 12.2 we have summarized our predictions about changes in community structure as a result of the trends we discern.

Table 12.2 Predicted changes in community structure as a result of an increased tendency towards woody encroachment and expanse of forests, and to increased availability of soluble carbohydrates

Predicted community trend		Typical for the tropics	Typical for the temperate zone
Browsers will increase		Greater kudu, lesser kudu	Moose
Concentrate selectors will increase	Browsers	Duiker antelopes	Mountain hare
	Grazers	Oribi; perhaps warthog and hippopotamus	Rabbits, picas, other hares
	Omnivores	Pigs, peccaries, tapirs, browsing rhinoceroses	Wild boar
Mixed feeders will increase or stay the same	More typical grazer will stay the same	Asian elephant, impala	White-tailed deer, red deer
	More typical browser will increase	African elephant	Roe deer
Bulk or roughage grazers will decrease or increase	Ruminants will decrease	Asian buffalo, African buffalo, wildebeest,	American bison, European wisent, fallow deer, Père David's deer
	Hind-gut fermenters will increase	Zebra species, white rhino	Kiang

12.6 Ways of Managing Browsers and Grazers

12.6.1 Managing for Variability Rather than Stability

Natural grazing systems are typified by dynamic herbivore populations (Prins and Douglas-Hamilton 1990 and Saether 1997; Clutton-Brock and Pemberton 2004). The present management ethos which advocates managing herbivore populations for stability (du Toit et al. 2003), often at levels well below carrying capacity, is, therefore, 'unnatural'. For example, for over 50 years the management of the Kruger National Park in South Africa was predicated on the basis of culling large herbivore populations to maintain predetermined levels. This management strategy does not reflect the natural dynamics of this semi-arid grazing system where climatic variability, predator/prey interactions, and disease would have meant dramatic annual, decadal, and centurial fluctuations in the numbers of large herbivores (Owen-Smith and Ogutu 2003). This would have lead to increased diversity of vegetation in the park, reflecting periods of low grazing pressure when tree recruitment, for example, would have been high, and periods of high grazing pressure when trees and shrubs would be been rare in the system and grasslands dominant (e.g., Prins and van der Jeugd 1993). In an area the size of the Kruger National Park it should be possible to restore this temporal variation in herbivore grazing pressure by using spatially variable population management policies.

However, in many other parklands, especially in Europe, this management opportunity does not exist. In the European context, the management of large herbivore populations will require cooperative management policies which link groups of (national) parks together allowing the build-up of large herbivore densities in some parks whilst reducing densities in others. This situation could be reversed at specific time intervals to allow ecosystem dynamics to mimic natural boom and bust cycles, with movement of animals between parks maintaining genetic integrity of the metapopulation (Hanski and Gilpin 1997). Whilst this strategy would require human intervention, it mimics the natural cycles of interactions between plants and herbivores, allowing plant adapted to both severe and lax grazing pressures to coexist.

12.6.2 Gardening Versus Laisser Faire

The European and American models of nature conservation differ in that the latter views a natural area as one that is left alone to reach some hypothesised wilderness equilibrium, whereas the former takes a much more interventionist approach in which nature has to be managed primarily for specific interests, e.g., birds, biodiversity, habitats of conservation importance, Scottishness! Each of the approaches has its benefits and drawbacks. In East Africa, the *laisser faire* policy has been for years the dominant one, while in southern Africa the interventionist 'gardening' approach was dominant for many decades. In West Africa, the *laisser faire* approach was a matter of fact, not of choice. It is poignant that with the abolishment of apartheid, the interventionist gardening approach has been imported into East Africa; specialists from South Africa and Zimbabwe now take up practice there. On the other hand, in Germany, France, and the Netherlands there is increased public interest in applying the American 'wilderness' concept to local nature management. Even in Japan, in the northeastern corner of Hokkaido Island, the management is keenly interested in cooperating with Yellowstone National Park to apply management principles for increased 'wilderness'—as is also happening in Russia's Far East.

With changes in land use occurring in both the developing and the developed world, there may be opportunities for both approaches since land abandonment in the developed world may allow large areas to be left to nature, whereas in the developing world, where the populations of rural areas are likely to expand dramatically over the next 50 years, there can be expected to be a requirement to strictly manage wildlife to reduce interactions with human agricultural interests. However, it may also be the case that in developed countries urban populations will call for strong control of wildlife as contact is made with predators in the countryside (on weekend visits), or as wildlife comes into conflict with people in peri-urban areas (e.g., traffic accidents, diseases such as Lyme disease). Incongruously, it may be that places such as Europe and North America will become the bastions for predators whilst in Africa and Asia predators are extirpated from much of their range as they come into conflict with ever expanding human populations.

12.7 Where Do We Go From Here?

As we have outlined in previous sections of this chapter, there are likely to be major changes in the structure and composition of the vegetation communities across large parts of the globe over the next 50 years. In turn this will lead to changes in the population densities and community composition of the large herbivores that rely on the vegetation. What can we do, as scientists to help managers to make decisions in this changing world? Firstly, we will have to make predictions about what is likely to happen. This can be at the qualitative level such as we have described above, however, we are likely to be asked to make more quantitative predictions in order to inform decisions about such things as what is the vegetation composition likely to be in a certain area in 20 years, what population levels of herbivores could these changed landscapes hold, what culling policies might need to be adopted in order to manage vegetation composition/structure to meet other biodiversity objectives?

A large number of models are available that predict different components of the system response, from those dealing with changes in plant growth in relation to CO_2 levels, through changes in tree/grass ratios, to herbivore population dynamics models (Illius and O'Connor 1999). However, given the complexity of the responses of these multi-dimensional systems it is our view that complex mechanistic models are likely to be too unwieldy to provide realistic quantitative outputs for management. We advocate the use of relatively simple Bayesian approaches that take into account the levels of understanding of system linkages and the strength of those linkages. This will allow scientists to work with managers to develop models of systems with the degree of certainty associated with the quantitative predictions.

12.7.1 Landscape Scale Experiments

Some exciting new initiatives have recently been taking place in Russia and the USA; parks are being established where the extinct large mammal communities are being replaced by functionally similar species (Zimov 2005; http://www.faculty.uaf.edu/fffsc/park.html). We have to use the kinds of predictions we provide in Tables 12.1 and 12.2 if we, as ecologists, are to help the developers of these parks decide how to introduce and manage species in a changing world. The predictions we have made about the changes in population density and community structure as a result of changes in CO_2 are not amenable to classic replicated experiments, however, the need remains to provide management advice at the landscape scale. This often forces a paradigm shift where the ecologist has to adopt a logical stance closer to that of the forensic scientist and address questions more closely aligned to the particular management issue at hand rather than any general scientific posture. We would suggest that model-based approaches, including Bayesian methods, offer powerful ways of addressing such questions, but the status of the conclusions is different from those that a truly replicated and scientifically controlled experiment would provide (Hobbs and Hilborn 2006). In cases where only observational data are available, some model-based approach is inescapable. We see no clear demarcation between model-based approaches and classical experimental approaches, particularly in ecology, since both rely on underlying assumptions about repeatability in the natural world, which in practice may not be entirely correct. In the end, the new "Pleistocene" Parks offer us a great opportunity to test hypotheses and provide guidance for adaptive learning management where scientists and managers work hand in hand to manage our natural resource for the future.

12.8 Conclusions

The world in which we live is changing rapidly. Rising levels of CO_2 and changes in land use patterns will lead to changes in vegetation productivity and vegetation community composition and structure. Herbivore populations and communities will respond to these changes. Unless society is going to adopt a *laisser faire* attitude to these changes (in which case there could be significant societal and economic consequences) then managers will be expected to make decisions which will affect landscapes in the long term. If science is to play a part in determining the decisions that are made, then we have to develop partnerships with managers in which we work in collaboration to use our knowledge to develop predictions of how systems will change and to gather data to improve our understanding of how these systems operate. This is a challenge facing us all—and scientists must meet the challenge if they want to remain relevant and valued by society.

References

Ainsworth EA, Long SP (2005) What have we learned from 15 years of free-air CO_2 enrichment (FACE)? A meta-analytic review of the responses of photosynthesis, canopy. New Phytol 165:351–371
Behnke RH, Scoones I (1992) Rethinking range ecology: implications for rangeland management in Africa. London: International Institute for Environment and Development
Bond WJ (2005) Large parts of the world are brown or black: a different view on the 'Green World' hypothesis. J Veg Sci 16:261–266
Bond WJ, Midgley GF, Woodward FI (2003) The importance of low atmospheric CO_2 and fire in promoting the spread of grasslands and savannas. Glob Change Biol 9:973–982
Bowes G (1993) Facing the inevitable - plants and increasing atmospheric CO(2). Annu Rev Plant Phys 44:309–332
Coe MJ, Cumming DH, Phillipson J (1976) Biomass and production of large African herbivores in relation to rainfall and primary production. Oecologia 22:341–354
Coley PD, Massa M, Lovelock CE, Winter K (2002) Effects of elevated CO2 on foliar chemistry of saplings of nine species of tropical tree. Oecologia 133:62–69

Convention on Biol Divers, Secretariat (2001) Handbook of the convention on biological diversity, section ecosystem approach. CBD/UNEP, Montreal

Drent RH, Prins HHT (1987) The herbivore as prisoner of its food supply. In: van Andel J, Bakker J, Snaydon RW (eds) Disturbance in grasslands; species and population responses. Junk Publishing, Dordrecht, pp 133–149

FAO (2005) http://www.fao.org/forestry/foris/data/fra2005/kf/common/GlobalForestA4-ENsmall.pdf

Ford J (1971) The role of trypanosomiases in African ecology: a study of the tsetse fly problem. Oxford University Press, Oxford

Fritz H, Duncan P (1994) On the carrying capacity for large ungulates of African savanna ecosystems. P Roy Soc Lond B Bio 256:77–82

Goverde M, Erhardt A, Stocklin J (2004) Genotype-specific response of a lycaenid herbivore to elevated carbon dioxide and phosphorus availability in calcareous grassland. Oecologia 139:383–391

Hattas D, Stock WD, Mabusela WT, Green IR (2005) Phytochemical changes in leaves of subtropical grasses and fynbos shrubs at elevated atmospheric CO_2 concentrations. Global Planet Change 47:181–192

Higgins SI, Bond WJ, Trollope WSW (2000) Fire, resprouting and variability: a recipe for grass–tree coexistence in savanna. J Ecol 88:213–229

Hirakawa, H (2001) Coprophagy in leporids and other mammalian herbivores. Mammal Rev 31:61–80

Hobbs NT, Hilborn R (2006) Alternatives to statistical hypothesis testing in ecology: a guide to self teaching. Ecol Appl 16:5–19

Illius AW, O'Connor TG (1999) On the relevance of nonequilibrium concepts to arid and semiarid grazing systems. Ecol Appl 9:798–813

IPCC (2001) Climate Change 2001: impacts, adaptation and vulnerability-contribution of Working Group II to the IPCC Third Assessment. Cambridge Univ Press, Cambridge

Kimball BA (1983) Carbon dioxide and agricultural yield – an assemblage and analysis of 430 prior observations. Agron J 75:779–788

King D (2005) Climate change: the science and the policy. J Appl Ecol 42:779–783

Kinney KK, Lindroth RL, Jung SM, Nordheim EV (1997) Effects of CO_2 and NO_3-availability on deciduous trees: phytochemistry and insect performance. Ecology 78:215–230

Kjekshus H (1977) Ecology control and economic development in East African history: the case of Tanganyika, 1850–1950. Heinemann, London

Lewis D (2002) Slower than the eye can see: environmental change in northern Australia's cattle lands – a case study from the Victoria River district, Northern Territory. Tropical Savannas CRC, Darwin

Mduma SAR, Sinclair ARE, Hilborn R (1999) Food regulates the Serengeti wildebeest: a 40-year record. J Anim Ecol 68:1101–1122

Nowak RS, Ellsworth DS, Smith SD (2004) Functional responses of plants to elevated atmospheric CO_2 – do photosynthetic and productivity data from FACE experiments support early predictions? New Phytol 162:253–280

Parton WJ (1995) Impact of climate change on grassland production and soil carbon worldwide. Global Change Biol 1:13–22

Prins HHT (1996) Ecology and behaviour of the African buffalo: social inequality and decision-making. Chapman and Hall, London

Prins HHT, Beekman JH (1989) A balanced diet as a goal of grazing: the food of the Manyara buffalo. Afr J Ecol 27:241–259

Prins HHT, Douglas-Hamilton I (1990) Stability in a multi-species assemblage of large herbivores in East Africa. Oecologia 83:392–400

Prins HHT, van der Jeugd HP (1993) Herbivore population crashes and woodland structure in East Africa. J Ecol 81:305–314

Sankaran M, Hanan NP, Scholes RJ et al (2005) Determinants of woody cover in African savannas. Nature 436:846–849

Scoones I (1994) Living with uncertainty: new directions for pastoral development in Africa. International Institute for Environment and Development, London

Siegenthaler U, Stocker TF, Monnin E et al (2005) Stable carbon cycle–climate relationship during the late Pleistocene. Science 310:1313–1317

Sinclair ARE (1977) The African Buffalo: a study in resource limitations of populations. Univ of Chicago Press, Chicago

Stuart-Hill GC, Tainton NM (1999) Savanna: geographical distribution and extent. In: Tainton NM (ed) Veld management in South Africa. Univ of Natal Press, Pietermaritzburg, pp 312–317

Van de Koppel J, Rietkerk M, van Langevelde F et al (2002) Spatial heterogeneity and irreversible vegetation change in semi-arid grazing systems. Am Nat 159:209–218

van Langevelde F, van de Vijver CADM, Kumar L et al (2003). Effects of fire and herbivory on the stability of savanna ecosystems. Ecology 84:337–350

Vucetich JA, Peterson RO (2004) The influence of top-down, bottom-up and abiotic factors on the moose (*Alces alces*) population of Isle Royale. P Roy Soc Lond B Bio 271:183–189

Wand SJE, Midgley GF, Jones MH, Curtis PS (1999) Responses of wild C4 and C3 grass (Poaceae) species to elevated atmospheric CO_2 concentration: a meta-analytic test of current theories and perceptions. Global Change Biol 5:723–741

Subject Index

A
Agricultural
   production, 5, 9
   subsidies, 7
Allometry, 134
Alternate stable state, 285
Annual rainfall, 151–153, 162, 170, 182–184, 313
Anthropogenic grassland, 273
Apical meristem, 89, 107, 219, 221, 231, 268
Avoidance
   animals, 139, 226
   plants, 225, 226, 229, 232, 272

B
Basal metabolic rate, 57
Biodiversity, 7, 10, 180, 183, 185, 191, 263–286, 317, 318
Biomass
   ratio, 271
Bite size, 119, 122, 123, 133, 136–138, 227, 231, 237
Body mass, 48, 52, 61, 62, 64, 66–68, 71, 72, 94, 95, 123, 124

C
Census, 5, 166, 167
Cheek teeth-crown height
   Brachydont, 27, 28
   Hypselodont/evergrowing, 25
   Hypsodont, 25–29
Cheek teeth–occlusal morphology
   Bilophodont, 26, 27, 33
   Bunodont, 26, 27, 33
   Bunolophodont, 24, 27
   Lophed, 26–28
   Lophodont, 27, 28
   Plagiolophodont, 27, 28
   Selenodont, 27, 28, 33
Chemical defences, 127, 227, 228
Chewing
   investment, 123, 124, 138
   rate, 123, 124, 138
Climate change, 8, 299, 310
Clone, 219
CO_2, 181, 299
Community
   diversity, 264, 265
   structure, 201, 206, 210, 213, 263, 271, 310, 312, 313, 315, 316, 318
Complementary grazing, 189
Concentrate selectors, 90, 312, 313, 315, 316
Convergent evolution, 63, 64, 67, 71, 77
Conversion efficiency, 180
Craniodental morphology, 26, 29, 36, 71
Cropping time, 120–123, 138
Cytochrome P450, 111

D
Defence
   chemical, 127, 224, 227, 228
   constitutive, 228
   induced, 228, 232
   qualitative, 228
   quantitative, 228
   structural, 120–122, 227
Demography, 7, 298
Density dependence, 164, 294, 302
Dental
   mesowear, 29, 57
   microwear, 29, 35, 38
Detoxification, 34, 53, 56, 107, 109–111

Subject Index

Diet
  optimisation, 139
  overlap, 186–189, 193
Digestibility, 49–51, 54, 57, 58, 71, 77, 91, 92, 95
Disease, 2, 15, 164, 227, 236, 316, 317
Dispersion, 118, 120, 121, 125, 127, 133

E
Ecosystem
  functioning, 183, 185, 189, 193, 195
  structure, 9
Enamel, 27–30, 50, 54, 56, 57, 59, 70, 73, 124
Encounter rate, 124, 125, 134
Equilibrium, 169, 244, 245, 317
Evapo-transpiration, 9
Evolutionary history, 21–41, 134, 186, 241, 268, 271, 272, 284

F
Facilitation, 124, 125, 190, 279–281
Feeding
  guild, 134, 183, 194
  niche, 52, 64, 77, 89, 93, 97, 108, 109, 186
  stations, 126, 135–138
Fibre
  cellulose, 41, 50, 51, 56, 90, 93, 118, 121
  lignin, 50, 53, 90, 95, 118, 121, 130, 131, 204
  neutral detergent fibre, 91
Fire, 2, 8, 15, 118, 119, 151, 153, 158, 168, 170
Flower, 218, 236, 266, 268, 275, 276
Food intake, 54, 55, 57, 58, 77
Forage
  production, 153
  quality, 128, 170, 204, 241, 242, 280
Foregut fermenter, 65
Functional
  group, 270, 273
  niche, 185, 191
  redundancy, 179
Functional response, 117–120, 122, 138

G
Gain function, 123, 137, 138
Game farms, 10
Genet, 218, 219, 221
Genotype, 106, 129–132, 221, 263, 268
Global change, 299, 309
Grass
  C3 vs. C4, 51, 54, 72, 73, 311–314
Grasslands, 3, 4, 21, 30–32, 34, 35, 40–42, 62, 101

Grazing history, 268, 269, 272, 286
Grazing lawn, 134, 232, 233, 235, 270, 280

H
Habitat heterogeneity, 264, 276
Herbage
  nutritive value, 21, 90, 117
  physical characteristics, 102
Hierarchy theory, 126
Hindgut fermenter, 26, 34, 41, 64, 312, 315
Home range, 126, 160, 162, 163, 170
Hypsodonty, 28–31, 34–36, 39, 54, 57, 66–68, 70–72

I
Ingesta retention, 54, 58, 62, 64–67, 70, 77
Insurance hypothesis, 196
Intake rate, 21, 117–119, 121, 122, 133, 134, 137, 227, 231, 237

K
Key resources, 238

L
Land abandonment, 7–11, 13–16, 317
Land use intensification, 8
Life history, 218, 295–297, 299, 303
Litter accumulation, 274
Litter size, 163, 295
Liver, 53, 56, 93, 107, 111, 228
Livestock, 3, 5, 56, 109, 111, 168, 169, 189, 227, 242, 244, 276, 278–282, 284, 309

M
Mandible, 26, 38, 54, 55, 57, 72
Marginal Value Theorem, 135
Masseter, 29, 54, 57, 66, 67, 71, 73
Mega herbivore, 33, 40, 168, 195, 232, 237, 243, 266, 267
Memory, 124, 125
Meristem, 89, 107, 118, 127, 132, 218–223, 229, 231, 232, 234, 236, 268
Migration, 35, 139
Mixed farming system, 189
Mixed-feeders, 35, 40, 182, 183, 193, 315
Mixed-species production systems, 196
Morphology, 26, 29, 30, 35, 36, 47, 50, 54, 56, 57
Muzzle width, 53, 55, 66–68

N

Neutral theory, 179
Niche
  complementarity, 185, 191, 195, 196
  concept, 179
  differentiation, 179, 185
  theory, 179
Nitrogen, 4, 49, 90, 99, 110, 128–132, 134
Non-linear dynamics, 285
Normalised vegetation index (NDVI), 181
Nutrient
  retention, 183, 185
  use efficiency, 185
Nutritive value, 28, 90, 91, 93, 96, 111, 117

O

Omasum, 60, 62, 74, 90

P

Passage rate, 41, 93, 94, 100–103
Pastoralism, 2
Patches
  optimisation models, 138
  perception, 130, 132–137
Pectin, 49, 50
Persistence, 219, 231, 233, 236, 240
Phenology, 129, 163, 268
Phenotypic variation, 264, 268
Photosynthesis, 30, 220, 222–224, 228, 229, 311
Photosynthetic rate, 223, 267
Phylogenetic control, 63, 64, 66, 67, 71, 72, 77
Phylogeny, 33, 64, 71, 100, 104, 295
Plant communities, 2, 16, 126, 128, 240–244, 270–275
Plant diversity, 32, 185, 272, 273, 277
Plant nutrient allocation
  long-term responses, 207–208
  short-term responses, 206–207
Plant secondary metabolites
  alkaloids, 107
  detoxification, 107
  mimosine, 109
  post absorptive metabolism, 110–111
Plant-herbivore interactions, 118, 119
Plasticity
  phenotypic, 221, 228
Population
  abundance, 150, 165, 293
  biomass, 129, 130, 132, 135, 136, 150, 153, 156, 179, 182, 183, 186, 189, 190, 193–195, 207, 211, 212, 220, 221, 227, 243, 270, 272, 312
  crashes, 150, 165, 166, 170, 171
  density, 150, 159–162, 164, 170, 294, 295, 301
  dynamics, 118, 119, 149–171, 181, 243, 312, 313, 318
  eruptions, 150
  explosion, 1, 239
  growth, 3, 8, 150, 163, 164, 166, 294, 299, 303, 312
  management, 302, 303, 316
  regulation, 163–165
  structure, 150, 164, 242
  synchrony, 150, 169, 171
Predation risk, 117, 238, 282
Predation, 139, 164–166, 168, 169, 171, 264, 271, 278, 283
Preferential grazing, 270
Primary production, 180, 181, 186, 190, 192, 193, 195, 201, 205, 240, 242, 273
Productivity gradient, 284
Protected areas, 167, 170, 194
Protein, 4, 49, 52, 53, 56, 67, 90, 91, 93, 103, 105, 107–109, 135, 138, 153, 159, 223, 228
Protozoa, 58, 90

R

Rainfall, 29, 129, 150, 151, 153, 155, 156, 162, 164, 166, 168–170, 182, 184, 194, 195, 245, 313
Ramet, 218, 221, 222, 230, 231, 233, 265
Regional species pool, 265, 280
Remote sensing, 181
Resilience, 183, 185
Resistence
  plants, 130, 225–229, 240, 241
Resource partitioning, 179, 280
Retention time, 58, 60, 70, 71, 77, 100–101, 104, 110
Reticulum, 77, 78
Root
  biomass, 206–208, 213
  exudation, 205–207, 213
Roughage feeders, 312
Ruderal species, 276
Rumen
  capacity, 64
  detoxification, 109–111
  fermentation rate, 91, 93, 96, 103, 104
  fibre digestibility, 103, 104
  fluid, 60, 98, 110
  function, 78, 98, 110
  microbes, 90, 98, 104

Rumen (continued)
  musculature, 96, 102
  N recycling, 104
  papillae, 60, 96
  stratification, 92
  surface area, 73, 102
Ruminant, 26, 28, 33, 34, 41, 42, 54, 58, 60–62, 64, 67, 71–78, 89, 90, 100, 102, 103, 105, 106, 108, 109, 124, 149, 153, 182, 237, 312, 313, 316

S

Saliva flow, 93, 103, 110
Salivary glands, 53, 56, 60, 71, 103
Salivary tannin-binding proteins, 67, 97, 107–109
Savanna, 31, 35, 36, 38, 40, 42, 129, 130, 150, 152, 156, 158, 168, 170, 182, 189, 195, 196, 219, 237, 243, 275, 299, 300, 310, 311
Seasonal variation, 63
Secondary compounds, 34, 49, 56, 107–111, 128, 210, 311, 312
Secondary metabolites, 89, 91–93, 107, 109, 112, 131, 202, 211, 213
Secondary productivity, 183, 185, 189–191, 193, 195
Selective foraging, 155, 208, 210, 211, 213, 217, 237
Sex ratio, 163, 240
Shrub, 9, 48, 89, 92, 107, 131, 136, 158, 163, 186, 217–220, 236, 244, 265, 278, 285, 310, 312, 316
Social facilitation, 124, 125
Soil community structure, 206, 210, 212
Soil nutrient cycling
  accelerating effects, 202
  decelerating effects, 202
Spatial heterogeneity, 126, 133, 181, 182, 222
Spatial variations
  browse, 130–132
  grass, 128–130
Species
  assemblage, 1, 263,
  diversity, 179–196, 263, 264, 271, 273
  richness, 10, 11, 15, 37, 179–185, 190, 193, 195, 264, 270, 273, 279, 280
Starch, 8, 50, 90, 105, 223, 224, 310
State-and-transition, 245, 246, 271
Steppe, 4, 31, 40, 41, 150, 209, 219, 313, 315
Structural defences, 120–122, 227, 288
Succession, 15, 204, 244–246, 270, 273, 274, 276, 285
Swards, 121, 122, 128, 129, 133, 134, 136, 138, 231, 279

T

Tannin degrading bacteria, 98, 99, 110
Tannins, 50, 91, 93, 107–110, 159, 228
Tensile strength, 121
Territory, 161–163, 238
Thornveld, 189
Tolerance
  animals, 182, 241
  plants, 229
Trait
  diversity, 263–265
  variation, 185, 268
Trampling, 212, 229, 270, 274, 276, 278

U

Ungulate, 4, 5, 21–26, 29, 30, 32–41, 57, 64, 65, 67
Urban areas, 5–7, 317

V

Vegetation dynamics, 170, 190, 217–247
Volatile fatty acid
  proportions, 105

W

Water infiltration, 9
Whole-ecosystem effect, 275
Wildlife sanctuaries, 6
Woody encroachment, 8–15, 316
Woody species, 8–11, 13, 218, 231, 233, 234, 244, 265, 309–311

Species Index

A
American bison, 4, 316
Argali, 188
Aurochs, 4, 265, 282, 314

B
Banteng, 2, 3
Bharal, 281
Bison, 4, 16, 40, 42, 136, 149, 162, 186, 206, 281, 282
Black rhinoceros, 56, 58, 108, 187
Black wildebeest, 4
Black-tailed deer, 186
Blue sheep, 188
Bluebuck, 4
Boar, 15, 281, 316
Buffalo, 2, 3, 5, 62, 162, 164, 167, 188, 239, 280, 316

C
Camel, 2–4, 24
Caribou, 165, 282, 283
Chevrotain, 188
Chiru, 281
Chital, 188

D
Deer, 3, 4, 7, 15, 16, 24, 40, 41, 78, 101, 108, 123, 132, 149, 158, 161–163, 165, 169
Dromedary, 4
Duiker, 188, 315, 316

E
Eland, 162, 163, 166, 187, 189
Elephant, 21, 23, 28, 41, 70, 77, 168, 182, 187, 193–195, 224, 234, 265, 280, 315, 316
Equids, 33–36, 39–42, 67, 70, 315

G
Gaur, 2
Gayal, 2, 4
Gerenuk, 280
Giraffe, 61, 62, 78, 122, 157, 164, 166–168, 187–189

H
Hare wallaby, 4
Hippopotamus, 188, 280, 316
Hyraxes, 22, 24, 26, 70, 124

I
Ibex, 188, 300
Impala, 149, 161, 164, 166, 188, 316

K
Kangaroo, 3, 27, 29, 106, 124
Kiang, 281, 316
Kob, 163, 188, 280
Kouprey, 4
Kudu, 56, 62, 122, 136, 137, 157, 164, 166, 168, 188, 189, 280, 316

L
Lama, 2, 13,
Llama, 40, 41, 187, 189

M
Macropods, 56, 59, 61, 64, 66, 67, 77, 121, 149
Moose, 15, 25, 78, 122, 132, 136, 137, 158, 162, 163, 165, 168, 186, 210, 211, 238, 243, 265, 283, 316

Mule deer, 108, 161, 162, 165, 186

N
Nilgai, 188

O
Onager, 2
Oryx, 280

P
Père david's deer, 4, 40, 316
Perissodactyls, 22, 24, 26, 28, 32–34, 57, 64, 65, 67
Pronghorn, 29, 36, 39, 40, 186, 205
Przewalski horse, 4

Q
Quagga, 4

R
Red deer, 15, 132, 137, 149, 158, 161, 165, 166, 169, 186, 187, 281, 283, 301, 316
Reindeer, 2, 3, 40, 105, 106, 135, 187, 205, 206, 208, 209, 212
Rhinoceroses, 67, 77, 108, 315, 316
Roe deer, 15, 16, 78, 123, 132, 137, 162, 163, 169, 187, 293, 295, 299–301, 316

S
Sambar, 188
Suids, 39, 41, 61, 71, 315
Suni, 188
Swamp deer, 4

T
Tapirs, 33, 39–41, 61, 316
Tarpan, 265
Thar, 3
Topi, 122, 163, 188, 280

W
Wapiti, 149, 161, 162, 281
Warthog, 27, 28, 41, 188, 315, 316
Water buffalo, 2, 3, 5
Waterbuck, 164, 166, 188, 280
White rhinoceros, 4, 108, 312, 315
White-tailed deer, 7, 16, 161, 163, 165, 166, 169, 186, 187, 316
Wildebeest, 4, 40, 122, 161–163, 166, 167, 169, 188, 235, 239, 280, 316
Wisent, 16, 316

Y
Yak, 2, 188, 281

Z
Zebra, 42, 149, 162–164, 166, 167, 169, 188, 239, 280, 315, 316

Ecological Studies
Volumes published since 2003

Volume 163
Fluxes of Carbon, Water and Energy of European Forests (2003)
R. Valentini (Ed.)

Volume 164
Herbivory of Leaf-Cutting Ants: A Case Study on *Atta colombica* in the Tropical Rainforest of Panama (2003)
R. Wirth, H. Herz, R.J. Ryel, W. Beyschlag, B. Hölldobler

Volume 165
Population Viability in Plants: Conservation, Management, and Modeling of Rare Plants (2003)
C.A. Brigham, M.W. Schwartz (Eds.)

Volume 166
North American Temperate Deciduous Forest Responses to Changing Precipitation Regimes (2003)
P. Hanson and S.D. Wullschleger (Eds.)

Volume 167
Alpine Biodiversity in Europe (2003)
L. Nagy, G. Grabherr, Ch. Körner, D. Thompson (Eds.)

Volume 168
Root Ecology (2003)
H. de Kroon and E.J.W. Visser (Eds.)

Volume 169
Fire in Tropical Savannas: The Kapalga Experiment (2003)
A.N. Andersen, G.D. Cook, and R.J. Williams (Eds.)

Volume 170
Molecular Ecotoxicology of Plants (2004)
H. Sandermann (Ed.)

Volume 171
Coastal Dunes: Ecology and Conservation (2004)
M.L. Martínez and N. Psuty (Eds.)

Volume 172
Biogeochemistry of Forested Catchments in a Changing Environment: A German Case Study (2004)
E. Matzner (Ed.)

Volume 173
Insects and Ecosystem Function (2004)
W.W. Weisser and E. Siemann (Eds.)

Volume 174
Pollination Ecology and the Rain Forest: Sarawak Studies (2005)
D. Roubik, S. Sakai, and A.A. Hamid (Eds.)

Volume 175
Antarctic Ecosystems: Environmental Contamination, Climate Change, and Human Impact (2005)
R. Bargagli

Volume 176
Forest Diversity and Function: Temperate and Boreal Systems (2005)
M. Scherer-Lorenzen, Ch. Körner, and E.-D. Schulze (Eds.)

Volume 177
A History of Atmospheric CO_2 and its Effects on Plants, Animals, and Ecosystems (2005)
J.R. Ehleringer, T.E. Cerling, and M.D. Dearing (Eds.)

Volume 178
Photosynthetic Adaptation: Chloroplast to Landscape (2005)
W.K. Smith, T.C. Vogelmann, and C. Chritchley (Eds.)

Volume 179
Lamto: Structure, Functioning, and Dynamics of a Savanna Ecosystem (2006)
L. Abbadie et al. (Eds.)

Volume 180
Plant Ecology, Herbivory, and Human Impact in Nordic Mountain Birch Forests (2005)
F.E. Wielgolaski (Ed.) and P.S. Karlsson, S. Neuvonen, D. Thannheiser (Ed. Board)

Volume 181
Nutrient Acquisition by Plants: An Ecological Perspective (2005)
H. BassiriRad (Ed.)

Volume 182
Human Ecology: Biocultural Adaptations in Human Cummunities (2006)
H. Schutkowski

Volume 183
Growth Dynamics of Conifer Tree Rings: Images of Past and Future Environments (2006)
E.A. Vaganov, M.K. Hughes, and A.V. Shashkin

Volume 184
Reindeer Management in Northernmost Europe: Linking Practical and Scientific Knowledge in Social-Ecological Systems (2006)
B.C. Forbes, M. Bölter, L. Müller-Wille, J. Hukkinen, F. Müller, N. Gunslay, and Y. Konstantinov (Eds.)

Volume 185
Ecology and Conservation of Neotropical Montane Oak Forests (2006)
M. Kappelle (Ed.)

Volume 186
Biological Invasions in New Zealand (2006)
R.B. Allen and W.G. Lee (Eds.)

Volume 187
Managed Ecosystems and CO_2: Case Studies, Processes, and Perspectives (2006)
J. Nösberger, S.P. Long, R. J. Norby, M. Stitt, G.R. Hendrey, and H. Blum (Eds.)

Volume 188
Boreal Peatland Ecosystem (2006)
R.K. Wieder and D.H. Vitt (Eds.)

Volume 189
Ecology of Harmful Algae (2006)
E. Granéli and J.T. Turner (Eds.)

Volume 190
Wetlands and Natural Resource Management (2006)
J.T.A. Verhoeven, B. Beltman, R. Bobbink, and D.F. Whigham (Eds.)

Volume 191
Wetlands: Functioning, Biodiversity Conservation, and Restoration (2006)
R. Bobbink, B. Beltman, J.T.A. Verhoeven, and D.F. Whigham (Eds.)

Volume 192
Geological Approaches to Coral Reef Ecology (2007)
R.B. Aronson (Ed.)

Volume 193
Biological Invasions (2007)
W. Nentwig (Ed.)

Volume 194
***Clusia*: A Woody Neotropical Genus of Remarkable Plasticity and Diversity** (2007)
U. Lüttge (Ed.)

Volume 195
The Ecology of Browsing and Grazing (2008)
I.J. Gordon and H.H.T. Prins (Eds.)

Printing: Krips bv, Meppel
Binding: Stürtz, Würzburg